T0313361

Onions
and
Allied Crops

Volume III
Biochemistry, Food Science,
and Minor Crops

Editors
James L. Brewster, Dr.Phil.
Principal Scientific Officer
National Vegetable Research Station
AFRC Institute for Horticultural Research
Wellesbourne, Warwick, U. K.

Haim D. Rabinowitch, Ph.D.
Chairman
Department of Field and Vegetable Crops
The Hebrew University of Jerusalem
Rehovot, Israel

CRC Press
Taylor & Francis Group
Boca Raton London New York

CRC Press is an imprint of the
Taylor & Francis Group, an **informa** business

CRC Press
Taylor & Francis Group
6000 Broken Sound Parkway NW, Suite 300
Boca Raton, FL 33487-2742

Reissued 2019 by CRC Press

© 1990 by Taylor & Francis Group, LLC
CRC Press is an imprint of Taylor & Francis Group, an Informa business

No claim to original U.S. Government works

A Library of Congress record exists under LC control number:

Publisher's Note
The publisher has gone to great lengths to ensure the quality of this reprint but points out that some imperfections in the original copies may be apparent.

Disclaimer
The publisher has made every effort to trace copyright holders and welcomes correspondence from those they have been unable to contact.

ISBN 13: 978-0-367-40376-8 (hbk)
ISBN 13: 978-0-429-35575-2 (ebk)

Visit the Taylor & Francis Web site at http://www.taylorandfrancis.com and the CRC Press Web site at http://www.crcpress.com

INTRODUCTION

The edible alliums are of major economic and dietary importance in all parts of the world. The common onion *(A. cepa)* and garlic (*A. sativum)* are grown, traded and consumed in most countries (Tables 1 to 3). Because of their economic importance, great efforts have been made in the selection and breeding of locally adapted cultivars, and in the development of cultural techniques. As a result these crops can be cultivated from the tropics to subarctic regions. Other edible alliums are more localized in economic significance although they are still of great importance.

It is estimated that the value of the world production of bulb onions alone approaches $5 billion annually, more than 90% of which is consumed within the countries of production. Bulb onions are an easily transportable commodity and between $400 and $500 millions worth (at 1984 prices) are traded internationally each year. The crop is a major export earner for some economies. In addition, the common onion is also an important salad crop when eaten green (scallions). Statistics on this crop are harder to come by, but by way of example, in the U.K., the monetary value of this crop is approximately 65% of that of the dry bulb crop. Leeks (*A. ampeloprasum)* are a valuable crop in Europe (Table 4). Although total European production is well below that for bulb onions, the price per unit weight of leeks is higher than for bulb onions, and in monetary terms the value of the leek crop approaches that of the dry bulb onion crop. Several other alliums are of major economic importance in far eastern countries (Table 5). The annual production of the Japanese bunching onion (*A. fistulosum)* in Japan is approximately 50% of that for dry bulb common onions. However the mean wholesale prices (yen per kilogram obtained during the period January 1983 to December 1985 in Tokyo for common onion, Japanese bunching onion, Chinese chive (*A. tuberosum)*, and rakkyo (*A. chinense)*, respectively, were 93, 216, 344 and 271.[6] Thus the monetary value of the Japanese bunching onion crop is probably similar to that of the common bulb onion in Japan. Japanese bunching onion, Chinese chive, and rakkyo are also important crops in China.[5] Hence, the alliaceous crops are clearly a major agricultural asset and have a great impact on many economies.

It is a quarter of a century since the last comprehensive account of the edible *Allium* crops was written by Jones and Mann.[7] Since then, continued and increased efforts have been made to better understand the physiology, chemistry, genetics, and pathology of these important crops. Some of these investigations have involved topics which were unexplored as regards alliums in 1963, for example, tissue culture, the effects of plant growth regulators, the biochemistry of carbohydrates and flavor compounds, and mycorrhizal associations, to name but a few. Scientific information concerning virtually all aspects of these crops has been accumulating at an accelerating rate in recent years. The extent of this knowledge means that it is no longer feasible for any individual to competently review the alliaceous crops. Consequently, in this book, we have brought together contributions from experts in many disciplines. It has been the policy of the editors to allow authors to freely express their views on matters on which there is still some scientific uncertainty or debate. We, therefore, make no apologies for any difference of emphasis or views which may be apparent between chapters. We are confident however, that these volumes consitute an up to date and soundly based summary of the current knowledge on the edible *Allium* crops.

TABLE 1
Total Production and Areas Harvested of Dry Bulb Onions in 1988[1] and Exports in 1987[2]

	Total production 1000 MT	Area harvested 1000 ha	Exports Amount 1000 MT	Exports Value 10⁶$
World	25496	1787	2009	467
Africa	2088	159	67	32
N. and Central America	2286	67	256	57
S. America	1990	136	29	7
Asia	12120	1003	520	93
Europe	4817	233	1063	260
Oceania	178	5	73	19

TABLE 2
The Leading Ten Producer, Exporter, and Importer Countries for Dry Bulb Onions in 1987[2,3]

Producer country	Total production 1000 MT	Area harvested 1000 ha
China	3600	237
India	2790	280
USSR	2000	184
U.S.	1993	49
Turkey	1500	84
Japan	1260	29
Spain	1104	32
Brazil	856	75
Iran	740	42
Poland	615	31

Exporter country	Quantity 1000 MT	Value 10⁶$
Netherlands	474	106
Spain	287	59
India	210	38
Mexico	141	23
Turkey	135	13
U.S.	94	26
Poland	110	20
Italy	60	29
Australia	44	11
Pakistan	49	3

Importer country	Quantity 1000 MT	Value 10⁶$
FR Germany	377	101
UK	242	62
U.S.	168	64
France	142	38
Malaysia	105	28
Canada	98	37
United Arab Emirates	85	15
Belgium	81	18
Singapore	53	18
Kuwait	44	12

TABLE 3
Garlic Production in 1987[4]

	Production 1000 MT	Area harvested 1000 ha
World	2662	421
Africa	237	8
N. and Central America	137	15
S. America	140	32
Asia	1612	264
Europe	488	90
Oceania	4	—

Leading Ten Countries

China	615	71
Korean Republic	280	40
Spain	230	40
India	217	60
Egypt	215	5
Turkey	98	14
Thailand	98	29
Brazil	76	18
USA	71	5
Pakistan	57	7

TABLE 4
Production of Leeks Within Countries of the European Economic Community[5]

	Production[a] 1000 MT	Area harvested 1000 ha
EEC Total	601	26.8

Major Producer Countries

France	225	9.8
Belgium	107	3.3
Netherlands	75	2.8
U.K.	61	2.7
Spain	55	2.9
FRG	43	1.7
Greece	37	1.8
Italy	30	1.3

[a] Data are for 1987.

TABLE 5
Production of Japanese Bunching Onion (A. *fistulosum*), in Japan and South Korea and Production of Rakkyo (A. *chinense*) and Chinese Chives (A. *tuberosum*) in Japan[6]

Crop	Production[a] 1000 MT	Area 1000 ha
Japan		
Japanese bunching onion	563	24.1
Rakkyo	31.5	2.3
Chinese chives	66.1	2.6
South Korea		
Japanese bunching onion	432	18.9

[a] Data are for the year 1984.

REFERENCES

1. **FAO,** *Quarterly Bulletin of Statistics,* Food and Agriculture Organization, Rome, 1988.
2. **FAO,** *Trade Yearbook,* Food and Agriculture Organization, Rome, 41, 149, 1988.
3. **FAO,** *Production Yearbook,* Food and Agriculture Organization, Rome, 41, 190, 1988.
4. **FAO,** *Production Yearbook,* Food and Agriculture Organization, Rome, 41, 192, 1988.
5. **Eurostat,** *Crop Production Quarterly Statistics 4-1987,* Statistical Office of the European Communities, Luxembourg, 1987, 35.
6. **Komochi, S.,** personal communication, 1987.
7. **Jones, H. A. and Mann, L. K.,** Onions and Their Allies, Leonand Hill Books, London, and Interscience Publishers, New York, 1963.

THE EDITORS

Haim D. Rabinowitch, Ph.D., is an Associate Professor at the Faculty of Agriculture, The Hebrew University of Jersalem, Rehovot, Israel and currently leads the Group of Vegetable Physiology and Breeding.

Prof. Rabinowitch graduated from the Hebrew University of Jersalem, in 1965, obtained his M.Sc. (*suma cum laude*) in 1967, and received his Ph.D. in 1973 from the same university. He was appointed Lecturer in Vegetable Production at his alma mater in 1976, promoted to Senior Lecturer in 1981, and to Associate Professor in 1986, at Rehovot.

He did post-doctoral work at the Department of Applied Genetics, the John Innes Institute, Norwich England, from 1974 to 1976. In 1981—1982 and in 1986—1987 he spent sabbatical leaves at the Department of Biochemistry, Duke University Medical Center Durham, North Carolina

Prof. Rabinowitch is a member of the American Society for Horticultural Science, The International Society for Horticultural Science, The Scandinavian Society for Plant Physiology, and EUCARPIA. He is a member of the National Vegetable Commodity Committee and of the Professional and Research Committees for Onion Breeding and Production, and for Seed Production, of the Israel Ministry of Agriculture; Member of the Working Group on *Allium*, the European Cooperative Group for the Conservation and Exchange of Crop Genetic Resources, the International Board for Plant Genetic Resources (IBPGR); served as a consultant to the IBPGR on Germplasm Conservation of *Allium* species; in charge of the IBPGR Global Field Bank for vegetatively propagated short-day *Allium* species; and recipient of the American Society for Horticultural Science, National Food Processing Association Award in raw products research for 1983.

Prof. Rabinowitch published more than 45 scientific papers, developed 2 tomato cultivars, and 3 short-day F_1 onion cultivars for fresh market and for the dehydration industry. His current research interests are in physiology and breeding of alliaceous and some *Solanaceae* crops, and in the protective mechanisms against oxygen toxicity in plants.

James L. Brewster, D.Phil is a Principal Scientific Officer in the Crop Production Division, Institute of Horticultural Research, Wellesbourne, Warwick, U.K.

Dr. Brewster graduated in 1966 from Wye College, University of London, with a B.Sc (Honors) degree in Agricultural Science and obtained his D.Phil degree from the University of Oxford in 1971. He remained at Oxford working on the quantitative modeling of nutrient uptake by plant roots until transferring to Wellesbourne in 1974.

Dr. Brewster is a member of the Society of Experimental Biology, Association of Applied Biologists, and Institute of Horticulture. He is an Associate Editor of the *Journal of Horticultural Science*. He is a member of the Onion panel for the National Institute of Agricultural Botany, U.K. He has frequently presented lectures on aspects of his work at scientific and technical meetings in the U.K. and Europe.

Throughout his research career he has been concerned with aspects of the physiology of alliums and has authored 33 research papers covering aspects of the nutrition, flowering, growth rate, control of bulbing, and agronomy of these crops. The main theme of his research is to obtain quantitative relationships for the environmental control of physiological processes and to apply these to understand and predict growth in the field.

ACKNOWLEDGMENTS

As editors and also as contributors, we would like to thank firstly the heads of our respective institutes, namely Professor T. J. Swinburne and Professor J. K. A. Bleasdale, of the Institute of Horticultural Research, Wellesbourne, U.K. for J. L. Brewster, and Professor Ilan Chet, the Dean of the Faculty of Agriculture, Hebrew University of Jerusalem, Rehovot, Israel, for H. D. Rabinowitch for their support for the production of this book. Prof. Rabinowitch also wishes to thank Professor Irwin Fridovich, Duke University Medical Center, Durham, North Carolina, with whom he spent a sabbatical. Without the help and facilities available at the above institutes, the task would not have been possible. We would also like to acknowledge the support of the contributing authors to these volumes, who have so generously given of their time and expertise. We thank them and the institutions in which they are based. We would also like to thank Dr. Suren S. Apte and other staff at the International Agricultural Centre, Wageningen, Netherlands for arranging, and the Government of the Netherlands for providing scholarships for 1 month of intensive joint editing work at Wageningen. Again without this generous support, it is difficult to imagine this project being completed. We would like to thank many individuals who contributed their assistance and expertise. In particular Miss Heather A. Butler of IHR Wellesbourne, for meticulously redrawing some figures, for checking reference abbreviations, and her careful and rapid work in many other respects. Also, we would like to thank Mrs. Margaret Jonas of IHR Wellesbourne for much secretarial assistance with the production of papers and with correspondence. We would like to acknowledge the help of our computer departments, particularly Mrs. Julie Rose at Wellesbourne and Mrs. Zehavit Regev at Rehovot. Without the BITNET and EARN communications between U.K. and Israel and U.K. and U.S., it is difficult to envisage how we as editors could have completed this task. We acknowledge the help of a number of colleagues in the reviewing of chapters on subjects on which our own expertise is limited. In particular we would like to thank Dr. Lesley Currah, Mr. B. D. Dowker, Dr. R. J. Whenham, Dr. D. J. Ockendon, and Professor Y. L. Bar-Sinai.

DEDICATION

Prof. Haim D. Rabinowitch would like to dedicate these volumes to his wife, Shoshie and to his mother, Sara. Prof. James L. Brewster would like to dedicate these volumes to his parents.

To my daughter Noa

— HDR

CONTRIBUTORS

Tadashi Asahira, Dr.Agr.
Professor
Laboratory of Vegetable and Ornamental
  Horticulture
Faculty of Agriculture
Kyoto University
Sakyoki, Kyoto, Japan

K. T. Augusti, Ph.D.
Reader
Department of Biochemistry
University of Kerala
Trivandrum, India

Michael Boland, Ph.D.
Applied Biochemistry Division
Department of Scientific and Industrial
  Research
Palmerston North, New Zealand

James L. Brewster, D.Phil.
Principal Scientific Officer
National Vegetable Research Station
AFRC Institute for Horticultural Research
Wellesbourne, Warwick, U. K.

Ben Darbyshire, Ph.D.
Warden
St. George's College
Crawley, Western Australia

G. R. Fenwick, Ph.D.
Molecular Sciences Department
Norwich Laboratory
AFRC Institute of Food Research
Norwich, Norfolk, U.K.

A. B. Hanley, B.Sc.
Department of Chemical Physics
AFRC Institute of Food Research
Norwich, Norfolk, U.K.

Jane Lancaster, B.Sc.
Applied Biochemistry Division
Department of Scientific and Industrial
  Research
Palmerston North, New Zealand

Haruhisa Inden, M.Agr.
Instructor
Laboratory of Vegetable and Ornamental
  Horticulture
Faculty of Agriculture
Kyoto University
Sakyoki, Kyoto, Japan

Niels Poulsen
Scientific Officer
Institute of Vegetables
Danish Research Service for Plant and
  Soil Science
Aarslev, Denmark

Haim D. Rabinowitch, Ph.D.
Chairman
Department of Field and Vegetable Crops
The Hebrew University of Jerusalem
Rehovot, Israel

Susumu Saito, Ph.D.
Professor
Department of Nutrition
Tokyo University of Agriculture
Setagaya, Tokyo, Japan

Barrie T. Steer, Ph.D.
Associate Professor
Tropical Crops Group
School of Agriculture
The University of Western Australia
Nedlands, Western Australia

Hideaki Takagi, Ph.D.
Associate Professor
Faculty of Agriculture
Yamagata University
Tsuruoka, Yamagata, Japan

Masao Toyama, Ph.D.
Associate Professor
Sand Dune Research Center
Tottori University
Tottori, Tottori, Japan

Izumi Wakamiya, M.Agr.
Research Associate
Sand Dune Research Center
Tottori University
Tottori, Tottori, Japan

Q. P. van der Meer
Engineer
Institute for Horticultural Plant Breeding
Wageningen, The Netherlands

TABLE OF CONTENTS

VOLUME III

Chapter 1

CARBOHYDRATE BIOCHEMISTRY

Ben Darbyshire and Barrie T. Steer

TABLE OF CONTENTS

I. INTRODUCTION

This chapter reviews the carbohydrate physiology and biochemistry of *Allium* species. We limit discussion mainly to issues peculiar to these species, and particular reference will be made to soluble carbohydrates, their distribution, and the synthesis of fructans. Much of our comment refers to work with onion (*A. cepa*). Some reference is made to investigations of other *Allium* species, but this is limited, since knowledge of interspecies differences is limited. No reference is made to the broader issues of carbohydrate metabolism, such as sucrose synthesis, which are beyond the scope of this chapter and for which readers are referred to other sources.[1-3]

II. COMPOUNDS AND THEIR DISTRIBUTION

The nonstructural carbohydrates in *Allium* spp. include glucose, fructose, and sucrose together with a series of oligosaccharides, the fructans.[4-10] Fructans occur in the Asteraceae,[11] but more widely in the monocotyledons, especially in the Amaryllidaceae, Iridaceae, Liliaceae, and Poaceae.[12] In all species fructans are considered to be fructose polymers containing a single glucose residue. Early reports[12-13] suggested only fructose was present, but later Palmer[14] demonstrated the presence of glucose in fructans from a number of sources. The mechanism of fructan synthesis, demonstrated by Edelman and Jefford[15] supports the presence of a glucose residue in fructans because fructose is transferred from one sucrose molecule to another sucrose molecule.

It is important to appreciate that in plants, three different trisaccharides are potentially possible, and all three have been identified in a variety of plants.[5,16,17] They are 1^F-fructosylsucrose, 6^F-fructosylsucrose, and 6^G-fructosylsucrose. These three trisaccharides arise as a result of the transfer of a fructosyl residue from one sucrose molecule to one of three primary hydroxyl groups of another sucrose molecule. The products of such a reaction are one of three possible trisaccharides and glucose. For a simple, diagrammatic representation of the structure of these trisaccharides readers are referred to Henry.[43] In onion the trisaccharides 1^F-fructosylsucrose and 6^G-fructosylsucrose have been identified,[5,9,18] and both have been reported in leek.[5] These isomers have different functions in fructan synthesis (see Section IV). The addition of further fructosyl groups to these trisaccharides leads to the production of tetrasaccharides and higher polymers, giving rise to many possible polymers. The degree of polymerization (DP) of fructans can be quite different in different species. Polymers up to a DP of 260 have been reported in grasses.[19] Figure 1 shows the separation of polymers from onion.

Starch is usually not detectable in onion, but Wilson et al.[20] have reported starch inclusions in the chloroplasts of the uniseriate layer of cells around the vascular bundles in onion leaves. Starch was not found in the chloroplasts of the other mesophyll cells. The plants used in their study were 10 weeks old, but it is not clear whether they were grown in bulb-inducing conditions or not. When there is a substantial translocation of carbohydrate to the leaf base, starch may not be present even in the vascular sheath cells.

The soluble carbohydrate concentration per unit of fresh weight of onion leaf is lowest at the top and highest in the basal part (Figure 2). This pattern does not vary substantially with leaf size or level of insertion. Sugar concentration increases in leaf blades, pseudostem, and bulbs with both increasing age and light intensity[7] (Figure 3).

Bacon[5] examined the distribution of the lower molecular weight fructans (up to DP 5) in onion bulbs and found these compounds to be absent from outer, older leaf bases and present in increasing amounts from the outer to the inner leaf bases. De Miniac[8] reported a similar distribution. These observations were extended by Darbyshire and Henry[9] who reported the distribution of fructans of DP 3 to DP 9 (Figures 4a and b). Free fructose

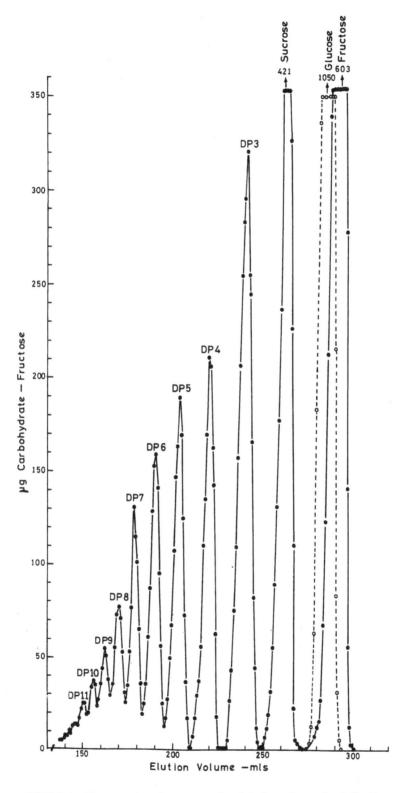

FIGURE 1. The separation of nonstructural carbohydrates from onion bulbs (*A. cepa* v. Creamgold). Using a 1.6 × 210-cm column of Bio-Gel P-2, minus 400 mesh; collecting 1-ml fractions; with a constant head pressure of 30 kPa and operating temperature of 25°C.[9]

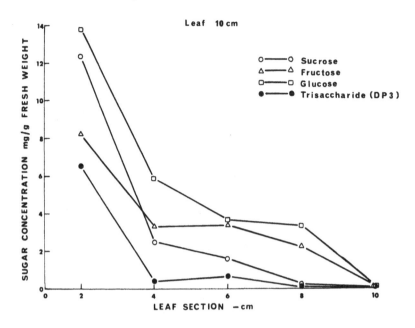

FIGURE 2.   The distribution of nonstructural carbohydrates in sections of a 10-cm onion leaf (*A. cepa* v. Creamgold), measured from the base: glucose (□); fructose (△); sucrose (○); trisaccharide fructan (●).[10]

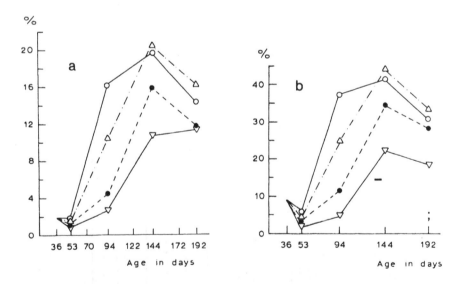

FIGURE 3.   Effect of light intensity during growth on soluble sugar (fructose + glucose + sucrose) % dry weight in the leaf blades (a) and neck (b) of *Allium cepa* ○ 100% light at 93.7 J m^{-2} s^{-1}; △ 78%; ● 35%; ▽ 11%.[7]

concentration was highest in the outer leaf bases and lowest in the innermost, an observation made also by Rutherford and Whittle.[9a] Glucose and sucrose did not change across the bulb.

Under controlled conditions Steer[21] examined the influence of day/night temperature on the accumulation of nonstructural carbohydrates within a cultivar (Figure 5). No significant differences in fructose, sucrose, or fructan concentrations occurred at the different temperatures. However, differences in the bulb dry-matter percentage may have been related to changes in water content influenced by these compounds. The concentration of monosac-

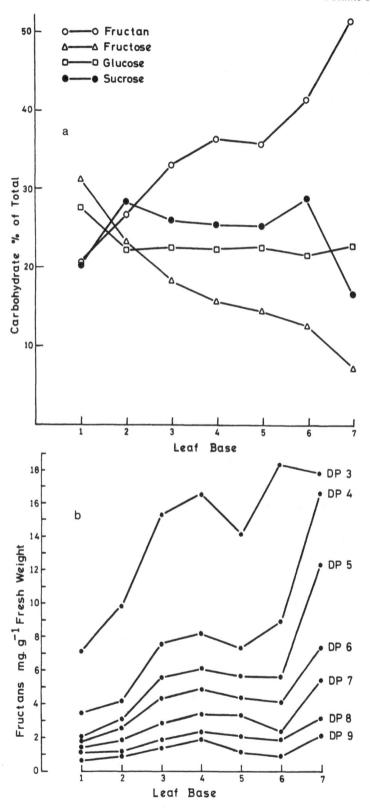

FIGURE 4. The distribution of nonstructural carbohydrates at harvest across an onion bulb (*A. cepa cv.* Creamgold) from the outer leaf base (1) to the inner (7). (a): glucose (□), fructose (△), sucrose (●), and total fructans (○) as a percentage of the total, and (b) the concentration of fructans DP 3 to DP 9.[9]

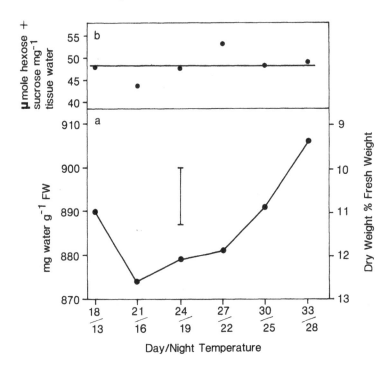

FIGURE 5.   (a) The water content or dry weight as a percentage of fresh weight and (b) the concentration of glucose + fructose + sucrose in the cell sap of onion bulbs grown under six different temperature regimes. The bar in a is LSD 0.05; the line in b is the grand mean; there were no significant differences between treatments.[21]

charides plus sucrose per unit of tissue water was constant in all temperature treatments, so that the osmolarity remained constant.

Soluble carbon compounds in onion leaves vary diurnally.[22] Glucose and fructose reach a maximum concentration during the 4th hour of the photoperiod, followed by a sucrose maximum in the 7th hour. The net accumulation rates are shown in Figure 6. The rates do not differ between the upper three fifths and the lower two fifths of the leaf, but the absolute concentrations are about twice as high in the lower as in the upper leaf. The maximum net accumulation rate of sucrose occurs when glucose and fructose have stopped accumulating. Net sucrose accumulation stops when the rates of loss of glucose and fructose are at their maximum. During the dark period, between the 15th and 23rd hour there are net losses of glucose, fructose, and sucrose, presumably due to both respiration and translocation of sucrose to younger leaves. While these differences may not have great cultural significance, it is important to consider them when experimentation is undertaken.

III. NONSTRUCTURAL CARBOHYDRATES DURING BULBING. DIFFERENCES BETWEEN CULTIVARS AND SPECIES

A. BULB PRODUCTION

Recent work examined the changes in nonstructural carbohydrates when onion plants, grown in a controlled environment, were transferred from 10- to 16-h daylengths, i.e., to bulb-inducing conditions. Figures 7a and b summarize these changes. As bulb fresh weight increased, so did the bulbing ratio (Figure 7a). However, before these morphological changes were detectable, there was a dramatic increase in fructan concentration. All fructan fractions showed a similar change (Figure 7b), but the increase in DP 3 preceded that of DP 4, which

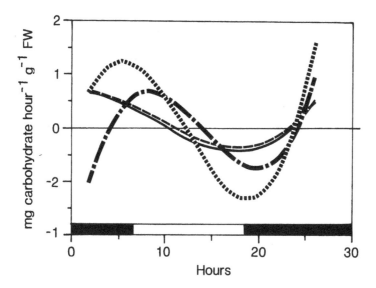

FIGURE 6. Net accumulation rates for fructose (--), glucose (—), sucrose (-·-) and total hexoses + sucrose (····) in onion leaves during a 24-hour cycle with a 12-h photoperiod. The open and closed horizontal bars represent the light and dark period, respectively. These are the first derivatives of regression equations fitted to the data of Darbyshire et al.[22]

in turn preceded that of DP 5 and above. There was no further increase in total fructan concentration after 13 d in bulb-inducing conditions, but it was 16 d before bulb fresh weight increased. These changes are discussed further in Section IV.

For bulb production to occur, not only are long photoperiods needed, but particular phytochrome photoequilibria (P_{fr}/P_{total}) are required during the photoperiod.[23] Furthermore, carbohydrates accumulated in the leaf blades and leaf bases only when both phytochrome photoequilibrium and photoperiod were suitable for bulbing.[24,25] Thus, the concentration of reducing sugars, sucrose, and fructans (measured after hydrolysis by exogenous invertase) increased in the leaf bases between 5 and 35 d after plants were transferred to an 18-h photoperiod with P_{fr}/P_{total} of 0.54.[24] Photoperiods of 10 h and/or P_{fr}/P_{total} of 0.78 did not support carbohydrate accumulation.[24] Under bulbing conditions reducing sugars accumulated in leaf blades, but oligosaccharides did not. When bulb-inducing conditions were terminated, by shortening the photoperiod to 10 h, the concentration of reducing sugars in leaf blades and bases immediately decreased. Sucrose and oligosaccharide concentrations decreased in leaf bases after a 3-d lag.[24] Lercari[26] has adduced other evidence that the high irradiance reaction of phytochrome is involved in bulb production and carbohydrate accumulation.

Lercari[24] used a low photon flux density in his experiments, and Mondal et al.[25] confirmed his findings on carbohydrate accumulation in response to photoperiod and phytochrome photoequilibria. Thus, bulb production and carbohydrate accumulation both need particular phytochrome photoequilibria and continuing long photoperiods. A number of inductive cycles are not sufficient.

Increases in carbohydrate concentration may occur independently of bulbing. Mondal et al.[25] demonstrated that carbohydrate was accumulated at high photon flux densities and high P_{fr}/P_{total} in the absence of bulb scale formation (see chapter on "Physiology of Crop Growth and Bulbing"). Carbohydrate accumulation in the leaf bases occurred at a high photon flux density and a P_{fr}/P_{total} of 0.8. As a result, the bulbing ratio increased up to about two. Thus, carbohydrate accumulation and bulb production can occur separately in some conditions. At P_{fr}/P_{total} of 0.5 and with long photoperiods, leaf development was modified

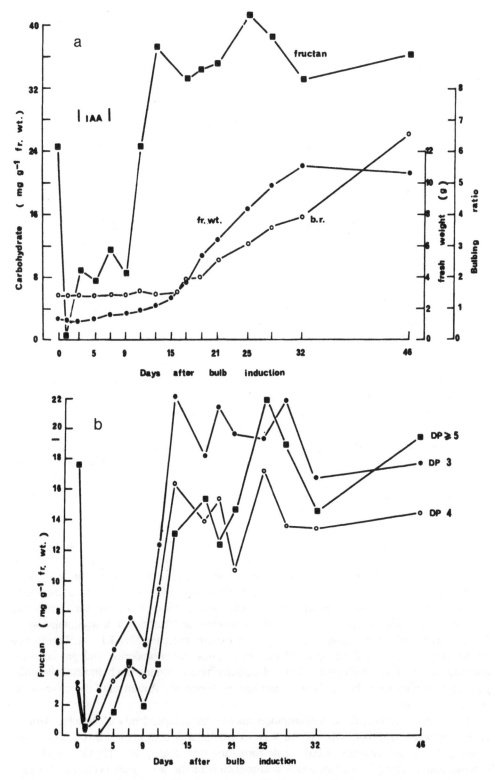

FIGURE 7.   (a) The change of bulb fresh weight, bulbing ratio, and total fructan concentration and (b) the concentrations of fructans DP 3 (●), DP 4 (○), and DP ≥5 (■) with time after transferring onion plants from 10- to 16-h daylengths. Figures a and b are from separate experiments. (Darbyshire and Henry, unpublished).

so that laminae were suppressed and bulb scales were formed. These became the major sinks for carbohydrates that accumulated at a P_{fr}/P_{total} of 0.5, and the bulbing ratio increased to a high value (>2). In the experiment illustrated by Figure 7a, the total fructan concentration reached a maximum 4 d before the bulbing ratio attained a value of two. From this we infer that nonstructural carbohydrate accumulation precedes the bulbing. Bulbing may involve hydrolysis of oligosaccharides previously accumulated.

Observations which offer support of this hypothesis are

1. Fructan hydrolase activity is minimal and fructan synthetic activity is at a maximum in the inner, unexpanded leaf bases, while hydrolytic activity is maximum and synthetic activity minimum in the outer, expanded leaves.[27]
2. As a result, levels of hexoses are lowest and levels of fructans are highest in the innermost, unexpanded leaf bases of bulbs, while hexoses are highest and fructans lowest in the outer, expanded leaf bases.[9]
3. Lercari[28] has demonstrated lowered invertase activities under conditions when leaf bases are accumulating hexoses, sucrose, and fructans, and has suggested a function of acid and neutral invertase in hydrolysis of sucrose when there is no accumulation of oligosaccharides.

These observations indicate that potentially the hydrolysis of fructans to monosaccharide fructose occurs increasingly as leaf bases expand during bulbing. Whether expansion of leaf bases causes the accumulation of monosaccharide, rather than fructans, or whether the accumulation of monosaccharides promotes bulbing, is not known.

B. CULTIVAR DIFFERENCES

High dry-matter onion cultivars have been developed specifically for processing purposes (see chapter on "Processing of *Alliums*: Use in Food Manufacture"). Onions used for the fresh market typically have a dry-matter percentage between seven and ten, while cultivars for processing have up to 20% dry matter. Differences occurring between a number of cultivars and changes in dry weights between leaf bases are shown in Figure 8a, and there is a close relationship between dry weight and nonstructural carbohydrate content (Figure 8b).

Further comparison between cultivars of low, medium, and high percentage dry matter[29] showed that low dry-matter cultivars contain high levels of glucose, fructose, and sucrose and only trace amounts of fructans whereas medium and, in particular, high dry-matter cultivars contain lower glucose and fructose concentrations, and substantial amounts of fructans (Figure 9). These differences are considered further under fructan synthesis (see Section IV).

C. SPECIES DIFFERENCES

Very little information is available for comparing the nonstructural carbohydrate content of different *Allium* species. Russian work[30] compared absolute amounts of "water-soluble polysaccharides" between a number of species, and some detail was offered for two species, namely, *A. longicuspis* and *A. coeruleum*.[31,32] Belval[34] mentions that Chevastelon[33] detected specific "sugar reserves" in shallot and onion and that these were similar to reserves from garlic that he classified as "garlic inulin". Belval[34] suggested that the fructoside of garlic is identical to the carbohydrate tuberose from the Amaryllidaceae, and he suggested the name "tuberoholoside" rather than "garlic inulin". More recently, Darbyshire and Henry[35] compared *A. cepa* (cv. Creamgold), *A. fistulosum, A. ampeloprasum v. porrum*, and *A. sativum*. Figure 10 shows clearly the difference between them in the molecular weights of the fructan polymers. Glucose, fructose, sucrose, and fructans were the only nonstructural

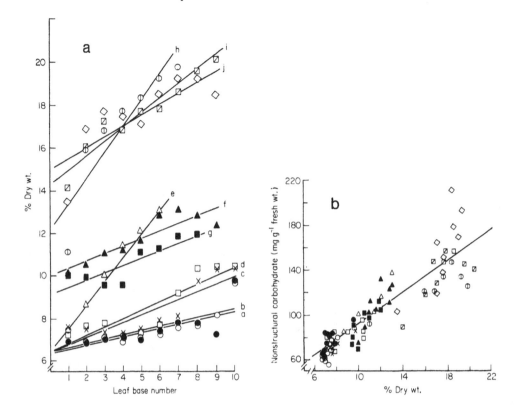

FIGURE 8.   (a) Change in dry-matter percentage of onion leaf bases from onion bulbs from the outer (1) to the inner ones (10); (b) relationship between nonstructural carbohydrates and dry weight % fresh weight. Cultivars: a (●) Dai-Maru; b (○) White Spanish Hybrid; c (×) Bronze Wonder; d (□) Valdez 3; e (△) Early Lockyer Brown; f (▲) Australian Brown; g (■) Lemon Skinned Creamgold; h ((∅) Dehyso; i (⊘) White Coreole × Southport White Globe; j (◇) Southport White Globe.[29]

carbohydrates found.[35] *A. cepa* and *A. ampeloprasum v. porrum* contained fructan polymers in the same range of lengths, but maximum concentrations occurred at DP 5 for *A. fistulosum* (Figure 11a), and at DP 12 for *A. ampeloprasum v. porrum* (Figure 11c). *A sativum* contained polymers of very high molecular weight (DP 50), and these were the most abundant (Figure 11b). The trisaccharides 1[F]- and 6[G]-fructosylsucrose were found in all the above species. Further consideration of species differences is in Section IV.

IV. FRUCTAN SYNTHESIS

Fructan synthesis in plants has been extensively and well reviewed.[12,15,36-42] In particular, the thesis of Henry[43] deals specifically with fructan synthesis in *Allium* species, with a brief comparison with fructans from wheat.

Earlier studies favored the involvement of nucleotide sugars in the biosynthesis of fructans,[44,45] but evidence now suggests synthesis from sucrose involving a number of glycosyl transferases.[15,37,41,46] The synthesis of β-(2-1)-linked fructans found in onion is effected by two enzymes. First, a sucrose: sucrose fructosyltransferase (SST) catalyzes the synthesis of the trisaccharide 1[F]-fructosylsucrose from sucrose. Sucrose acts as both a donor and an acceptor molecule, liberating free glucose.

$$G\text{-}F + G\text{-}F \rightarrow G\text{-}F\text{-}F + G \tag{1}$$
$$\text{sucrose sucrose trisaccharide glucose}$$

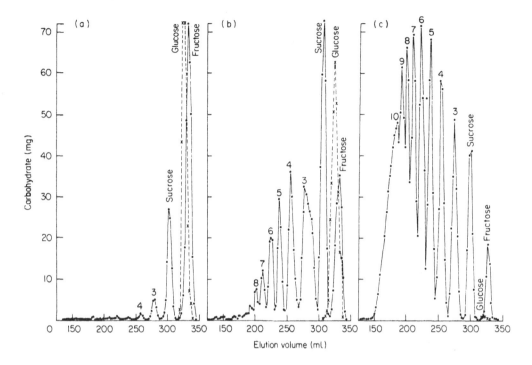

FIGURE 9. Separation of nonstructural carbohydrates extracted from bulbs of cultivars with (a) low (White Spanish Hybrid, 7.5%,) (b) medium (Australian Brown, 10.6%), and (c) high (White Creole × Southport White Globe, 17.2%) dry-matter percentage. Using a 1.6 × 200-cm column of Bio-Gel P-2 with a void volume of 150 ml; collecting 2.5-ml fractions; other conditions as in Figure 1. Numbers indicate degree of polymerization of fructans.[29]

Second, a fructan: fructan fructosyltransferase (FFT) catalyzes the synthesis of fructans of longer chain length from two molecules of trisaccharide yielding sucrose and a tetrasaccharide:[46]

$$\text{G-F-F} + \text{G-F-F} \rightarrow \text{G-F} + \text{G-F-F-F} \qquad (2)$$
$$\text{trisaccharide} \quad \text{trisaccharide} \quad \text{sucrose} \quad \text{DP4}$$

The second trisaccharide found in onions, 6^G-fructosylsucrose[5,9] is not synthesized by SST but by the action of FFT which facilitates self-transfer of fructose residues from the 6^F- to the 6^G-position.[46] A fructan hydrolase (2, 1-β-D-fructan fructanofuranosidase, E.C. 3.2.1.,[47]) has also been detected in onions.[27] While all SST, FFT, and fructan hydrolase activity is present in all parts of onion plants, most acitivty is detected in leaf bases.[27] The synthetic activities of SST and FFT increase from the outer to the inner leaf bases, while fructan hydrolase activity decreased from the outer to the center. This distribution is consistent with the suggestion that fructan synthesis predominates in the inner-leaf bases and fructan hydrolysis in the outer-leaf bases.

No information is available relating fructan synthesis to the bulbing process. However, reference was made earlier to the dramatic increases in fructan content, following bulb induction (Figures 7a and b). This increase in bulb fructan content was preceded by substantial increases in sucrose content (Figure 12). This relationship is consistent with an increase in SST and FFT activity starting 3 d after transfer to bulb-inducing conditions. Between 3 and 7 d there was a temporary halt to sucrose accumulation while fructans accumulated. After 9 d there was a massive increase in both, followed after 15 d by small net losses of both fructans and sucrose.

While *A. cepa* and *A. sativum* have radically different fructan complements (Figures 10

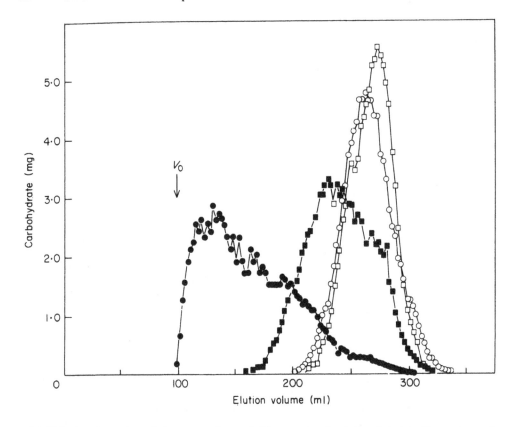

FIGURE 10.   Comparison of the content of several *Allium* species eluted through 2.6 × 45-cm column of Sephadex G-50. ● *A. sativum;* ■ *A. ampeloprasum v. porrum;* ○ *A. fistulosum;* □ *A. cepa vs.* Creamgold. The void volume (Vo) was measured with dextran blue, MW = 2,000,000; 2.5-ml fractions.[35]

and 11), the fructan:fructan fructosyl transferases isolated from these species are similar. Enzymes from both species transfer fructosyl residues from trisaccharides to form a tetrasaccharide and sucrose as the major products.[35] This transfer between trisaccharides predominates even in the presence of higher molecular weight acceptor molecules. Because of this Darbyshire and Henry[35] suggested that the supply of the substrate sucrose may control the ability of plants to synthesize fructan polymers. It was further suggested that the DP of polymers may be related to the total concentration of nonstructural carbohydrate. This was based on the finding that *A. sativum* which has much higher DP polymers than *A. cepa* contains much higher concentrations of nonstructural carbohydrate (252 mg g^{-1} fresh weight compared with 42 mg g^{-1} fresh weight (unpublished data). Furthermore, within *A. cepa*, cultivars of higher nonstructural carbohydrate concentration have a higher average DP (see Figure 9).

V. CONCLUDING REMARKS

Our knowledge of carbohydrate biochemistry in *Allium* species remains incomplete. Two approaches will rapidly add to our understanding: first, comparative studies of species of *Allium* and second, advances in understanding synthesis and function in general.

Recently, Hendry[48] assessed the ecological significance of fructans in plants. He emphasized a number of important points. First, geographic distribution is not related to whether or not plants contain fructans. Second, closely related species, or species within the same family, do not necessarily have similar fructan profiles.[35] Third, conditions favoring CO_2

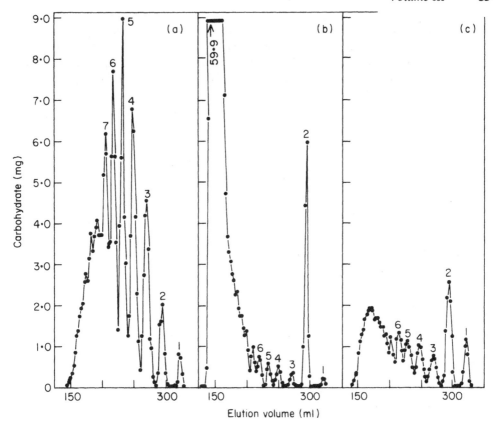

FIGURE 11.   Nonstructural carbohydrates extracted from (a) *A. fistulosum,* (b) *A. sativum,* and (c) *A. ampeloprasum v. porrum.* 1.6 × 210-cm column of Bio-Gel P-2 minus 400 mesh with void volume of 125 ml; 2-ml fractions; other conditions as in Figure 1. Numbers indicate the degree of polymerization of fructans.[35]

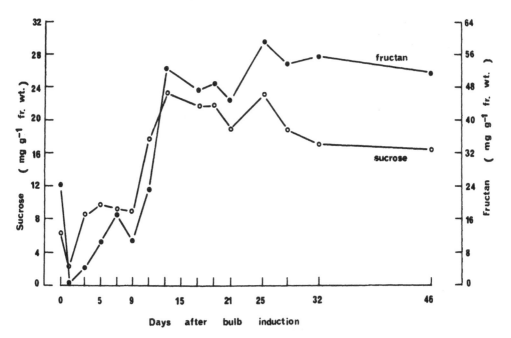

FIGURE 12.   Changes in sucrose (○) and total fructan (●) concentrations in leaf bases with time from transfer of onion plants from 10- to 16-h daylengths. (Darbyshire and Henry, unpublished.)

fixation tend to favor the accumulation of fructans. This is probably related primarily to the accumulation of sucrose which then promotes fructan accumulation. Finally, it is unlikely that fructans make a significant contribution to the tolerance of low temperature by plants. It is to be hoped that these ideas will stimulate further research with *Allium* species.

Additional comparisons between species and cultivars with respect to nonstructural carbohydrates are required. Relationships between the fructan content and the geographic origin of *Allium* species may suggest why *Allium* species contain different amounts of nonstructural carbohydrates although they have similar enzyme systems. Further understanding of the controlling role of sucrose accumulation in the synthesis of fructans is needed.

The bulbing process of onion requires further investigation. The interaction of light quality, photoperiod length, and photon flux density with the activities of SST, FFT, and fructan hydrolase is of particular interest. The activities of these enzymes and the flux rate of carbon through the carbohydrate pools should be measured in individual leaf bases from the stage of bulb initiation until maturity. It would be of interest to know why morphologically related species do not bulb and whether carbohydrate synthesis plays a part in these differences.

Nothing is known about how bulb reserves are utilized in the sprouting of bulbs or in the flowering process and in what way fructans are involved.

Our comments in this chapter are confined to aspects of carbohydrate biochemistry that we considered to be specifically related to *Allium*. We deliberately excluded the broader issues of plant carbohydrate biochemistry. This topic is a fascinating area that should involve research on both starch and fructan accumulators. It is an area that we recommend for attention because it will give both academic and applied benefits.

REFERENCES

1. **Loewus, F. A. and Tanner, W., Eds.,** *Plant Carbohydrates. I. Intracellular Carbohydrates,* Vol. 13A, Springer-Verlag, Berlin, 1982.
2. **Hawker, J. S.,** Sucrose, in *Biochemistry of Storage Carbohydrates in Green Plants,* Dey, P. M. and Dixon, R. A., Eds., Academic Press, London, 1985, chap. 1.
3. **Pontis, H. G.,** Riddle of sucrose, in *International Review of Biochemistry II,* Northcote, D. H., Ed., University Park Press, London, 1977, chap. 3.
4. **Bacon, J. S. D.,** The water soluble carbohydrates of the onion, *Allium cepa* L., *Biochem. J.,* 67, 5p, 1957.
5. **Bacon, J. S. D.,** The trisaccharide fraction of some monocotyledons, *Biochem. J.,* 73, 507, 1959.
6. **Bose, S. and Shrivastava, A. N.,** Soluble carbohydrates from onion (*Allium cepa,* Linn.), *Sci. Cult.,* 27, 253, 1961.
7. **Butt, A. M.,** Vegetative growth, morphogenesis and carbohydrate content of the onion plant as a function of light and temperature under field and controlled conditions, *Meded. Landbouwhogesch. Wageningen,* 68-10, 29, 1968.
8. **De Miniac, M.,** Application of gas chromatography to the study of *Allium cepa* (onion) bulb carbohydrates, *C. R. Acad. Sci. Paris,* 270(12), 1583, 1970.
9. **Darbyshire, B. and Henry, R. J.,** The distribution of fructans in onions, *New Phytol.,* 81, 29, 1978.
9a. **Rutherford, P. P. and Whittle, R.,** Methods of predicting long-term storage of onions, *J. Hortic. Sci.,* 59, 537, 1984.
10. **Steer, B. T. and Darbyshire, B.,** Some aspects of carbon metabolism and translocation in onions, *New Phytol.,* 82, 59, 1979.
11. **Bacon, J. S. D. and Edelman, J.,** The carbohydrates of the Jerusalem artichoke and other compositae, *Biochem. J.,* 48, 114, 1951.
12. **Archbold, H. K.,** Fructosans in the monocotyledons. A Review, *New Phytol.,* 39, 185, 1940.
13. **Challinar, S. W., Haworth, W. N., and Hirst, E. L.,** Carbohydrates of grass. Isolation of a polysaccharide of the levan type, *J. Chem. Soc.,* p. 1560, 1934.
14. **Palmer, A.,** The specific determination and detection of glucose as a probable constituent radical of certain fructosans by means of notation, *Biochem. J.,* 48, 389, 1951.

15. **Edelman, J. and Jefford, T. G.**, The mechanism of fructosan metabolism in higher plants as exemplified in *Helianthus tuberosus*, *New Phytol.*, 67, 517, 1968.
16. **Hammer, H.**, The trisaccharide fraction of some plants belonging to the Amaryllidaceae, *Acta Chem. Scand.*, 22, 197, 1968.
17. **Hammer, H.**, The trisaccharide fraction of the bulbs of some Liliaceous species, *Acta Chem. Scand.*, 23, 3268, 1969.
18. **Shiomi, N.**, Isolation and identification of 1-kestose and neokestose from onion bulbs, *J. Fac. Agric. Hokkaido Univ.*, 58, 548, 1978.
19. **Groteloeschen, R. D. and Smith, D.**, Carbohydrates in grasses. III. Estimations of the degree of polymerization of the fructosans in the stem bases of Timothy and Bromegrass near seed maturity, *Crop. Sci.*, 8, 210, 1968.
20. **Wilson, C., Oross, J. W., and Lucas, W. J.**, Sugar uptake into *Allium cepa* leaf tissue: an integrated approach, *Planta*, 164, 227, 1985.
21. **Steer, B. T.**, The effect of growth temperature on dry weight and carbohydrate content of Onion (*Allium cepa* L. cv. Creamgold) bulbs, *Aust. J. Agric. Res.*, 33, 559, 1982.
22. **Darbyshire, B., Henry, R. J., Melhuish, F. M., and Hewett, R. K.**, Diurnal variations in non-structural carbohydrates, leaf extension, and leaf cavity carbon dioxide concentrations in *Allium cepa* L., *J. Exp. Bot.*, 30, 109, 1979.
23. **Lercari, B.**, Bulb formation in the long day plant *Allium cepa* L.: the effect of simultaneous irradiations with red and far-red light, *Acta Hortic.*, 128, 55, 1981.
24. **Lercari, B.**, The effect of far-red light on the photoperiodic regulation of carbohydrate accumulation in *Allium cepa* L., *Physiol. Plant.*, 54, 475, 1982.
25. **Mondal, M. F., Brewster, J. L., Morris, G. E. L., and Butler, H. A.**, Bulb development in onion (*Allium cepa* L.). II. The influence of red:far-red spectral ratio and of photon flux density, *Ann. Bot. (London)*, 58, 197, 1986.
26. **Lercari, B.**, The promoting effect of far-red light on bulb formation in the long day plant *Allium cepa* L., *Plant Sci. Lett.*, 27, 243, 1982.
27. **Henry, R. J. and Darbyshire, B.**, The distribution of fructan metabolising enzymes in the onion plant, *Plant Sci. Lett.*, 14, 155, 1979.
28. **Lercari, B.**, Changes in invertase activities during the photoperiodically induced bulb formation of onion (*Allium cepa* L.), *Physiol. Plant.*, 54, 480, 1982.
29. **Darbyshire, B. and Henry, R. J.**, The association of fructans with high percentage dry weight in onion cultivars suitable for dehydrating, *J. Sci. Food Agric.*, 30, 1035, 1979.
30. **Khodzhaeva, M. A. and Ismailov, Z. F.**, *Allium* carbohydrates. I. Isolation and characterization of polysaccharides (transl.), *Khim. Prir. Soedin.*, 15, 114, 1979.
31. **Khodzhaeva, M. A. and Kondratenko, E. S.**, Carbohydrates of *Allium*. IV. Glucofructans in *Allium longicuspis* (transl.), *Khim. Prir. Soedin.*, 0(6), 703, 1984.
32. **Khodzhaeva, M. A. and Kondratenko, E. S.**, Carbohydrates of *Allium*. VIII. Polysaccharides in *Allium coeruleum* (transl.), *Khim. Prir. Soedin.*, 0(1), 17, 1985.
33. **Chevastelon, -,** Contribution a l'etude des hydrates de carbone contenus dans l'Ail, l'Echalotte et l'Oignon, *These Doct.*, Paris, 1894.
34. **Belval, H.**, The glucide reserves of the garlic and tuberose, *Bull. Soc. Chim. Biol.*, 21, 294, 1939.
35. **Darbyshire, B. and Henry, R. J.**, Differences in fructan content and synthesis in some *Allium* species, *New Phytol.*, 87, 249, 1981.
36. **Hirst, E. L.**, Some aspects of the chemistry of the fructosans, *Proc. Chem. Soc. London*, p. 193, 1957.
37. **Scott, R. W.**, *Transfructosylation in higher plants containing fructose polymers*, Ph.D. thesis, University of London, 1968.
38. **Smith, D.**, The non-structural carbohydrates, in *Chemistry and Biochemistry of Herbage*, Vol. 1, Butler, G. W. and Dailey, R. W., Eds., Academic Press, New York, 1973, 105.
39. **Meier, H. and Reid, J. S. G.**, Reserve polysaccharides other than starch in higher plants, in *Encyclopaedia of Plant Physiology, New Series*, Loewus, F. A. and Tanner, W., Eds., Springer-Verlag, Berlin, 1982, 418.
40. **Pontis, H. G. and Del Campillo, E.**, Fructans, in *Biochemistry of Storage Carbohydrates in Green Plants*, Dey, P. M. and Dixon, R. A., Eds., Academic Press, New York, 1985, 205.
41. **Pollock, C. J.**, Physiology and metabolism of sucrosyl-fructans, in *Storage Carbohydrates in Vascular Plants*, Lewis, D. H., Ed., Cambridge University Press, Cambridge, 1984, 97.
42. **Pollock, C. J.**, Fructans and the metabolism of sucrose in vascular plants, *New Phytol.*, 104, 1, 1986.
43. **Henry, R. J.**, *Fructan metabolism in Allium species*, M.Sc. thesis, Macquarie University, North Ryde, New South Wales, Australia, 1980.
44. **Pontis, H. G. and Fischer, C. L.**, Synthesis of D-fructopyranose 2-phosphate and D-fructofuranose 2-phosphate, *Biochem. J.*, 89, 452, 1963.

45. **Umemura, Y., Nakamura, M., and Funahashi, S.,** Isolation and characterisation of uridine diphosphate fructose from tubers of Jerusalem artichoke *(Helianthus tuberosus), Arch. Biochem. Biophys.,* 199, 240, 1967.
46. **Henry, R. J. and Darbyshire, B.,** Sucrose:sucrose fructosyltransferase and fructan:fructan fructosyltransferase from *Allium cepa, Phytochemistry,* 19, 1017, 1980.
47. **Edelman, J. and Jefford, T. G.,** The metabolism of fructose polymers in plants. IV. β-Fructofuranosidases of tubers of *Helianthus tuberosus* L., *Biochem. J.,* 93, 148, 1964.
48. **Hendry, G.,** The ecological significance of fructan in a contemporary flora, *New Phytol.,* 106 (Suppl), 201, 1987.

Chapter 2

CHEMICAL COMPOSITION

G. R. Fenwick and A. B. Hanley

TABLE OF CONTENTS

I. INTRODUCTION

Elsewhere in this volume are chapters dealing with flavor biochemistry and carbohydrates. These reflect two aspects of the composition of alliums being important in relation to their flavor and pharmacological effects and harvesting/processing, respectively. It is only comparatively recently that the overall chemical composition has been considered,[1] and, hence, this chapter will rely heavily on references contained therein. In Chapter 4 consideration will be given to the processing of alliums, mainly onion and garlic, and the effect of processing on composition will be discussed therein.

II. PROXIMATE COMPOSITION

The proximate composition of onions, garlic, and other alliums is shown in Table 1, data being assembled mainly from the U.K.,[2] North American,[3] Middle Eastern,[4] and East Asian[5] food tables. It should be emphasized that the composition of individual cultivars varies considerably.[9-11] For example, for three cultivars of Nigerian onion,[10] crude protein ranged from 0.49 to 0.79 g 100 g^{-1} fresh weight, crude fiber from 0.20 to 0.41 g 100 g^{-1}, and ash content from 0.25 to 0.41 g 100 g^{-1}. Among 20 differing varieties of garlic examined by Treutner et al.,[12] dry matter contents varied between 31 to 56 g 100 g^{-1}. Additional factors which affect gross composition include growing conditions, time of harvest, and length and nature of subsequent storage.

III. TRACE ELEMENTS AND VITAMINS

The trace element content of raw alliums is shown in Table 2. There are considerable differences between the mineral contents reported by Christensen et al.[13] and those included in later composition and dietary tables. Claims have been made for high selenium levels (37 to 101 μg 100 g^{-1} dry weight) in onion bulbs, but these have been disputed.[14] Garlic was found to contain 14 μg germanium 100 g^{-1}, making it a relatively rich source of this element.[15] In constrast, germanium was not detected in onions. Garlic has been found to contain 42 μg molybdenum 100 g^{-1} fresh weight, ranking below only beans as a dietary source of this mineral.[16] Roshania and Agrawal[17] have reported onions to be richest in vanadium (680 to 750 μg 100 g^{-1} fresh weight) among a variety of foodstuffs examined. Levels of Cr, Co, Ni, Mo, and Sn in onions were all below the detection limits of dry ashing atomic absorption (approximately 7, 9, 9, 28, and 56 μg 100 g^{-1}, respectively).[18] Among 20 varieties of garlic examined,[12] phosphorus levels varied between 300 to 620 mg 100 g^{-1}, and similar variations were found in levels of potassium (990 to 1910 mg 100 g^{-1}), calcium (40 to 230 mg 100 g^{-1}), copper (0.6 to 1.5 mg 100 g^{-1}), and manganese (0.6 to 1.7 mg 100 g^{-1}).

Vitamin contents are listed in Table 3. While vitamin C levels are highest in leafy vegetables, reference has recently been made to high levels (~150 mg 100 g^{-1} fresh weight) in the Asian species *Allium obliquum* and this has been suggested as an early spring food plant in the U.S.S.R.[19] Saleh et al.[20] have reported the contents of selected vitamins in immature bulb/green leaves of Egyptian onion and of kurrat (*A. ampeloprasum*) pseudostems. They are, respectively, carotene (7000, 4160 μg 100 g^{-1}), riboflavin (0.02, 0.02 mg 100 g^{-1}), nicotinic acid (0.2, 0.2 mg 100 g^{-1}), and ascorbic acid (67, 69 mg 100 g^{-1}). Yamauchi and Matsushita[21] have examined the tocopherol levels of onion, garlic, Welsh onion (*A. fistulosum*), and Chinese leek (*A. odorum*, nira). Total (α, β, γ, δ) tocopherol levels were 0.1 mg, 0.1 mg, 0.42 mg, and 0.81 mg 100 g^{-1} fresh weight, respectively. If per capita consumption was taken into account, onions ranked ninth among common U.S. vegetables as a dietary source of the ten major minerals and vitamins.[22]

TABLE 1
Proximate Composition (%) and the Energy Values (cal 100 g⁻¹) of some
Alliums and their Products

	Moisture	Protein	Fat	Carbohydrate	Ash	Energy	Ref.
Onion (*A. cepa*)	92.8	0.9	tr	5.2		23	2
	89.1	1.5	0.1	8.7	0.6	38	3
	88.6	1.6	0.2	9.0	0.6	38	5
Immature bulbs/tops	91.6	1.6	0.4	5.8	0.6	28	5
Green onion	86.8	0.9	tr	8.5		35	2
Bulb + top	89.4	1.5	0.2	8.2	0.7	36	3
Bulb + white top	87.6	1.1	0.2	10.5	0.6	45	3
White top	91.8	1.6	0.4	5.5	0.7	27	3
Welsh onion	90.5	1.9	0.4	6.5	0.7	34	3
(*A. fistulosum*)	91.4	1.6	0.3	6.1	0.6	30	5
Leaf	90.4	1.5	0.5	6.3	0.2	36	6
Bulb	85.4	1.1	0.2	12.3	0.6	57	6
Wild Welsh onion							
Leaf	93.1	1.9	0.3	3.5	0.7	25	6
Bulb	86.1	3.1	0.2	8.2	0.9	49	6
Garlic	61.3	6.2	0.2	30.8	1.5	137	3
(*A. sativum*)	63.8	5.3	0.2		1.4	140	4
	67.8	3.5	0.3	27.4	1.0	117	5
Leaves, stems	86.4	2.6	0.5	9.5	1.0	44	5
Shoots	77.7	1.2	0.3	20.1	0.7	76	5
Flowers	88.4	1.4	0.2	9.4	0.6	39	5
Great-headed garlic (*A. ampeloprasum*)	86.3	2.2	0.3	10.3	0.9	45	5
Wild garlic	87.9	3.3	0.4	7.5	0.9	41	5
Leek (*A. ampeloprasum*)	83.0	1.8	0.2	11.2	0.8	66	4
	85.4	2.2	0.3	6.0	0.9	52	3
	86.0	1.9	tr	5.0		31	2
	89.4	1.6	0.4		0.6	35	5
Blanched	94.8	1.9	0.3	2.6	0.4	16	5
Flowers	83.1	5.5	0.5	10.5	0.4	55	5
Chive	92.6	1.1	0.2		0.6	28	4
(*A. schoenoprasum*)	92.0	2.7	0.6	4.3	0.4	27	5
Chinese chives (*A. tuberosum*)							
Leaf	92.2	1.4	0.6	3.4	0.9	26	6
Bulb	84.6	2.6	0.2	7.6	0.7	30	6
Chinese leek (*A. odorum*)	88.4	3.0	0.4	6.8	1.4	34	5
Rakkyo (*A. chinense*, Syn. *A. bakeri*)		2.2	0.3	13.1	3.8		7
Suckochee (*A. carolinianum*)	86.4	2.3	0.7	6.4	1.7		8

IV. FATTY ACIDS AND LIPIDS

Table 1 shows alliums to contain <0.7 g 100 g⁻¹ fresh weight of fat. Abdel-Fattah and Edrees[23,24] have reported pigmented onion skins to contain tenfold more lipids than garlic skins (4 vs. 0.43%). Palmitic, oleic, and linoleic acids accounted for more than three quarters of the total fatty acid content of the Egyptian cultivars, the ratio of total unsaturated/saturated fatty acids being ~1.9.[25] Recently, Kamanna and Chandrasekhara[26] and Yang and Shin[27] have examined the lipid content and composition of garlic skins. Levels of 300 to 600 mg

TABLE 2
Elemental Contents of Alliums (mg 100 g^{-1})

	Ca	P	K	Na	Mg	Al	Ba	Fe	Ref.
Onion (*A. cepa*)	190—540	200—430	80—110	31—50	81—150	0.5—1	0.1—1	1.8—2.6	13
	31	30	140	10	8			0.3	2
	27	36	157	10				0.5	3
	30	44	166	9				1.0	5
Immature bulb & tops	43	36	178	4				1.2	5
Welsh onion (*A. fistulosum*)	18	49							3
	55	41	192	15				1.1	5
Spring onion (*A. cepa*) whole	51	39	231	5				1	3
	140	24	231	5	11			1.2	5
Garlic (*A. sativum*)	50—90	390—460	100—120	10.22	43—77	0.5—1	0.2—1	2.8—3.9	13
	29	202	529	19				1.5	3
	2.7		411	8.5	17.8			1.9	2
	18	88							
Leaves, stems	58	76	326	4				0.6	5
Shoots	12	52	273					1.7	5
Flowers	25	46						0.9	5
Great-headed garlic (*A. ampeloprasum*)	52	50						1.1	5
Wild garlic	169	64						2.2	5
Leek (*A. ampeloprasum*)	63	43	310	9	10			1.1	2
	58	48	316	5				2.7	5
	52	50	347	5				1.1	3
Blanched leek	16	20	62					0.6	5
Chinese leek (*A. odorum*)	59	66	234	6				2.6	5
Chive (*A. schoenoprasum*)	83	41	373	18				0.8	5
Suckochee (*A. carolinianum*)	18	27						40	8

	Sr	B	Cu	Zn	Mn	Cr	S	Cl	Ref.
Onion (*A. cepa*)	0.8—7	0.6—1	0.05—0.64	1.5—2.8	0.5—1	<0.5			13
			0.08	0.1			51		2
Spring onion (*A. cepa*)			0.13				50	36	3
Garlic (*A. sativum*)	0.1—0.7	0.3—0.6	0.02—0.03	1.8—3.1	0.2—0.6	0.3—0.5			13
			0.17	3.9		0.02			3
Wild garlic								43	5

100 g^{-1} fresh weight were obtained, neutral lipid predominating over phospholipid and glycolipid. The main fatty acids in leek parenchymatous tissue were palmitic, linoleic, and linolenic acids.[1]

V. AMINO ACIDS

The free amino acid contents of onions vary considerably (Table 4), with both arginine and glutamic acid being abundant. Schuphan and Schwerdtfeger[29] have suggested that these amino acids function as nitrogen sources. During the growth and physiological development of the onion, arginine and glutamic acid contents were increased by 29 and 7%, respectively; in contrast, all other free amino acids were relatively little changed. Within the bulb, highest

TABLE 3

Vitamin Content of Alliums (100 g^{-1})

	Retinol (µg)	Vit E (mg)	Carotene (µg)	Vit B$_6$ (mg)	Vit B$_{12}$ (mg)	Vit. D (µg)	Thiamin (mg)	Folic acid (µg)	Riboflavin (mg)	Biotin (µg)	Nicotinic acid (mg)	Ref.
Onion (A. cepa)	0	tr	0	0.1	0	0	0.3	16	0.05	0.9	0.2	2
							0.3		0.04		0.2	3
			tr				0.06		0.04		0.2	3
Immature, bulbs and tops			890				0.06		0.11		0.5	5
Welsh onion (A. fistulosum)	0		0	0.22	0	0	0.05	20.7	0.09		0.4	3
			630				0.06		0.08		0.5	5
Spring onion (A. cepa)	0		tr			0	0.03		0.05		0.2	2
							0.05		0.04		0.4	3
Garlic (A. sativum)			tr				0.25		0.08		0.5	2
							0.24		0.05		0.4	3
Leaves, stems			920				0.11		0.14		0.6	5
Shoots			200				0.14		0.06		0.5	5
Flowers			60				0.11		0.06		0.4	5
Great-headed garlic (A. ampeloprasum)			25				0.11					5
Wild garlic			486				0.06		0.10		56	5
Leek (A. ampeloprasum)	0		40		0		0.1		0.05		0.6	2
			2435				0.09		0.10		0.6	5
							0.11		0.06		0.5	4
Blanched leek			15				0.04		0.06		0.08	3
Flowers			2550				0.14		0.19		0.9	5
Chinese leek (A. odorum)			1020				0.8		0.15		0.9	5
Chive (A. schoenoprasum)							0.10		0.06		0.5	5

	Vit C (mg)	Pantothenic acid (mg)	Ref.
Onion (A. cepa)	10	0.14	2
	10		3
	9		5
Immature bulbs and tops	29		
Welsh onion (A. fistulosum)	27		3
	19	0.17	5

TABLE 3 (continued)
Vitamin Content of Alliums (100 g^{-1})

	Retinol (µg)	Carotene (µg)	Vit. D (µg)	Thiamin (mg)	Riboflavin (mg)	Nicotinic acid (mg)	Ref.
Spring onion (A. cepa)	25	0.1	0	40	0.9	0.14	2
Garlic (A. sativum)	15	tr					3
	10						5
Leaves, stems	39	0.96		6.2			5
Shoots	42	0.2	0	123			5
Flowers	44						5
Great-headed garlic (A. ampeloprasum)	17						5
Wild garlic	11						5
Leek (A. ampeloprasum)	15—30	0.8	0	—	1.4	0.12	2
	32	0.25	0	58		0.12	5
	17	0.15					
Blanched leek	12						3
Flowers	40						5
Chinese leek (A. odorum)	36						5
Chive (A. schoenoprasum)	32						5

TABLE 4
Free Amino Acid Composition (mg 100 g^{-1}) of Onion Bulbs

	Onion (A. cepa)			Welsh onion (A. fistulosum)
Isoleucine	82[a]	2.5[a]	1.9—13.1[b]	37[c]
Leucine	74	7.9	1.9—15.9	62
Lysine	67	10.5	4.2—18.8	121
Methionine	8	0.5	<1.1	35
Cystine				24
Phenylalanine	36	8.9	2.4—10.6	46
Tyrosine		16.2	2.6—6.5	22
Threonine	22	154		39
Tryptophan	10	tr	0.8—3.6	15
Valine	38	6.5	1.7—7.6	31
Arginine		144.2	18—68	115
Histidine		11.6	1.1—8.1	19
Alanine		6.1	1.9—3.8	78
Aspartic acid				83
Glutamic acid	}	391	18.6—24.9	300
Glycine		—	1.1—2.2	70
Proline		2.8	<0.8	45
Serine		16.6		40

[a]   Reference 10.
[b]   Reference 28.
[c]   Reference 5.

arginine contents were measured in the inner scales surrounding the bud, outer scales and buds having 27 and 35% less arginine, respectively. Sulfur-containing compounds are important in alliums, but the concentrations of L-cysteine, L-cystine, and L-methionine are all relatively low, indicative of their rapid metabolism. A range of other involatile sulfur compounds are found in onion (Table 5). The S-alk(en)yl cysteines are readily oxidized to the corresponding (+)-sulfoxide,[30] the involatile precursors (alliins) of the flavoring and pharmacologically active substances (allicins) of onions and other alliums.

A wide range of γ-glutamyl derivatives of amino acids, predominantly those containing sulfur or their derivatives (Table 6) have been reported. Of these 14 have been identified in onion, 11 in chives, with many fewer in garlic.[31,32] It is considered that these compounds, which are present in dormant seeds and resting bulbs, exhibit a storage function.

Whitaker[32] has recorded levels of γ-glutamyl derivatives of L-methionine, S-methyl-L-cysteine and trans-S-(1-propenyl)-L-cysteine sulfoxide of 5 to 13 mg, 5 to 19 mg, and 130 to 200 mg 100 g^{-1} fresh weight, respectively. Such compounds are rapidly hydrolyzed during germination or sprouting; γ-glutamyl transpeptidase has been found in germinating chives and may also be present in sprouting onions.

Discussion of other sulfur compounds in alliums is presented elsewhere in this volume (see chapter on "Flavor Biochemistry").

VI. CARBOHYDRATES

According to Bajaj et al.[11] the reducing sugar contents of 12 varieties of onion ranged from 12 to 22 g 100 g^{-1} dry weight, while nonreducing sugars varied between 25 to 62 g 100 g^{-1}. No differences were noted between red and white onions. In general, such classification of sugar content is of limited usefulness. Nonstructural carbohydrate accounts for

TABLE 5
Involatile Sulfur Compounds of Onion[1]

L-Cysteine	*trans*-*S*-(1-propenyl)-L-cysteine
L-Cystine	*S*-(Carboxymethyl)-L-cysteine
L-Methionine	*S*-(Carboxyethyl-L-cysteine
L-Methionine sulfoxide	*S*-(Carboxypropyl)-L-cysteine
S-Methyl-L-cysteine	*S*-(Carboxy*iso*propyl)-L-cysteine
S-Propyl-L-cysteine	*S*-(2-Propenyl)-L-cysteine

TABLE 6
γ-Glutamyl Peptides Isolated from Alliums[1]

Valine	Onion
Isoleucine	Onion, chive
Leucine	Onion
Phenylalanine	Onion
Tyrosine	Onion
Cysteine	Chive
Methionine	Onion, chive
S-Methylcysteine	Onion, chive
S-Methylcysteine sulfoxide	Onion, garlic (?)
S-Propylcysteine	Garlic, chive
S-(2-Propenyl)cysteine	Garlic, chive
S-(1-Propenyl)cysteine	Chive
S-(1-Propenyl)cysteine sulfoxide	Onion, chive
S-(1-Propenyl)cysteine-*S*-(1-propenyl)-cysteine sulfoxide	Chive
S-(2-Carboxypropyl)cysteinylglycine	Onion, garlic
Glutathione	Onion
Glutathione cysteine disulfide	Onion
Glutathione glutamylcysteine disulfide	Onion
S-Sulfoglutathione	Onion
N,N'-*bis*(-Glutamyl)cysteine	Chive
N,N' bis(-Glutamyl)3,3' (2-methylethylene-1,2-dithio)-dialanine	Chive

the major portion of the dry weight of onions and comprises free sugar, trisaccharides, and polysaccharides (fructans). These are covered elsewhere in this volume (see chapter on "Carbohydrate Biochemistry").

Onion and garlic skins, and probably those of other alliums, are good sources of pectin. Pectins have been isolated in 33% yield from the skins of two Egyptian onion varieties[23] and in slightly lower yield (27%) from garlic skins by Abdel-Fattah and Edrees,[24] while Alexander and Sulebele[33] isolated pectins in 12, 9.8, and 7.8% yield from the skins of white and red onions and garlic, respectively. Details of the properties of these pectins are shown in Table 7. According to Sen and co-workers[34,35] onion pectin contained galactose (30%), arabinose (18%), and galacturonic acid (45%), and consisted of 1→3 linked galactopyranose units linked to hexose residues at C-1, C-4, and C-6. The average repeating unit was considered to be ~10 residues. Recently, Redgwell and Selvendran[36] have reported the results of an exhaustive investigation into the structural features of the cell-wall polysaccharides of the onion. Pectic substances from the primary cell walls (as opposed to the middle lamellae) contained more highly branched rhamnogalacturonan backbones. The isolated rhamnogalacturonans contained mainly (1→4)-linked galactose branches, with smaller amounts of (1→4, 1→6)- and (1→2, 1→6)-linked galactose, (1→2)-linked galactose, and (1→5)-linked arabinose. Most of the branched residues were terminated by galactopyranosyl and arabinofuranosyl residues.

In contrast, Abdel-Fattah and Edrees[24] considered garlic-skin pectin to comprise (1→3)-linked galacturonic residues with little or no cross-linking.

TABLE 7
Properties of Pectic Substances from Onion and Garlic Skins[33]

Characteristic	Red onion	White onion	Garlic
Moisture (%)	12.4	9.0	10.2
Ash (%)	1.8	1.4	1.2
Jelly grade	200	235	210
Setting temp. (C)	70—80	70—80	30—40
Setting time (min)	1—2	1—2	13—15
Equiv. weight	758	568	470
Methoxyl content (%)	8.6	8.3	7.2
Esterification (%)	60.2	59.7	51.3
Intrinsic viscosity (CP)	2.60	4.15	2.47
Mol. weight	5.55×10^4	8.8×10^4	5.25×10^4
Viscosity of 0.5% soln. (CP)	3.18	5.61	11.42

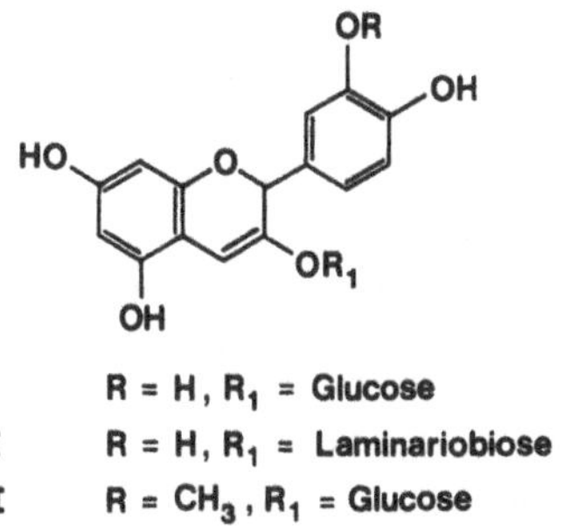

I	R = H, R_1 = Glucose
II	R = H, R_1 = Laminariobiose
III	R = CH_3, R_1 = Glucose

FIGURE 1. Anthocyanins found in red onion cultivars. The full names of the compounds denoted by the
roman numerals are given in the text where they are
also denoted by the same numerals.

VII. ANTHOCYANINS

Despite the fact that the anthocyanins of red onions were first investigated more than
50 years ago, there is still relatively little known about their composition. Fuleki[37] has
separated eight anthocyanins from extracts of cvs. Ruby and Southport Red Globe. Seven
of these compounds contained cyanidin as the aglycone,[38,39] two being characterized as
cyanidin-3-glucoside (I) and cyanidin-3-diglucoside (later shown to be cyanidin-3-lamina-
riobioside, II,[40] Figure 1). The remaining compound was identified as peonidin-3-glucoside
(III). Du and Francis[41] have separated seven pigments from an extract of garlic, one of
which was positively identified as cyanidin-3-glucoside. Of the remaining major components,
both yielded cyanidin and glucose, and the authors suggested the possibility that acyl groups
were also present.

VIII. FLAVONOLS

Quercetin (IV, Figures 2A and B) was first isolated from the yellow outer skins of dry
onions at the end of the last century and has subsequently been obtained from leaves, peelings,
or processed onions.[1] The compound is found in the free form in the outer skins but is bound
to sugar residues in the epidermal tissues.[42] In contrast to the presence of quercetin in the
skins of colored onions (2.5 to 6.5 g 100 g^{-1} dry weight), only traces have been found ($\sim$1
mg 100 g^{-1}) in the skins of white onion varieties.[43] Kiviranta et al.[44] have found the total

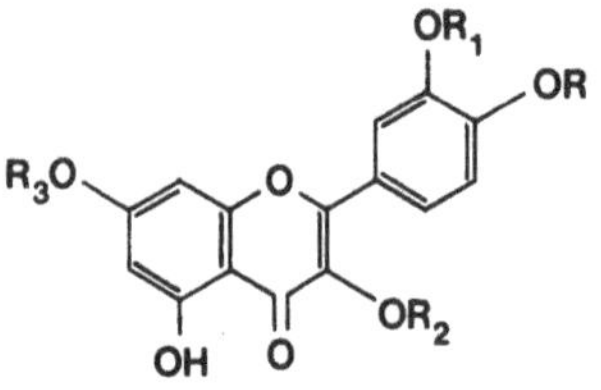

	R	R₁	R₂	R₃
IV	H	H	H	H
VIII	Glucose	H	H	H
IX	H	H	H	Glucose
X	H	H	Glucose	Glucose
XI	H	H	Glucose	H
XII	Glucose	H	Glucose	H
XIII	Glucose	H	H	Glucose
XVI	H	CH₃	Glucose	H

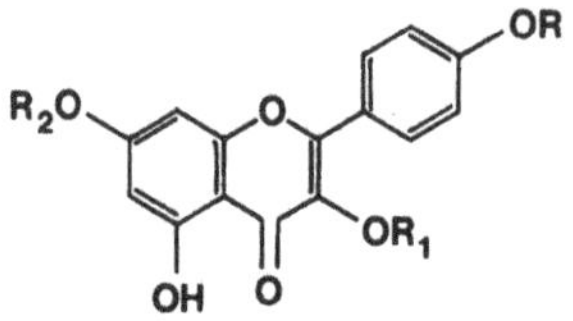

	R	R₁	R₂
V	H	H	H
VI	Glucose	H	H
VII	Glucose	H	Glucose
XIV	H	Glucose	H
XV	H	Xylosylglucose	H
XVII	H	Rhamnosylgalactose	Rhamnose

FIGURE 2A and B.   Flavonols found in edible alliums. The full names of the compounds denoted by the roman numerals are given in the text where they are also denoted by the same numerals.

content of flavonols in the edible portion of a red onion to be double that of a yellow variety. Recent work by Bilyk et al.[45] has identified variations in the distribution of quercetin and kaempferol (V) in the edible portions of eight cultivars. Chive, leek, and garlic chive contained greater amounts of kaempferol than quercetin,[46] with the two former species containing higher levels of flavonols in the green leaves than in the white leaf portions. In contrast, garlic chive contained 1 and 2 mg flavonols per 100 g fresh weight in the green and white parts, respectively. Panisset and Tissut[47] have found kaempferol and its glycosides in all eight onion cultivars examined and subsequently identified kaempferol-4'-glucoside (VI) and kaempferol-7,4'-diglucoside (VII). The same authors also confirmed the presence of spiraeoside (quercetin-4'-glucoside, VIII), first found by Herrmann[48] and thereafter by Koeppin and Vander Spuy[49] as well as quercetin-7-glucoside (IX) and -3,7-diglucoside (X). Brandwein[50] found quercetin-3-glucoside (XI) in Southport Red, Yellow, and White Globe onions and, independently of Harborne,[51] reported two additional quercetin glycosides, the 3,4'- and 7,4'-diglucosides (XII and XIII, respectively), the presence of which had been strongly suggested earlier by Herrmann.[48,52]

According to French workers,[53] quercetin its 4'-glucoside, and 3,4'- and 7,4'-diglucosides are all found in shallots while quercetin and kaempferol glycosides including quercetin-3-glycoside, kaempferol-3-glycoside (XIV), and -3-xylosyl-3-glucoside (XV), but apparently not spiraeoside, occur in leeks. Bruising of leek leaf tissue causes a significant change in flavonol composition with a reduction in quercetin glycosides and the appearance of free and bound forms of kaempferol. While mono- and diglycosides predominated in leek, in chive leaves it is the di- and triglycosides which are more important; of the monoglycosides present the 3-glucosides of kaempferol, quercetin, and isorhamnetin (XVI) have all been identified.[54] Traces of quercetin and kaempferol glycosides were reported in garlic bulbs and in addition of kaempferol-3-0-rhamnosylgalactoside-7-0-rhamnoside (XVIII),[55] Yoshida et al.[56] have recently identified kaempferol-3-0-sophoroside, -3,4'-diglucoside, -3-(2-0-feruloyl) glucoside-7,4'-diglucoside, -3,4'-diglucoside-7-(2-0-feruloyl) glucoside, -3-sulforoside-7-(2-0-feruloyl) glucoside, and quercetin-3,4'-glucoside in *A. tuberosum*.

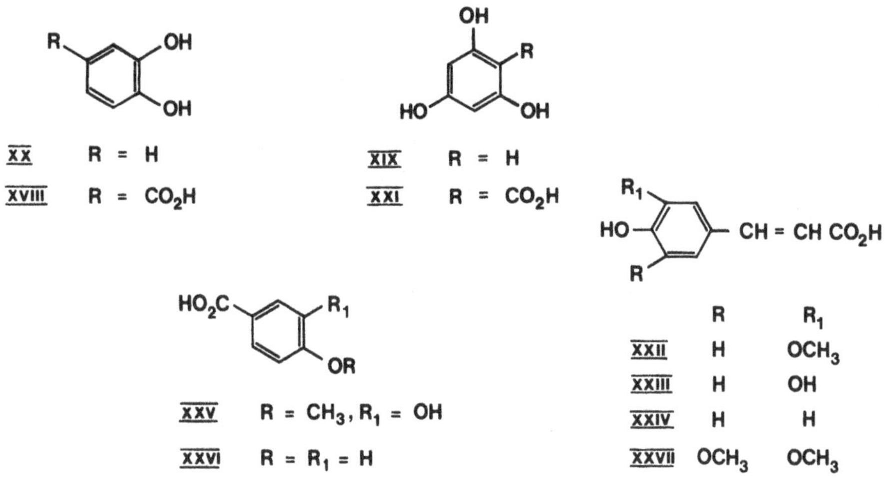

FIGURE 3A and B.   Phenolic compounds found in edible alliums. The full names of the compounds denoted by the roman numerals are given in the text where they are also denoted by the same numerals.

IX. PHENOLICS

Extracts of the colored outer skins of onions have been shown to contain protocatechuic acid (XVIII, Figure 3), phloroglucinol (XIX), pyrocatechol (XX), and methyl esters of protocatechuic acid and phloroglucinol carboxylic acid (XXI).[43,52] Fleshy onion scales contained protocatechuic acid, phloroglucinol, pyrocatechol, and ferulic acid (XXII), while caffeic acid (XXIII) and esterified forms of *p*-coumaric (XXIV) and ferulic acids were found in the leaves. According to Schmidtlein and Herrmann[57] white onions (cv. Weisse Frühlingszwiebeln) contained ~4 mg protocatechuic acid per 100 g dry weight in the outer scales, with only traces 50 μg 100 g^{-1} being found in the remaining tissues. In contrast colored onions contained 1 to 2 g protocatechuic acid per 100 g. The dry, outer skins of cvs. Braunschweiger-Dunkelblot, Stuttgarten Riesen, and Weisse Frühlingszwiebeln contained 26, 65, and 9 mg vanillic acid (XXV) per 100 g and 11, 7, and 4 mg *p*-hydroxybenzoic acid (XXVI) per 100 g, respectively. In agreement with the above, higher levels of phenolic acids were found in the inner scales of colored onions (300 μg 100 g^{-1}) than in white onions, although the difference was less than tenfold. Leeks contained ferulic acid (1 mg 100 g^{-1} fresh weight), methyl caffeate (200 to 700 μg 100 g^{-1}), caffeic acid (300 μg 100 g^{-1}), and traces of both *p*-coumaric acid and sinapic acids (XXVII.)[43]

X. STEROLS AND SAPONINS

Eichenberger and Grob[58] have found onion leaves to contain ~160 mg free sterols and sterol esters per 100 g dry weight, ~100 mg steryl glycosides per 100 g, and ~50 mg acylated steryl glycosides per 100 g. Much lower levels were found in the onion bulb, 30, 10, 10 mg 100 g^{-1} dry weight, respectively. Itoh et al.[59] have identified 15 sterols in onion bulbs (total content 262 mg 100 g^{-1} dry weight) and 16 sterols (72 mg 100 g^{-1}) in shallots, with β-sitosterol predominant in both cases.

Welsh onion (*A. fistulosum*) greens, Chinese garlic, *A. bakeri*, and garlic all contained cholesterol, campesterol, and β-sitosterol. In addition, garlic and Welsh onion greens contained stigmasterol, Chinese garlic (possibly) brassicasterol, and *A. bakeri* both these sterols.[60-62] The concentrations of all these compounds decreased from the outer scales inward. Smoczkiewicz et al.[63] have examined onion, garlic, and leek for saponins and steryl glycosides and found combined levels of 95, 21, and 100 mg 100 g^{-1} fresh weight. Onions contained bound β-sitosterol, oleanolic acid, and amyrin; garlic contained glycosides of β-sitosterol; and leek possessed oleanolic acid- and gitogen-containing saponins. Recently, Kravets et al.[64,65] have reported two new steroid saponins from the onion and Harmatha et al.[66] have isolated aginosid, a steroidal saponin from leek leaves (30 mg 100 g^{-1} fresh weight). This compound, present at much greater concentrations in the flowers, has been shown to possess growth inhibitory activity against the larvae of the leek moth (*Acrolepiopsis assectella*). Russian workers[1] have examined a number of wild alliums from Central Asia for saponin composition. A number of steroidal saponins have been isolated, including some possessing aglycones (e.g., diosgenin) of importance as raw materials in the pharmaceutical industry.

XI. OTHERS

Erazo and Concha[67] have reported the presence of oxalic, citric, malic, and pyruvic acids in *A. cepa* cv. Valencia Sintetica 14, in agreement with the earlier findings of Russian workers,[68] who additionally reported succinic acid. In common with many other species, garlic, onion, and other alliums possess well-defined antioxidant activity.[69,70] The active principles were considered to be *S*-alk(en)yl cysteine sulfoxides[70] and quercetin and its flavone aglycone analogues.[71] Tannins isolated from the skins of red onions[72,73] also exhibit antioxidant properties. The presence of prostaglandin A_1 or a compound possessing similar blood pressure-reducing and chemical properties was first reported by Attrep et al.[74] While Panosyan[75] could not confirm these findings, the American workers later positively identified this compound in extracts of yellow onions, although they were unable to exclude the possibility of its being formed from prostaglandin E_1 during isolation and purification.[76] Subsequently, Al-Nagdy et al.[77] have reported the presence of four prostaglandins ($F_{1\alpha}$, E_1, B_1, and A_2) in green onions. Claeys et al.[78] isolated lipoxygenase metabolites of linoleic acid, notably 9,10,13-trihydroxy-11-octadecenoic- and 9,12,13-trihydroxy-10-octadecenoic acids, from onion bulbs and found these to have a similar biological effect to prostaglandin E. Garlic contains a group of structurally related, biologically active thioglycosides, scordinins, which have been used as tonics in the orient.[79,80]

ACKNOWLEDGMENT

The patience and persistence of Mrs. D. Davies in typing this chapter is gratefully acknowledged.

REFERENCES

1. **Fenwick, G. R. and Hanley, A. B.**, The genus *Allium*. II, *CRC Crit. Rev. Food Sci. Nutr.*, 23, 273, 1985.
2. **Widdowson, E. M. and McCance, R. A.**, *The Composition of Foods*, 3rd ed., Special Report, Ser. No. 297, Her Majesty's Stationery Office, London, 1960.
3. **Watt, B. K. and Merrill, A. L.**, *Composition of Foods: Raw, Processed, Prepared*, Agricultural Handbook No. 8, U.S. Department of Agriculture, Washington, D.C., 1963.
4. **Pellet, P. L. and Shadarevian, S.**, *Food Composition Tables for Use in the Middle East*, 2nd ed., American University, Beirut, 1970.
5. Food Composition Tables for Use in East Asia, Food and Agriculture Organization, Rome, 1972.
6. **Hahn, S.-J.-K., Sang, S.-D., and Chae, G.**, Agronomic characteristics and food value of wild Welsh onion (*Allium schoenoprasum* L.) in Korea, *J. Korean Soc. Hortic. Sci.*, 18, 40, 1977.
7. **Mann, L. K. and Stearn, W. T.**, Rakkyo or ch'iao t'ou (*Allium chinense* G. Don, Syn. *A. bakeri* Regel), a little known vegetable crop, *Econ. Bot.*, 14, 69, 1960.

8. **Katiyar, S. K., Kumar, N., Bhatia, A. K., and Atal, C. K.**, Nutritional quality of edible leaves of some wild plants of Himalayas and culinary practices adopted for their processing, *J. Food Sci. Technol.*, 22, 438, 1985.
9. **Qureshi, E. I., Chaudry, M. I., and Malik, M. A.**, Chemical composition and nutritive value of some varieties of onion, *Pak. J. Med. Res.*, 7, 155, 1968.
10. **Ketiku, A. O.**, The chemical composition of Nigerian onions (*Allium cepa* Linn.), *Food Chem.*, 1, 41, 1976.
11. **Bajaj, K. L., Kaur, G., Singh, J., and Gill, S. P. S.**, Chemical evaluation of some important varieties of onion (*Allium cepa* L.), *Qual. Plant. Plant Foods Hum. Nutr.*, 30, 117, 1980.
12. **Treutner, R., Jankovsky, M., and Hubacek, J.**, The content of some substances in selected varieties of garlic (*Allium sativum* L.) *Rostl. Vyroba*, 24, 1003, 1978.
13. **Christensen, R. E., Beckman, R. M., and Birdsall, J. J.**, Some mineral elements of commercial spices and herbs as determined by direct reading emission spectroscopy, *J. Assoc. Off. Anal. Chem.*, 51, 1003, 1968.
14. **Noda, T., Taniguchi, H., Suzuki, A., and Hirai, S.**, Comparison of the selenium contents of the genus *Allium* measured by fluorimetry and neutron activation analysis, *Agric. Biol. Chem.*, 47, 613, 1983.
15. **Ninomiya, T., Okada, T., and Hosogai, Y.**, Germanium contents in foods, *Shokuhin Eiseigaku Zasshi*, 17, 481, 1976.
16. **Lupea, V. and Vranceanu, C.**, Beitrag zur Untersuchung des Molybdangehaltes einiger Lebensmittel and Gebieten mit endemischer Nephritis, *Nahrung*, 16, 637, 1972.
17. **Roshania, R. D. and Agrawal, Y. K.**, *N-p*-Chlorophenyl substituted hydroxamic acids and analytical reagents for the trace determination of vanadium (V), *Chem. Anal. (Warsaw)*, 26, 191, 1981.
18. **Lopez, A. and Williams, H. L.**, Essential elements in fresh and canned onions, *J. Food Sci.*, 44, 887, 1979.
19. **Kucherov, E. V. and Khairetdinov, S. S.**, *Allium obliquum* productivity and its ascorbic acid content, *Rastit. Resur.*, 19, 185, 1983; *Chem. Abstr.*, 99, 19658m, 1983.
20. **Saleh, N., El-Hawary, Z., El-Shokabi, F. A., Abbassy, M., and Morcos, S. R.**, Vitamin content of fruits and vegetables in common use in Egypt, *Z. Ernaehrungswiss.*, 16, 158, 1977.
21. **Yamauchi, R. and Matsushita, S.**, Tocopherol levels of vegetables and fruits, *Nippon Nogei Kagaku Kaishi*, 50, 569, 1976.
22. **Rich, R. C.**, The tomato, *Sci. A.*, 239, 66, 1978.
23. **Abdel-Fattah, A. F. and Edrees, M.**, Chemical investigations on some constituents of pigmented onion skins, *J. Sci. Food Agric.*, 22, 298, 1971.
24. **Abdel-Fattah, A. F. and Edrees, M.**, A study on the composition of garlic skins and the structural features of the isolated pectic acid, *J. Sci. Food Agric.*, 23, 871, 1972.
25. **Farag, R. S., Shabana, M. K., and Shallam, H. A.**, Biochemical studies on some chemical characteristics of sliced Egyptian onions, *Fette, Seifen, Anstrichm.*, 82, 233, 1981.
26. **Kamanna, V. S. and Chandrasekhara, N.**, Fatty acid composition of garlic (*Allium sativum* Linnaeus) lipids, *J. Am. Chem. Soc.*, 57, 175, 1980.
27. **Yang, K.-Y. and Shin, H.-S.**, Lipids and fatty acid composition of garlic (*Allium sativum* Linnaeus), *Korean J. Food Sci. Technol.*, 14, 388, 1982.
28. **Matikkala, E. J. and Virtanen, A. I.**, On the quantitative determination of the amino acids and gamma glutamyl peptides of onions, *Acta Chem. Scand.*, 21, 2891, 1967.
29. **Schuphan, W. and Schwerdtfeger, E.**, Biochemistry of the vegetative development of *Allium cepa* with special emphasis upon protein and amino acid pattern. The role of arginine as an N-pool, *Qual. Plant. Plant Foods Hum. Nutr.*, 21, 141, 1972.
30. **Granroth, B.**, Biosynthesis and decomposition of cysteine derivatives in onion and other *Allium* species, *Ann. Acad. Sci. Fenn.* Ser. A, 154,(11), 9, 1970.
31. **Virtanen, A. I.**, Studies on organic sulphur compounds and other labile substances in plants—a review, *Phytochemistry*, 4, 207, 1965.
32. **Whitaker, J. R.**, Development of flavour, odour and pungency in onion and garlic, *Adv. Food Res.*, 22, 73, 1976.
33. **Alexander, M. M. and Sulebele, G. A.**, Pectic substances in onion and garlic skins, *J. Sci. Food Agric.*, 24, 611, 1973.
34. **Sen, S. K. and Rao, C. V. N.**, Studies on pectic substances in onion (*Allium cepa* Linn.), *Indian J. Appl. Chem.*, 29, 127, 1966.
35. **Sen, S. K., Chatterjee, B. P., and Rao, C. V. N.**, A galactan from onion (*Allium cepa* Linn.), pectic substance, *J. Chem. Soc. (Sect. C)* 1788, 1971.
36. **Redgwell, R. J. and Selvendran, R. R.**, Structural features of cell-wall polysaccharides of onion, *Allium cepa, Carbohydr. Res.*, 157, 183, 1986.
37. **Fuleki, T.**, The anthocyanins of strawberry, rhubarb, radish and onion, *J. Food Sci.*, 34, 365, 1969.
38. **Fuleki, T.**, Pigments responsible for the colour of red onion, *Rep. Hortic. Res. Inst. Ont.*, 124, 1969.

39. **Fuleki, T.**, Anthocyanins in red onion, *Allium cepa, J. Food Sci.*, 36, 101, 1971.

40. **Du, C. T., Wang, P. L., and Francis, F. J.**, Cyanidin-3-laminariobioside in Spanish red onion (*Allium cepa* L.)., *J. Food Sci.*, 39, 1265, 1974.

41. **Du, C. T. and Francis, F. J.**, Anthocyanins of garlic (*Allium sativum* L.), *J. Food Sci.*, 40, 1101, 1975.

42. **Starke, H. and Herrmann, K.**, Flavonols and flavones of vegetables. VI. On the changes of the flavonols of onions, *Z. Lebensm. Unters. Forsch.*, 161, 137, 1976.

43. **Herrmann, K.**, On the contents and localization of phenolics in vegetables, *Qual. Plant. Plant Foods Hum. Nutr.*, 25, 231, 1976.

44. **Kiviranta, J., Huovinen, K., and Hiltunen, R.**, Variation of flavonoids in *Allium cepa, Planta Med.*, 517, 1986.

45. **Bilyk, A., Cooper, P. L., and Sapers, G. M.**, Varietal differences in the distribution of quercetin and kaempferol in onion (*Allium cepa* L.) tissue, *J. Agric. Food Chem.*, 32, 274, 1984.

46. **Bilyk, A. and Sapers, G. M.**, Distribution of quercetin and kaempferol in lettuce, kale, chive, garlic chive, leek, horseradish, red radish and red cabbage tissues, *J. Agric. Food Chem.*, 33, 226, 1985.

47. **Panisset, B. and Tissut, M.**, Metabolisme des flavonols dans les funiques isolees de huit varietes d'oignon: effect d'un traitement par l'acide gibberelique et l'amonozide, *Physiol. Veg.*, 21, 49, 1983.

48. **Herrmann, K.**, Quercetin glycosides of onions (*Allium cepa*), *Naturwissenschaften*, 43, 158, 1956.

49. **Koeppin, B. H. and Vander Spuy, J. E.**, Microbial hydrolysis of quercetin glycosides from the inner scales of onions, *S. Afr. J. Agric. Sci.*, 4, 557, 1961.

50. **Brandwein, B. J.**, The pigments in three cultivars of the common onion (*Allium cepa*), *J. Food Sci.*, 30, 680, 1965.

51. **Harborne, J. B.**, Characterization of flavonoid glycosides by acidic and enzymic hydrolysis, *Phytochemistry*, 4, 107, 1965.

52. **Herrmann, K.**, Flavonols and phenols of the onion (*Allium cepa*), *Arch. Pharm.*, 291, 238, 1958.

53. **Bezanger-Beauquesne, L. and Delelis, A.**, Sur les flavonoides du bulbe d'*Allium ascalonicum* (Liliacees), *C. R. Acad. Sci. Ser. D*, 265, 2118, 1967.

54. **Starke, H. and Herrmann, K.**, Flavonols and flavones of vegetables. VII. Flavonols of leek, chive and garlic, *Z. Lebensm. Unters. Forsch.*, 161, 25, 1976.

55. **Kaneta, M., Hikichi, H., Endo, S., and Sugiyama, N.**, Identification of flavones in thirteen Liliaceae species, *Agric. Biol. Chem.*, 44, 1405, 1980.

56. **Yoshida, T., Saito, T., and Kadoya, S.**, New acylated flavanol glucosides in *Allium tuberosum* Rottler, *Chem. Pharm. Bull.*, 35, 97, 1987.

57. **Schmidtlein, H. and Herrmann, K.**, Über Phenolsauren des Gemusen. IV. Hydroxyzimtsauren und Hydroxybenzolsauren weiterer Gemusearten und der Kartoffeln, *Z. Lebensm. Unters. Forsch.*, 159, 257, 1975.

58. **Eichenberger, W. and Grob. E. C.**, Über die quantitative Bestimmung von Sterinderivaten in Pflanzen und die Intrazellulare Verteilung der Steringlycoside in Blattern, *FEBS Lett.*, 11, 177, 1970.

59. **Itoh, T., Tomura, T., Mitsuhashi, T., and Matsumoto, T.**, Sterols of the *Liliaceae, Phytochemistry*, 16, 140, 1977.

60. **Oka, Y., Kiriyama, S., and Yoshida, A.**, Sterol composition of vegetables, *J. Jpn. Soc. Food Nutr.*, 26, 121, 1973.

61. **Oka, Y., Kiriyama, S., and Yoshida, A.**, Sterol composition of spices and cholesterol in vegetable foodstuffs, *J. Jpn. Soc. Food Nutr.*, 27, 347, 1974.

62. **Stoianova-Ivanova, B., Tzutzulova, A., and Caputto, R.**, On the hydrocarbon and sterol composition in the scales and fleshy part of *Allium sativum* Linnaeus bulbs, *Riv. Ital. Essenze Profumi Piante Off. Saponi*, 62, 373, 1980.

63. **Smoczkiewicz, M. A., Nitschke, D., and Wieladek, H.**, Microdetermination of steroid and triterpene saponin glycosides in various plant materials. I. *Allium* species,. *Mikrochim. Acta*, 11, 42, 1982.

64. **Kravets, S. D., Vollerner, Yu. S., Gorovits, M. B., Shashkov, A. S., and Abubakirov, N. K.**, Spirostan and furostan type steroids from plants of the *Allium* genus. XXI. Structure of alliospiroside A and alliofuroside A from *Allium cepa, Khim. Prir. Soedin.*, 188, 1986.

65. **Kravets, S. D., Vollerner, Y., Gorovits, M. B., Shashkov, A. S., and Abubakirov, N. K.**, Steroids of the spirostane and furostane series from plants of the genus *Allium*. XXII. The structure of alliospiroside B from *Allium cepa, Khim. Prir. Soedin.*, 589, 1986.

66. **Harmatha, J., Mauchamp, B., Arnault, C., and Slama, K.**, Identification of a spirostane-type saponin in the flowers of leek with inhibitory effects on growth of leek-moth larvae, *Biochem. Syst. Ecol.*, 15, 113, 1987.

67. **Erazo, G. S. and Concha, C. M. I.**, Effects of storage conditions on some chemical properties of onions of the *cv* Valencia Sintetica 14, *Alimentos*, 6, 13, 1981.

68. **Soldatenkov, S. V., Mazurova, T. A., and Panteleev, A. N.**, Organic acids in onion and spinach, *Tr. Peterqof. Biol. Inst., Leningr. Gos. Univ.*, 18, 55, 1960; *Chem Abstr.*, 55, 10739b, 1961.

69. **Naito, S., Yamaguchi, N., and Yokoo, Y.,** Studies on natural anti-oxidants. II. Antioxidative activities of vegetables of *Allium* species, *J. Jpn. Soc. Food Sci. Technol.,* 28, 291, 1981.

70. **Naito, S., Yamaguchi, N., and Yokoo, Y.,** Studies on natural anti-oxidants. III. Fractionation of the antioxidant extracted from garlic, *J. Jpn. Soc. Food Sci. Technol.,* 28, 465, 1981.

71. **Pratt, D. E. and Watts, B. M.,** The antioxidant activity of vegetable extracts. I. Flavone aglycones, *J. Food Sci.,* 29, 27, 1964.

72. **Odozi, T. O. and Agiri, G. O.,** Wood adhesives from modified red onion skin tannin extracts, *Agric. Waste,* 17, 59, 1986.

73. **Akarata, O. and Odozi, T. O.,** Antioxidant properties for red onion skin tannin extracts, *Agric. Waste,* 18, 299, 1986.

74. **Attrep, K. A., Mariuani, J. M., Jr., and Attrep, M., Jr.,** Search for prostaglandin A_1 in onion, *Lipids,* 8, 484, 1973.

75. **Panosyan, A. G.,** The search for prostaglandins and prostaglandin-like compounds in plants, *Khim. Prir. Soedin.,* 17, 102, 1981.

76. **Attrep, K. A., Bellman, W. R., Sr., Attrep, M., Jr., Lee, J. B., and Braselton, W. E., Jr.,** Separation and identification of prostaglandin A_1 in onion, *Lipids,* 15, 292, 1980.

77. **Al-Nagdy, S. A., Abdel Rahman, M. O., and Heiba, H. I.,** Extraction and identification of different prostaglandins in *Allium cepa, Comp. Biochem. Physiol.,* 85C, 163, 1986.

78. **Claeys, M., Üstünes, L., Laekeman, G., Herman, A. G., Vlietinck, A. J., and Özer, A.,** Characterisation of prostaglandin E-like activity isolated from plant source *(Allium cepa), Prog. Lipid Res.,* 25, 53, 1986.

79. **Komimato, K.,** Studies on biologically active component in garlic (*Allium scorodoprasum* L. or *Allium sativum*). I. Thioglycoside, *Chem. Pharm. Bull. (Tokyo),* 17, 2193, 1969.

80. **Komimato, K.,** Studies on biologically active component in garlic (*Allium scorodoprasum* L. or *Allium sativum*). II. Chemical structure of scordinin A_1, *Chem. Pharm. Bull. (Tokyo),* 17, 2198, 1969.

Chapter 3

FLAVOR BIOCHEMISTRY

Jane E. Lancaster and Michael J. Boland

TABLE OF CONTENTS

I. INTRODUCTION

A distinguishing feature of the 500 or so *Allium* species is the metabolic network of sulfur compounds.[1] One class of sulfur compounds, the *S*-alk(en)yl cysteine sulfoxides, gives rise to the numerous volatile sulfur compounds which have caused *Allium* species to be valued as herbal medicines and as a food.

Intact allium cells have no odor, but when cells are disrupted the enzyme alliinase hydrolyzes the *S*-alk(en)yl cysteine sulfoxides to produce pyruvate, ammonia, and the many volatile sulfur compounds associated with flavor and odor.[2,3] The reaction of enzyme and substrate to produce sulfur volatiles is the central point of allium flavor biochemistry. In this chapter we will outline the chemistry of the substrate, the flavor precursors, and potential substrate, the γ-glutamyl peptides. In the last 10 years considerable progress has been made on characterizing the enzyme alliinase, and the latest work on alliinase will be reviewed.

An account of the chemistry of the volatile products of the flavor reaction leads in to an evaluation of the wide range of methods used for determining and comparing flavor in *Allium* species.

We will then look at the flavor precursors, the *S*-alk(en)yl cysteine sulfoxides from the plants' point of view. Their biosynthesis, localization within the plant, and their developmental and environmental regualtion are areas which have seen renewed research interest within the last decade.

Research in the area of biochemical ecology has thrown light on that most speculative and interesting question of why *Allium* species use so much metabolic energy producing this network of sulfur compounds. The chapter finishes with some possible functions of these compounds in alliums. The reader is also referred to the reviews by Whitaker and by Fenwick and Hanley for further information on the subject of flavor in alliums.[4,5]

II. FLAVOR PRECURSORS

A. STRUCTURE

The nonvolatile sulfur compounds in the cell, which give rise to flavor and pungency when the tissue is disrupted, are nonprotein sulfur amino acids collectively referred to as *S*-alk(en)yl cysteine sulfoxides. There are four which occur naturally:

I

$$CH_3-\overset{\overset{O}{\uparrow}}{S}-CH_2-CH(NH_2)COOH$$

II

$$CH_3-CH_2-CH_2-\overset{\overset{O}{\uparrow}}{S}-CH_2-CH(NH_2)COOH$$

III

$$CH_3-CH=CH-\overset{\overset{O}{\uparrow}}{S}-CH_2-CH(NH_2)COOH$$

IV

$$CH_2=CH-CH_2-\overset{\overset{O}{\uparrow}}{S}-CH_2-CH(NH_2)COOH$$

S-(2-Propenyl)-L-cysteine sulfoxide (IV), commonly called alliin or *S*-allyl cysteine sulfoxide, was the first of these compounds isolated. It was isolated from *Allium sativum*, (garlic) by Stoll and Seebeck during work on the antibiotic action of garlic.[6,7] The remaining alkyl and alkenyl sulfoxides were isolated by Virtanen's group in Finland during work on the sulfur compounds of onions and garlic. (+)*S*-Methyl-L-cysteine sulfoxide (I) and (+)-*S*-propyl-L cysteine sulfoxide (II) were isolated first.[8] *Trans*-(+)-*S*-(1-propenyl)-L-cysteine sulfoxide (III) was isolated 2 years later and subsequently shown to differ from alliin (IV) by the position of the double bond (C1-C2, vs. C2-C3).[8,9] The double bond of III has the *trans* configuration.[10] The sulfoxide bond can be diasteromeric, but the naturally occurring compounds are all (+) isomers.[11]

B. FLAVOR PRECURSORS IN DIFFERENT SPECIES

There are quantitative and qualitative differences in flavor precursor content in *Allium* species. So far 37 species have been analyzed,[12-15] including the main alliums of economic importance, e.g., onion, garlic, chives, leeks, and many ornamental alliums.

The flavor precursor composition was estimated by the following method. Volatile sulfur compounds were produced either by chopping fresh onions and trapping the volatiles in a closed container or by steam distillation.[12-15] The volatiles were separated and characterized by gas liquid chromatography (GLC) and the proportion of *S*-alk(en)yl radicals (e.g., methyl, propyl, propenyl, or allyl) estimated (see Section VI.A for further details of this method). Artifacts are possible with this method: 1-propenyl radicals can be misidentified as 2-propenyl or allyl radicals,[13] or can be modified to propyl radicals, thereby underestimating the proportion of 1-propenyl cysteine sulfoxide in the original *Allium*.[12] The results of *S*-alk(en)yl radicals given by this method have been interpreted in the light of possible artifacts, and Table 1 contains estimates of the relative proportions (high + + +, medium + +, low +, and absent 0) of *S*-alk(en)yl-L-cysteine sulfoxides (flavor precursors) within a given species.

All the species analyzed contained *S*-methyl cysteine sulfoxide and *S*-propyl cysteine sulfoxide. *S*-Methyl cysteine sulfoxide was a minor constituent of onions, but the main constituent of many of the ornamental alliums. The species can be divided into four groups based on the presence or absence of propenyl and allyl flavor precursors. Group 1 contains onions, leeks, and many ornamental alliums and is characterized by the presence of *S*-methyl, *S*-propyl, and *S*-propenyl cysteine sulfoxides and the absence of *S*-allyl cysteine sulfoxides. Onions have a higher proportion of *S*-propenyl cysteine sulfoxide than other species in this group.

Group 2 contains garlic and is characterized by *S*-methyl cysteine sulfoxide, *S*-propyl cysteine sulfoxide, *S*-allyl cysteine sulfoxide, *S*-propenyl cysteine sulfoxide is absent. In garlic *S*-allyl cysteine sulfoxide is the dominant flavor precursor. Some species contain all four flavor precursors, e.g., Group 4, while others contain neither propenyl nor allyl, e.g., Group 3.

Freeman and Whenham grouped the alliums they analyzed into three: high proportions of propyl/propenyl (e.g., onions), high proportions of allyl, (e.g., garlic), and high methyl proportions, (e.g., ornamental alliums).[12] Bernhard used an existing taxonomic classification (L. K. Mann's) and compared alliums within each section for presence of flavor precursors.[15] The grouping we have used in Table 1 is based on presence or absence of each flavor precursor. Existing taxonomic classifications tend to be based on morphological characteristics, e.g., Moore.[1] Flavor precursor composition could be used as an additional characteristic in taxonomic classification.

Flavor precursors are also found in genera other than alliums. *S*-Methyl cysteine sulfoxide is found in several members of the Liliaceae and Cruciferae (where it contributed substantially to the sulfurous odor of cooked cabbage) and in some of the Compositae, Umbelliferae, and Leguminosae.[16,17] *S*-Propyl cysteine sulfoxide has been detected in *Ipheion uniflorum* Raf.[18]

The quantitative and qualitative differences in flavor precursor content result in the different flavors of many of the alliums, particularly the presence or absence of *S*-propenyl cysteine sulfoxide and *S*-allyl cysteine sulfoxide. This topic will be discussed more fully in Section V.

III. OTHER NONVOLATILE SULFUR COMPOUNDS

Allium cells contain other nonvolatile sulfur compounds which are not acted upon by alliinase and do not give rise to flavor and pungency. They include the γ-glutamyl peptides, the S-substituted cysteines, and cycloalliiinin. Their biosynthetic relationship to the flavor precursors will be discussed in Section IX.

A.   γ-GLUTAMYL PEPTIDES

γ-Glutamyl peptides of amino acids and amines are very widely distributed in the plant kingdom, particularly in the Leguminosae and Alliaceae; more than 70 such peptides have been isolated.[19] In alliums there are a total of 24 γ-glutamyl peptides, 18 of which contain sulfur. Most of the peptides were isolated and characterized by Virtanen and his co-workers, beginning with γ-glutamyl phenylalanine and γ-glutamyl S-2-carboxy propyl cysteinyl

TABLE 1
An Estimate of the Composition of Flavor Precursors in Alliums

Species - common name	S-Methyl C.S.[a]	S-Propyl C.S.	S-Propenyl C.S.	S-Allyl C.S.	Ref.
1. *A. cepa* L. - common onion	+	+ +	+ + +	0	12,13,15
A. ascalonicum hort. - shallott	+ +	+ +	+	0	12
A. chinense G. Don - rakkyo	+ +	+	+ +	0	12
A. fistulosum L. - Japanese bunching onion	+	+ +	+ +	0	12,15
A. porrum L. - leek	+ +	+ +	+	0	12,13
A. schoenoprasum L. - chives	+	+	+ +	0	12,13
A. scorodoprasum L. - sand leek	+ +	+ +	+	0	12,15
A. rotundum L.	+	+ +	+ +	0	15
A. galanthum Kar + Kir	+	+	+ +	0	15
A. pskemense(Alma Ata)	+	+ +	+ +	0	15
A. christophii Trautv.	+ + +	+	+	0	12,15
A. monophyllum Vved.	+ +	+ +	+	0	15
A. altaicum	+	+ +	+ +	0	15
A. nutans L.	+ +	+	+ +	0	15
A. scabriscapum Boiss. et Kotschy	+	+ +	+ +	0	15
A. senescens L.	+ +	+ +	+ +	0	15
A. flavum L.	+ +	+	+	0	12
A. pulchellum Don	+ +	+ +	+	0	12
A. karataviense Regel	+ + +	+	+	0	12
A. oleraceum L.	+ + +	+	+	0	15
A. globosum Marsch-Bieb	+ + +	+	+	0	15
A. caesium Schrenk	+ + +	+	+	0	15
A. canadense L.	+ + +	+ +	+	0	15
A. plummerae S. Wats	+ + +	+ +	+	0	15
A. platyspathum Schrenk	+ + +	+ +	?	0	15
2. *A. sativum* L. - garlic	+ +	+	0	+ + +	12,13,15
A. moly L.	+ + +	+	0	+ +	12
A. ampeloprasum L. - great headed garlic (ransoms)	+ +	+	0	+ + +	13
A. ursinum L. - wild garlic	+ +	+	0	+ + +	12,15
A. grayi Regal - Nobiru	+ +	+ +	0	+	14
3. *A. aflatunense* B. Fedtschenko	+ + +	+	0	0	12
A. ostrawskianum Regel	+ + +	+	0	0	12
A. siculum Ucria	+ + +	+	0	0	12
4. *A. tuberosum* Rottler ex Sprengel - Chinese chives	+ +	+	+	+ + +	12,13,15
A. vineale L. - wild onion or crow garlic	+ + +	+ +	+	+ + +	12
A. triquetrum L. - garlic chives	+ + +	+ +	+	+ + +	15
A. roylei Stearn	+	+ +	+ +	+	15

Note: + + + = high; + + = medium; + = low; 0 = absent.

[a] C.S. = cysteine sulfoxide.

TABLE 2
Sulfur-Containing γ-Glutamyl Peptides Found in Alliums

Compound	Formula	Source	Ref.
γ-L-Glutamyl-S-methyl-L-cysteine	$CH_3-S-CH_2-CH-COOH$ | NH | γ-L-glu	Garlic bulbs, onion bulbs	22
γ-L-Glutamyl-S-methyl-L-cysteine sulfoxide	$\overset{O}{\overset{\uparrow}{CH_3-S}}-CH_2-CH-COOH$ | NH | γ-L-glu	Garlic bulbs	22
γ-L-Glutamyl-S-(prop-1-enyl)L-cysteine	$CH_3-CH{=}CH-S-CH_2-CH-COOH$ | NH | γ-L-glu	Chives seeds	25
γ-L-Glutamyl-S-(prop-1-enyl)-L-cysteine sulfoxide	$CH_3-CH{=}CH-\overset{O}{\overset{\uparrow}{S}}-CH_2-CH-COOH$ | NH | γ-L-glu	Onion bulbs Chives seeds	26,27 28
γ-L-Glutamyl-S-allyl-L-cysteine	$CH_2{=}CH-CH_2-S-CH_2-CH-COOH$ | NH | γ-L-glu	Chives seeds Garlic bulbs	23 28,29
γ-L-Glutamyl-S-propyl-L-cysteine	$CH_3-CH_2-CH_2-S-CH_2-CH-COOH$ | NH | γ-L-glu	Garlic bulbs Chives seeds	30 31
γ-L-Glutamyl-S-(2-carboxy-propyl)-L-cysteine	$\overset{COOH}{\underset{}{\vert}}$ $CH_3-CH-CH_2-S-CH_2-CH-COOH$ | NH | γ-L-glu	Onion bulbs	21
γ-L-Glutamyl-S-(2-carboxy propyl)-cysteinyl glycine or S-(2-carboxy propyl glutathione	$\overset{COOH}{\underset{}{\vert}}$ $CH_3-CH-CH_2-S-CH_2-CH-CO-NH-CH_2-COOH$ | NH | γ-L-glu	Garlic bulbs Onion bulbs	22 20
γ-L-Glutamyl-S-allyl-mercapto-L-cysteine	$CH_2{=}CH-CH_2-S-S-CH_2-CH-COOH$ | NH | γ-L-glu	Garlic bulbs	24

N-N-bis (γ-L, Glutamyl)-3,3-(1-methyl-ethylene -1,2-dithio)-di-alanine	COOH-CH-CH$_2$-S-CH$_2$-CH-S-CH$_2$-CH-COOH 　　　|　　　　　　|　　　　　| 　　　NH　　　　　　CH$_3$　　　NH 　　　|　　　　　　　　　　　| 　　γ-L-glu　　　　　　　　γ-L-glu	Chives seeds	31,32
γ-L-Glutamyl-*S*-(prop-1-enyl)-cysteinyl-*S*(prop-1-eny)-cysteine sulfoxide	O 　　　　　　　|| CH$_3$-CH=CH-S-CH$_2$-CH-COOH 　　　　　　　　　| 　　　　　　　　　NH CH$_3$-CH=CH-S-CH$_2$-CH-C=O 　　　　　　　　| 　　　　　　　　NH 　　　　　　　γ-L-glu	Chives seeds	33
Glutathione	HS-CH$_2$-CH-CO-NH-CH$_2$-COOH 　　　| 　　　NH 　　γ-L-glu	Onion	28
Glutathione-cysteine disulfide	COOH-CH-CH$_2$-S-S-CH$_2$-CH-CO-NH-CH$_2$-COOH 　　　|　　　　　　　| 　　　NH$_2$　　　　　NH 　　　　　　　　　| 　　　　　　　　γ-L-glu	Onion	28
Glutathione-γ-glutamyl-cysteine-disulfide	COOH-CH-CH$_2$-S-S-CH$_2$-CH-CO-NH-CH$_2$-COOH 　　　|　　　　　　　| 　　　NH　　　　　　NH 　　　|　　　　　　　| 　γ-L-glu　　　　γ-L-glu	Onion	28
S-Sulfoglutathione	HSO$_3$-S-CH$_2$-CH-CO-NH-CH$_2$-COOH 　　　　　　| 　　　　　　NH 　　　　γ-L-glu	Onion	28
γ-Glutamyl-L-methionine	CH$_3$-S-CH$_2$-CH$_2$-CH-COOH 　　　　　　　| 　　　　　　　NH 　　　　　γ-L-glu	Onion, chives seeds	28
γ-L-Glutamyl-L-cysteine	H-S-CH$_2$-CH-COOH 　　　　　| 　　　　　NH 　　　γ-L-glu	Chives seeds	28
N,N^1-*bis*(γ-L-glutamyl)-cystine	COOH-CH-CH$_2$-S-S-CH$_2$-CH-COOH 　　　|　　　　　　　| 　　　NH　　　　　　NH 　　　|　　　　　　　| 　γ-L-glu　　　　γ-L-glu	Chives seeds	28

glycine in 1960,[20] and continuing until the isolation of γ-glutamyl arginine and γ-glutamyl S-2-carboxy propyl cysteine in 1970.[21] At the same time Suzuki and his co-workers in Japan were isolating and characterizing the γ-glutamyl peptides found in garlic.[22-24] Table 2 shows the structure and occurrence of the sulfur-containing γ-glutamyl peptides found in alliums. γ-Glutamyl peptides of 2-S-alk(en)yl cysteine sulfoxides, and all the S-substituted cysteines are present, as well as glutathione and several (four) glutathione derivatives. γ-Glutamyl peptides of the nonsulfur protein amino acids valine, leucine, isoleucine, tyrosine, arginine, and phenylalanine have also been isolated.[28] The concentrations of the peptides vary widely from the abundant γ-glutamyl S-propenyl cysteine sulfoxide (1.3 g isolated from 1 kg onions)[26] to the sparse γ-glutamyl S-2-carboxy-propyl cysteine (21 mg isolated from 13.5 kg onions)[21] γ-Glutamyl peptides of S-propyl cysteine sulfoxide and S-allyl cysteine sulfoxide have not yet been isolated, and one may speculate that this is because of their low concentration in rather than their absence from the tissue. Other anomalies also exist. γ-Glutamyl-S-methyl cysteine has been found in onions, but not γ-glutamyl-S-methyl cysteine sulfoxide; conversely γ-glutamyl-S-propenyl cysteine sulfoxide has been found in onions but not γ-glutamyl-S-propenyl cysteine.

The significance of γ-glutamyl peptides in the metabolism of the plant is unclear. They occur in highest concentrations in seeds and bulbs, and are rapidly hydrolyzed by γ-glutamyl transpeptidase during seed germination and sprouting.[34,35] It is possible that they serve as reserves of nitrogen and sulfur during quiescent times of the onion's life cycle. Kasai and Larson raised the possibiltity that γ-glutamyl peptides may play a role in the transport of amino acids across cell membranes,[19] although the information of this putative function is at present limited.[36,37] The measurement of peptides is discussed in Section VI.B.3, and their relationship to the flavor precursors is discussed in Section IX.

1. γ-Glutamyl Transpeptidase

Since a significant proportion of the flavor precursors exists as γ-glutamyl peptides and these peptides are not acted upon by alliinase, the enzymes which release the free amino acid derivative from the peptide have a considerable bearing on any discussion of onion flavor.

An enzyme, γ-glutamyl peptidase, which catalyzed the hydrolysis of γ-glutamyl peptide bond was prepared by Matikkala and Virtanen from germinating chive seeds and sprouting onion bulbs.[34,35] Glutamate is produced as a reaction product. Thus:

$$\gamma\text{-Glutamyl-R} + H_2O \rightarrow \text{Glutamate} + R \tag{1}$$

where R is an amino acid or amino acid derivative and in alliums is S-alk(en)yl cysteine sulfoxide. The enzyme was purified about 60-fold from germinating chive seeds, had a pH optimum of 8.0, and was inhibited by borate. A transpeptidase effect could not be found in the preparation.

Schwimmer and Austin purified the enzyme γ-glutamyl transpeptidase [E.C.2.3.2.2] from sprouting onion bulbs.[38] This enzyme acts as a transferase by displacing the amino acid of the peptide with another:

$$\gamma\text{-Glutamyl-R} + R^1 \rightarrow \gamma\text{-Glutamyl-R}^1 + R \tag{2}$$

In this case another amino acid or derivative must be available for the reaction to proceed. The enzyme from sprouting onion bulbs was purified 800-fold.[38] The purified enzyme was found to have a pH optimum of 9.0, was activated by amino acids (consistent with a transpeptidase function), and was inhibited by borate and competitively by some γ-glutamyl derivatives. Kinetic studies using the synthetic substrate γ-glutamyl-p-nitroanilide gave a K_m of 14.3 mM, and K_i values were 0.20 mM for glutathione and 2.14 mM for γ-glutamyl S-methyl cysteine.

Neither of the enzymes was present in dormant bulbs. Schwimmer and Austin also reported possible γ-glutamyl peptidase activity that lysed γ-glutamyl peptides in preparations from sprouting onions and which could not be accounted for by the transpeptidase activity.[38]

Transpeptidase-type enzymes, capable of releasing amino acids from the γ-glutamyl peptide bond have also been reported in yeast,[39] in the ackee plant (*Blighia sapida*),[40] in tobacco suspension cultures,[41] and in legumes, asparagus, radish, and iris bulbs.[42,43]

A dual role of γ-glutamyl transpeptidase as both transferase and hydrolase has been established *in vitro*. Using the synthetic substrate γ-glutamyl-p-nitroanilide, transpeptidase may act as a transferase at higher pH, e.g., 9.5, and a hydrolase at lower pH such as 6.5.[40,44,45] The pH required for optimum transferase activity is unphysiologically high, and it has been suggested that relatively high concentrations of amino acid *in vitro* favor the transfer function and thus dipeptide synthesis.[46] In yeast γ-glutamyl transpeptidase is thought to function as both a hydrolase and transferase of glutathione depending on the nitrogen source supplied to the cells.[47]

The mode of action of the *Allium* enzyme *in vivo* remains to be clarified. It may then be established whether two transpeptidase enzymes are present in sprouting onions or one enzyme with a dual role.

Enzymatic enhancement of the production of pyruvate and flavor in dehydrated onion and garlic powders has been demonstrated using transpeptidase from onion or mammalian kidney,[48] sometimes in conjunction with alliinase, or cysteine-CS-lyase from *Albizzia lophanta*. A patent on the use of enzymes to enhance onion flavor has been granted.[49]

B. S-SUBSTITUTED CYSTEINES

Low levels of some S-substituted cysteines have been shown to occur in onion and garlic. Matikkala and Virtanen reported the occurrence of S-methyl cysteine and S-propenyl cysteine in onion extracts.[50] The latter compound was also isolated from garlic.[51] Very small amounts of S-2-carboxy propyl cysteine have been shown to occur in onion and garlic during isotope labeling experiments.[52,53] None of these compounds is hydrolyzed by alliinase. Labeled S-substituted cysteines (methyl cysteine, ethyl cysteine, propyl cysteine, propenyl cysteine) fed to onion leaf tissue were rapidly oxidized to the corresponding sulfoxide. Within 4 h 50% of S-methyl cysteine was oxidized to S-methyl cysteine sulfoxide.[54]

C. CYCLOALLIINN

This compound, 3-ethyl-1,4-thiazine-S-carboxylic acid-S-oxide, was first isolated in the reduced, thioether form by Virtanen and Matikkala in 1956.[55] It is not hydrolyzed by alliinase, and does not contribute to onion flavor.

Cycloalliinin

The oxidized form, cycloalliinin, is formed from *trans*-(+)-*S*-(1-propenyl)-L-cysteine sulfoxide during alkaline conditions *in vitro*.[11,56] The high levels of cycloalliinin found in some plant extracts may result from alkaline conditions used in the extraction and purification.[57] However, Virtanen and Matikkala presented evidence that it does exist as an endogenous compound in onions.[58]

Cycloalliinin does not appear to be formed biosynthetically from *trans*-(+)-*S*-(1-propenyl)-L-cysteine sulfoxide since injections of labeled compound into onion bulb did not produce any labeled cycloalliinin.[59]

IV. ALLIINASE

Alliinase (E.C. 4.4.1.4, *S*-alk(en)yl-L-cysteine sulfoxide lyase) is the enzyme ultimately responsible for the development of the flavor compounds.[60] The reaction catalyzed by the enzyme is a beta elimination of the *S*-alk(en)yl sulfoxide group from the substrate:

$$\text{R-}\overset{\overset{\text{O}}{\uparrow}}{\text{S}}\text{-CH}_2\text{-}\underset{\underset{\text{NH}_2}{|}}{\text{CH}}\text{-COOH} \rightarrow \text{R-S=O:} + \text{CH}_3\text{-}\overset{\overset{\text{O}}{|}}{\underset{\underset{\text{NH}}{||}}{\text{C}}}\text{-COOH}$$

(3)

Both products of the reaction are chemically unstable: the ketimine product spontaneously hydrolyzes to pyruvate and ammonia, while the reactive sulfur species will combine with any of a number of coreactants, most often another of the same species, to give a range of flavor components. Thus, the reaction is commonly described as:

$$2\ \text{R-}\overset{\overset{\text{O}}{\uparrow}}{\text{S}}\text{-CH}_2\text{-}\underset{\underset{\text{NH}_2}{|}}{\text{CH}}\text{-COOH} + \text{H}_2\text{O} \longrightarrow \text{R-}\overset{\overset{\text{O}}{\uparrow}}{\text{S}}\text{-S-R} + 2\ \text{CH}_3\text{-}\overset{\overset{\text{O}}{||}}{\text{C}}\text{-COO} + 2\ \text{NH}_4$$

(4)

A. OCCURRENCE

Alliinase occurs in most, if not all, the members of the genus *Allium*.[61] It has also been reported in related genera of the Liliaceae, such as *Ipheion*,[18] and *Thulbaghia*.[62] Alliinase-like activity has also been found in *Brassica* species[63] and *Albizzia lophanta*,[64] as well as a variety of bacterial species.[65] This chapter is only concerned with alliinase from the genus *Allium*.

B. SOLUBILITY AND STABILITY

Solubility and stability of alliinase have been the cause of considerable frustration in studies of alliinase. The requirement for a cosolvent to maintain enzyme activity during purification, noted below, is an indication of the difficulty of maintaining this enzyme in aqueous solution.

Studies carried out in our laboratory have indicated that alliinase from onions may be considerably less soluble than previously supposed. A series of extractions were performed using 0.1 M phosphate, 30% ethanediol, 0.1% 2-mercaptoethanol, 2.5 mg/l pyridoxal phosphate, with different ratios of extraction medium to onion tissue. Results indicated that the amount of enzyme extracted per gram of onion tissue increased with increasing amount of buffer until a ratio of 5 ml/g was reached. This is in contrast with methods for preparation of the enzyme which typically use 0.5 ml/g to extract alliinase. We have observed with purified, concentrated solutions of alliinase that losses of activity are often correlated with the formation of turbidity. The turbidity appears to be due to precipitated enzyme, as it can

TABLE 3
Purification of Alliinase from Onion[a]

Step	Relative purity	Recovery (%)
Homogenize 500 g onions + 250 ml 0.1 *M* PEGM[b]	1	100
Add 15% (v/v) 1% protamine sulfate and centrifuge	1.1	102
Add 1 ml 10% SDS[c]; centrifuge	1.3	97
Precipitate with 65% ammonium sulfate; redissolve in 0.01 *M* PEGM	0.9	31
Gel filtration (Sephadex-G-150)	6.0	42
Absorb on C γ-alumina; reextract with 0.5 *M* PEGM	11	42
Chromatography on CM-cellulose[d]	64	36

[a] Data from Tobkin and Maxelis.[69]
[b] PEGM is phosphate buffer pH 7.5 containing 30% ethane diol and 0.05% 2-mercaptoethanol.
[c] Sodium dodecyl sulfate.
[d] Carboxymethylcellulose.

sometimes be recovered and redissolved to recover some of the lost activity.[154] This leads us to suggest that the mechanism of instability of the enzyme is through a tendency to aggregate and precipitate.

An interesting finding, of great practical significance, is that alliinase is totally inactivated by freezing. Onions that have been frozen are noticeable in their lack of flavor and pungency. This was investigated by Schwimmer and Guadagni, who treated extracts of frozen onion with L-cysteine C-S lyase from *Albizzia lophanta* and were able to restore both onion odor and pyruvate production.[66]

C. PURIFICATION

Alliinase from garlic was first prepared by Stoll and Seebeck in 1947; however, the preparation was relatively crude.[6] Attempts to purify the enzyme to homogeneity appear to have been unsuccessful until the late 1970s when purifications and properties were independently reported for alliinase from both garlic and onion.[67-69] A common feature of the two preparations is the use of a polyalcohol cosolvent, 10% glycerol in the case of the garlic enzyme and 30% ethane diol for the enzyme from onion. The purification of the garlic enzyme also required the presence of 10 μ*M* pyridoxal phosphate in the buffers. Purification of both enzymes used standard precipitation and chromatographic separation methods.

The purification of the onion enzyme is summarized in Table 3. We have found that this method works equally well for the garlic enzyme if pyridoxal phosphate is added to all buffers and the final chromatographic separation is carried out on DEAE Sephadex A-50. Addition of pyridoxal phosphate will improve the yield of the onion enzyme.

D. PHYSICAL PROPERTIES

There exists considerable confusion concerning the physical state of the alliinase molecule. The garlic enzyme has been described as having a molecular weight of 130,000, with two subunits of 65,000,[67] while the onion enzyme is described as having a molecular weight of 150,000 and three subunits of 50,000 each.[69] Attempts to resolve this discrepancy by determination of molecular weights of both native enzyme and subunits using a variety of methods are summarized in Table 4. Further attempts to solve this problem by using chemical cross-linkers followed by sodium dodecylsulfate (SDS)-gel electrophoresis were unsuccessful.[154]

There appears to be little doubt that the subunit molecular weight is close to 50,000; however, the weight of the native enzyme molecule is unresolved. The most likely expla-

TABLE 4
Molecular Weight of Alliinase and its Subunits

Method of determination	Molecular weight (000)			
	Onion enzyme		Garlic enzyme	
	Native	Subunit	Native	Subunit
Method of determination				
Analytical ultracentrifugation	127[a]	85		47
Density gradient ultracentrifugation	150[b]			
Gel chromatography	140		130[c]	65[c]
			144	
SDS-gel electrophoresis		50[b]		
		53		51
		54[d]		49[d]

[a] All data except where otherwise indicated are from Boland.[154] Partial specific volume of 0.728 calculated from amino acid composition of the onion enzyme. Subunit molecular weight was determined following reduction and derivatization of sulfhydryls with iodoacetate, in 6 M guanidine-HCl.

[b] Data from Tobkin.[68]

[c] Data from Kazaryan and Goryachenkova.[67] The subunit molecular weight was determined in the presence of 6 M urea.

[d] Periodic acid-Schiff base stain for glycoprotein.

nation for the observed behavior is that rather than existing as a native molecule of a set number of subunits, that the subunits aggregate into dimers and tetramers (possibly also trimers) and that what is seen as the "native" enzyme is a time-averaged rapid equilibrium between these forms.

E. CHEMICAL PROPERTIES

Spectral studies on alliinase from both onion and garlic revealed an absorption peak at 420 nm, characteristic of pyridoxal phosphate.[68,69] Quantitative analysis indicated that the onion enzyme contained three pyridoxal phosphate moieties per 150,000 molecular weight, i.e., one per equivalent subunit.[68,69] Determinations on the garlic enzyme yielded a somewhat higher content, which was attributed to nonspecific binding.[67]

When the periodic acid-Schiff base stain was used following gel electrophoresis of the purified onion enzyme, a positive stain suggested the presence of a carbohydrate constitutent, and carbohydrate analysis on the purified enzyme revealed that the enzyme contains 5.8% (w/w) of carbohydrate, including 1.5% hexosamine and 0.5% methyl pentose.[68,69]

Garlic alliinase has been reported to contain no carbohydrate;[70] however, the presence of a subunit band giving a periodic acid-Schiff base stain on SDS gels of the purified enzyme indicates that the enzyme is a glycoprotein.[154]

F. CATALYTIC PROPERTIES

The general catalytic properties of alliinase were well described long before the enzyme was purified to homogeneity. A broad pH optimum of 5 to 8 was described by Stoll and Seebeck for the garlic enzyme, with a temperature optimum under the conditions of assay of 37°C.[60] They also made the following observations on specificity of the reaction: the substrate must be a derivative of (−)L-cysteine, the sulfur atom must be linked to an aliphatic group, the amino group of cysteine must not be substituted, and the sulfur atom must be in the sulfoxide form, with the (+) and (−) configurations both being substrates. Subsequent work of several groups has established that onion alliinase has a similar specificity,[71,72] with the one difference that it can also lyse molecules with an aryl substitutent on the sulfur atom.[74] The catalytic behavior with different substrates is summarized in Table 5.

TABLE 5
Substrate Specificity of Alliinase

	Onion enzyme		Garlic enzyme	
S-substituent	K_m (mM)	Ref.	K_m (mM)	Ref.
(+) Methyl	16.6	71	15	75
	9.1	72		
(+) Ethyl	5.7	71	6	75
	12.5	72		
(+) Propyl	3.8	71	3	75
	6.3	72		
(+)Propyl	2.3	71		
(−) Propyl	11.7	71		
Butyl	4.7	71	5	75
	5.7	72		
Allyl	7.1	72		
Propenyl	6	73	6	75
			0.5	67
Benzyl	Lysed	74	Not lysed	

TABLE 6
Pyridoxal Phosphate Antagonist Inhibition of Alliinase

Onion enzyme[69]	Garlic enzyme[76]
Hydroxylamine (reversible)	Hydroxylamine (reversible)
Aminooxyacetate	Aminooxyacetate
Sodium cyanide	Aminooxypropionate
	D-cycloserine *(partial activity restored by replacement of pyridoxal phosphate)*
	L-cycloserine
	3-aminooxy-DL-alanine
	3-cyano-L-alanine (irreversible)

Inhibitor studies with alliinase fall into two classes: those involving pyridoxal phosphate antagonists and those using substrate analogues as competitive inhibitors. In the former group, reversible inhibition by hydroxylamine has been demonstrated by several authors.[69,76] Other inhibitors are listed in Table 6. It is likely that inhibition by cycloserine is due to the action of its product of hydrolysis, 2-aminooxy alanine. Taken together, these various results indicate unequivocally that pyridoxal phosphate is required for catalytic activity.

The second group of inhibitor studies indicates the nature of the binding site for the substrate using analogues that are not capable of undergoing the reaction, mostly derivatives of unoxidized cysteine. These results, collected in Table 7 show interesting parallels with the substrate specificity in Table 5.

G. CATALYTIC MECHANISM
The dependence of catalysis on pyridoxal phosphate, taken together with the specific requirements for the substrate molecule and the nature of the products leaves little doubt that the reaction mechanism of alliinase is via a pyridoxal phosphate-Schiff base derivative which undergoes beta elimination. Attemps to trap a Schiff base intermediate using ^{14}C-labeled S-methyl cysteine sulfoxide and cyanoborohydride (a reagent that will reduce a Schiff base but will not react with free pyridoxal phosphate) were unsuccessful; however, further studies with this reagent gave evidence that in the absence of substrate the enzyme is mostly in a resting form, with an internal Schiff base. Labeling studies with tritiated borohydride followed by hydrolysis have indicated that the internal Schiff base is formed with the 4-

TABLE 7
Substrate Analogue Inhibitors of
Alliinase by S-Derivatives of Cysteine
with *S*-Ethyl Cysteine Sulfoxide as
Reference Substrate

Derivative	Onion		Garlic	
	K_i (m*M*)	Ref.	K_i (m*M*)	Ref.
Methyl	2.6	71	33	75
			25	76
Ethyl	1.9	71	8	75
Propyl	1.5	71	5	75
Butyl	1.3	71		
Allyl			20	76
Benzyl			10	76

amino group of a lysine residue. When the experiment was repeated with the substrate analogue *S*-propyl cysteine present, the formation of a *S*-propyl cysteine-pyridoxal phosphate derivative was observed.[155]

V. FORMATION OF FLAVOR IN ALLIUMS

The chemistry of the flavor precursors and the structure and mode of action of alliinase have been discussed in previous sections. This section will outline the chemistry of the volatile sulfur compounds which are formed when the tissue is disrupted and alliinase hydrolyzes the flavor precursors.

A. ENZYMICALLY FORMED VOLATILES
1. Lacrimatory Factor in Onions
The lacrimatory or tear-producing characteristic is the hallmark of onions. Studies by Virtanen's group showed that the lacrimator was formed enzymically during the hydrolysis of *S*-propenyl cysteine sulfoxide, and they tentatively identified it as 1-propenyl sulfenic acid.[2,3,26] Wilkins proposed thiopropanal *S*-oxide as a possible structure for the lacrimator,[77] and this structure was confirmed by Brodnitz and Pascale who synthesized the identical compound.[78] The stereochemistry was determined, and the lacrimatory factor was shown to be a 19 to 1 mixture of (Z)- and (E)-thiopropanal *S*-oxide.[79]

Block et al. suggested that the lacrimatory factor was formed via the transient 1-propenyl sulfenic acid.[80] The lacrimatory effect of synthetic thioalkanal *S*-oxides was determined.[78-80]

Lacrimatory activity was present, but as the "R" group in RCH=SO became more bulky the lacrimatory activity diminished to the point where R=*t*-butyl was devoid of lacrimatory activity. Synthetic lacrimators were found to be stable at room temperature for less than 1 h.[78]

2. Sulfenic Acids
Alliinase cleaves the *S*-methyl, *S*-propyl, and *S*-allyl cysteine sulfoxides to produce

methyl, propyl, and allyl sulfenic acids. Sulfenic acids are very unstable, with a half-life of about 90s.[3] Methyl, propyl, and allyl sulfenic acids do not themselves possess lacrimatory activity; neither do they rearrange to form thioalkanal *S*-oxide which does have lacrimatory activity. Thus, garlic, which has mainly *S*-allyl cysteine sulfoxide, does not produce tears when it is cut.

B. NONENZYMICALLY FORMED VOLATILES

The products of the reaction between alliinase and flavor precursors are unstable and themselves undergo nonenzymic rearrangements to produce a wide range of volatiles which give the characteristic aroma of alliums.

Over 80 volatiles have been reported in fresh, cut, and steam-distilled extracts of alliums, particularly onion and garlic.[81,82] The sulfur volatiles and the carbonyl-type volatiles formed directly and indirectly from enzymic cleavage of flavor precursors are shown in Table 8.

1. Volatiles in Onion

The lacrimator factor, thiopropanal *S*-oxide, rearranges to produce propanal, carbonyl derivatives of propanal, and elemental sulfur.[81,83]

Freeman and Whenham used a synthetic *in vitro* system to demonstrate the importance of thiosulfinates as intermediates in the formation of onion volatiles.[84] Thiosulfinates are formed via propyl and methyl sulfenic acids and via thiopropanal *S*-oxide. Rearrangement of thiosulfinates leads mainly to monosulfides, disulfides and sulfur dioxide. A thiosulfinate rearrangement, of lesser significance, leads to thiosulfonates and disulfides.[81] Unsaturated disulfides, e.g., methyl, propenyl disulfide or propyl propenyl disulfide are converted into dimethyl thiophenes and saturated disulfides by the action of heat and UV light.[81] Disulfides rearrange to form trisulfides, tetrasulfides, and polysulfides.

Figure 1 shows the genesis of volatile sulfur compounds in onion.

2. Volatiles in Garlic

In garlic the main flavor precursor is *S*-allyl cysteine sulfoxide. Enzymic cleavage produces allyl sulfenic acid which rearranges to form alliicin or diallyl thiosulfinate. Alliicin is the main volatile in fresh garlic extract.[85] It undergoes nonenzymic rearrangement to mono-, di-, trisulfides, thiosulfonates, and sulfur dioxide, in a similar fashion to that outlined for onions in Figure 1.

C. FLAVOR NOTES OF VOLATILES

To the person eating them the tastes of fresh, boiled, or fried onions and fresh or fried garlic are very different and very distinctive. The particular complement of flavor precursors in each *Allium* determines potential flavor. Thus, only alliums with *S*-propenyl cysteine sulfoxide have a tear-producing effect, and alliums with *S*-allyl cysteine sulfoxide resemble the taste of garlic. The method of preparation, however, determines the makeup of volatiles, and the exact relationship between a particular volatile and a particular flavor note is still unclear.

Thiosulfinates and thiosulfonates are thought to be characteristic of freshly cut onions.[81,84,86] They are present in the headspace of cut onions, but not in steam-distilled onion oil. Synthetic thiosulfonates resemble the odor of freshly cut onions.[81] The propyl- and propenyl-containing di- and trisulfides possess the flavor of cooked onions.[81,87] They are also present in steam-distilled onion oil.[81] Dimethyl thiophenes are found in steam-distilled onion oil, in fresh onion,[81] and have been described as having a "fried-onion flavor". However, Galetto and Hoffman used pure samples of dimethyl thiophenes in organoleptic tests and found no resemblance to fried onions. If anything the compounds had a solvent, plastic-like flavor.[88]

TABLE 8
Volatile Compounds Identified from Alliums

Thiosulfinates

Dimethyl thiosulfinate
Dipropyl thiosulfinate
Diallyl thiosulfinate
Methyl methane thiosulfinate
Propyl methane thiosulfinate
Propyl propane thiosulfinate
Methyl propyl thiosulfinate
Methyl allyl thiosulfinate
Allyl propenyl thiosulfinate
Propyl propenyl thiosulfinate

Monosulfides

Dimethyl sulfide
Diallyl sulfide
Methyl allyl sulfide
Dipropenyl sulphide (3 isomers)
Allyl propyl sulfide
Methyl propenyl sulfide (2 isomers)
Propyl propenyl sulfide (2 isomers)

Disulfides

Dimethyl disulfide
Dipropyl disulfide
Diallyl disulfide
Dipropenyl disulfide
Methyl propyl disulfide
Allyl propyl disulfide
Methyl allyl disulfide
Isopropyl propyl disulfide
Methyl propenyl disulfide
Propyl propenyl disulfide
  (*cis* and *trans*)
Allyl propenyl disulfide

Sulfenic acids

Propenyl sulfenic acid
Allyl sulfenic acid
Propyl sulfenic acid
Methyl sulfenic acid
Thiopropanal *S*-oxide

Trisulfides

Dimethyl trisulfide
Dipropyl trisulfide
Diallyl trisulfide
Methyl allyl trisulfide
Propyl allyl trisulfide
Methyl propenyl trisulfide (2 isomers)
Methyl propyl trisulfide
Propyl propenyl trisulfide (2 isomers)
Diisopropyl trisulfide
Isopropyl propyl trisulfide

Tetrasulfides

Dimethyl tetrasulfide
Diallyl tetrasulfide

Thiosulfonates

Methyl methane thiosulfonate
Propyl methane thiosulfonate
Propyl propane thiosulfonate

Thiophene derivatives

2,5-Dimethyl thiophene
2,4-Dimethyl thiophene
3,4-Dimethyl thiophene
3,4-Dimethyl-2,5-dihydrothiophen-2-one

Thiols

Methanethiol
Ethanethiol
Propanethiol
2-Propene-1-thiol
2-Hydroxy propanethiol

Carbonyl compounds

Propanal
2-Methylpentanal
2-Methyl-pent-2-enal
Butanal
2 Methyl butanal
2 Methyl but-2-enal

The aldehyde, propanal, formed from thiopropanal *S*-oxide was found in steam-distilled onion oil, and in the headspace of cut onions and was described as being "one of the most important flavor compounds in raw onions", although its flavor note was not specified.[81]

Boiled onions have a characteristically sweet taste. Yamanishi and Orioka found that *n*-propanethiol increased in concentration during boiling, and that it was 50 to 70 times as sweet as sucrose.[89] They concluded that the sweet taste developed in boiled onions was due, at least partly, to the development of the thiol.

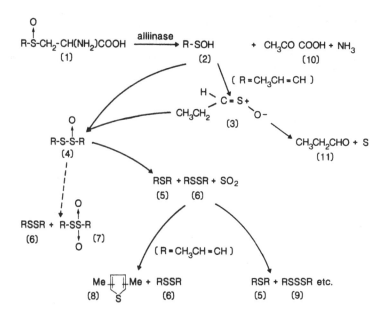

FIGURE 1. Formation of volatile sulfur compounds in onion. (1) *S*-alk(en)yl cysteine sulfoxide; (2) *S*-alk(en)yl sulfenic acid; (3) thiopropanal *S*-oxide; (4) thiosulfinate; (5) monosulfide; (6) disulfide; (7) thiosulfonate; (8) dimethyl thiophene; (9) trisulfide; (10) pyruvate; (11) propanal. R = methyl [CH_3], propyl [C_3H_7], or propenyl [CH_3–CH=CH].

D. DEVELOPMENT OF BITTERNESS

Freshly prepared onion juice develops a bitter, alkaloid-like taste within a few minutes and after an hour or so will turn pink.[90,91] The role of the *S*-alk(en)yl cysteine sulfoxides and/or alliinase in the development of this bitterness is not clear. The development of the bitterness required an enzyme, since heat-treated, or acid-treated onions did not develop bitterness; the addition of a crude enzyme preparation restored the development of bitterness. However, onion odor and pungency develop at a faster rate than bitterness. Schwimmer concluded that the bitter taste was probably due to secondary reactions following the initial action of alliinase on *S*-propenyl cysteine sulfoxide.[92] The exact compound responsible for the bitterness remains unknown.

The pinking of onion extracts is unrelated to the bitterness. It is thought to be caused by carbonyl compounds reacting with free amino acids to produce pink pigments.[91]

VI. METHODS OF MEASUREMENT OF FLAVOR

The assessment and comparison of flavor in onions requires a suitable method for measuring flavor. Numerous methods have been published in the literature, and these range from simple testing to the use of complex physicochemical methods. The number and diversity of methods can be attributed to three facets of flavor biochemistry:

1. The net flavor depends upon the presence and action of both enzyme and substrate.
2. The net flavor is produced by numerous, chemically diverse volatiles.
3. The sulfur compounds within the plant are chemically complex, and some of them are unstable *in vitro*.

Outlined in this section are some of the most widely used methods for estimating flavor in onions.

A. MEASUREMENT OF REACTION PRODUCTS

The cleavage of the flavor precursors by alliinase produces thiopropanal S-oxide, sulfenic acids, sulfur volatiles, pyruvate, and ammonia. Measurements of these reaction products have been used as estimates of flavor.

1. Thiopropanal S-Oxide (Lacrimator)

Three different methods have been reported for measuring the amounts of lacrimator and thiosulfinates in crushed onions. As early as 1944 Cavallito et al. observed that thiosulfinates would react with cysteine to form cysteine derivatives:[93]

$$2\,HSCH_2\text{-}CHNH_2\text{-}COOH + R\text{-}S\overset{O}{\overset{\uparrow}{\text{-}S}}\text{-}R^I \longrightarrow R\text{-}S\text{-}S\text{-}CH_2CHNH_2COOH + R^I\text{-}S\text{-}S\text{-}CH_2CHNH_2COOH \quad (5)$$

Lukes developed a method whereby crushed fresh onion was extracted with diisopropyl ether.[94] Cysteine was added to the ether, and the mixture chromatographed on thin layer chromatography (TLC) plates in t-butanol:formic acid (85%):concentrated HCl:water (95:15:18:2 V/V). The cysteine derivatives of lacrimator and thiosulfinates were visualized with ninhydrin and measured at 570 nm wavelength.

Tewari and Bandyopadhyay extracted the lacrimator and thiosulfinates from crushed onion into ether and reacted them with glycine-formaldehyde.[95] The resultant compounds have a pink color (E max = 520 nm) and may be determined in total by spectrophotometry or separated and determined by TLC.

Freeman and Whenham extracted the lacrimator and thiosulfinates from crushed onion into hexane at 0°C and within 5 to 10 min of comminution.[96] The absorbance of the extract at 254 nm gave a measure of the amount of lacrimator and thiosulfinates present. Synthetic thiopropanal S-oxide was used as a standard.

There are difficulties associated with the measurement of lacrimator (thiopropanal S-oxide) and thiosulfinates. The lacrimator is unstable, and, thus, the reproducibility of the results is affected by the temperature and the timing involved in the method. Reproducibility is also dependent on the technique used for macerating the tissue. Because of the instability of the lacrimator, standards should be stored under nitrogen at 0°C.

2. Pyruvate

Pyruvate, formed enzymatically when onion tissue is crushed, reacts with 2,4, dinitrophenyl hydrazine to produce a yellow-colored derivative which is measured spectrophotometrically. This is the basis of a robust, much-used method developed by Schwimmer and Weston.[97] Measurements of pyruvate may be used as measures of flavor strength: a correlation of r = 0.97 was shown between the pyruvate formed enzymatically and the olfactory threshold.[98]

This is a rapid method which has been used for screening large numbers of samples as in a breeding or selection program. It provides a measurement of total flavor, but does not provide any information about relative amounts of individual flavor precursors or final flavor volatiles. Because 2,4, dinitrophenyl hydrazine is nonspecific and reacts with any carbonyl-type compounds, it is necessary to have a heat-treated or protein-denatured sample as a blank to inactivate alliinase and to enable the detection of background carbonyl-type compounds. This method cannot be used with dehydrated onion, as carbonyls other than pyruvate form during storage.[99]

3. Volatile Compounds

The volatile sulfur and carbonyl compounds formed from crushing onions may be measured by GLC. The volatile mixture is separated on a liquid phase, usually carbowax,

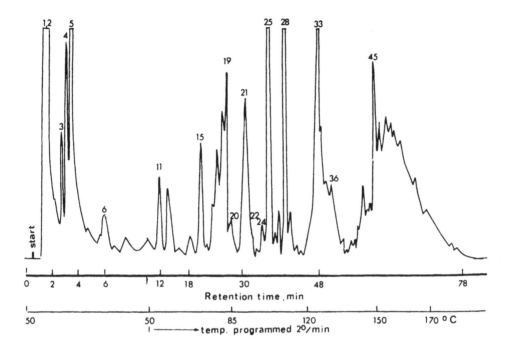

Peak No	Compound[a]
1	Methanethiol + ethanal
2	Propanal
3	Propanethiol + 2-methylbutanal
4	Methanol + 2-methylpentanal
5	Ethanol + propanol
11	Dimethyl disulfide
15	2-Methylpent-2-enal
19	Methyl propyl disulfide
20	3,4-Dimethylthiophene
21	Methyl cis-propenyl disulfide
22	Methyl trans-propenyl disulfide
23	Dimethyl trisulfide
24	Isopropyl propyl disulfide
25	Di-n-propyl disulfide
28	Propyl cis-propenyl disulfide
33	Propyl trans-propenyl disulfide
45	Di-n-propyl trisulfide

[a] Identification was based on retention times, co-chromatography and MS-data.

FIGURE 2. Gas liquid chromatogram of onion volatiles from onion extracted with dichloromethane. Separation column used: Carbowax 1500. For peak designation see table underneath. (Data from Mazza, G., LeMaguer, M., and Hadziyev, D., *Can. Inst. Food Sci. Technol. J.*, 13, 87, 1980. With permission.)

and the components identified by retention time, and sometimes also by mass spectrometry. The peak area is used to determine the amount of a volatile. Several reports have been published on this method.[81,87,88,100]

The main types of volatile compounds isolated from onions are shown in Figure 2 (a more detailed list may be found in Table 8). Calculations of total peak area may be used to compare onion varieties, products, and agronomic practices. However, in order for this method to be reproducible it is important to standardize the procedure used. Mazza et al.[100] outlined the main variables in the method.

1.   Incubation time: peak areas of most onion volatiles increased with increasing incubation time of the sliced onions, reaching a maximum at about 3 h, but sharply decreasing thereafter.

2.   Method of sample disintegration: the composition of volatiles depended on the extent of disintegration of the onion tissue. Total volatiles were enhanced 25% by slicing, rather than dicing, although a few volatiles, notably di-*n*-propyl disulfide and 2-methyl pentanal decreased in amount. Extensive disintegration, as in blending, produced a 7% increase in volatiles over slicing. Addition of water to onion slices decreased the amount of volatiles collected.

3.   At all temperatures between 20 and 70°C the number of volatiles produced was the same, but at the higher temperatures the amount greatly increased. The rise in volatiles at higher temperatures results from their solubility decrease in water and their overall increase in vapor pressure, not from increased enzyme activity (since enzyme activity decreases after 40°C).

4.   Dehydrated onions, fresh onion slices, and solvent-extracted sliced, fresh onions had similar chromatographs, but differed in the amounts of volatiles. Preliminary concentration of headspace volatiles in a Tenax trap provided a more sensitive method for small amounts of tissue and for dehydrated onions.

B.   MEASUREMENT OF FLAVOR PRECURSORS

Measurements of the amounts of flavor precursors have been used as indicators of flavor strength in alliums. The flavor precursors are potential flavor, and the presence of alliinase is needed to realize that flavor. Any methods for quantifying flavor precursors require good standards, and although the flavor precursors are homologous in structure, preparation of pure standards is not chemically straightforward.

1.   Preparation of S-alk(en)yl-L-Cysteine Sulfoxide Standards
a.   *Preparation of Thioether*
The thioethers *S*-2-propenyl cysteine and *S*-propyl cysteine may be prepared by either of two methods. L-Cysteine, formed *in situ* by reduction of L-cystine with sodium in liquid ammonia, was alkylated by means of 2-propenyl bromide or 1-propyl bromide. This method was first described by Grenby and Young[101] and used subsequently by Freeman and Whenham.[102] Alternatively, the method of Stoll and Seebeck may be used.[103] L-Cysteine hydrochloride is dissolved in a solution of sodium hydroxide in ethanol, and 2-propenyl bromide or 1-propyl bromide is added with stirring and the solution left overnight. A modified method of Stoll and Seebeck was used by Lancaster and Kelly.[104] The thioether *cis-S*-1-propenyl-cysteine is prepared from the 2-propenyl cysteine by base isomerization with potassium *t*-butoxide in dimethyl sulfoxide.[11] *S*-Methyl cysteine is available commercially.

b.   *Oxidation to Sulfoxides*
Thioethers were oxidized to sulfoxides by oxidation with 30% hydrogen peroxide.[102-105] Carson and Boggs outlined procedures for separating optical isomers of the sulfoxides.[11]

Synthesis of the natural isomer, (+) *trans-S*-1-propenyl-L-cysteine sulfoxide has been carried out but in very small amounts.[106]

2.   Measurement of S-Alk(en)yl-L-Cysteine Sulfoxides
In order to measure these compounds alliinase must first be inactivated. Steeping the intact tissue in solvent such as ethanol or a mixture of methanol:chloroform:water (12:5:3 v/v) at −70°C inactivates alliinase and extracts the flavor precursors.

The earliest method of direct measurement of flavor precursors made use of an amino acid analyzer.[50] An ethanol extract of onions was fractionated into neutral/basic and acidic fractions on an ion-exchange column, and the amino acids and peptides quantified by ninhydrin. A disadvantage of this method was the cyclization of S-propenyl cysteine sulfoxide to cycloalliinin which occurred in alkaline fractions from the ion exchange.

In order to circumvent this problem Granroth developed a method which used thin layer electrophoresis and chromatography in nonalkaline conditions to isolate and separate flavor precursors, amino acids, and peptides.[107] The method described as "rapid, sensitive, and nonartifact producing" was based on Bieleski and Turner's technique for extracting plant tissue in methanol:chloroform:water (12:5:3 v/v), and purifying and separating the amino acids on two-dimensional electrophoresis and TLC.[108]

Lancaster and Kelly extended Granroth's method, and by using the nonendogenous analogue S-butyl cysteine sulfoxide as an internal standard made the method quantitative.[104] Sulfoxides are labile components, and an internal standard in each extract allows losses during extraction, purification, and spearation to be accounted for.

TLC on cellulose, with butanol:acetic acid:water (12:3:5 v/v) as a solvent system is useful as an additional separation system for flavor precursors. The preparation of n-butyl trifluoro-acetyl derivatives of flavor precursors and their quantitative measurement by GLC was attempted, but was not achieved because of the breakdown of these compounds, particularly the S-propenyl cysteine sulfoxide derivative, during high temperatures and acidic conditions.[156]

3. Measurement of Peptides
Studies on the biosynthesis and interrelationships of nonprotein sulfur amino acids in alliums require a technique which separates and measures most of the flavor precursors and peptides.

In the method developed recently by Shaw et al.[109] methanol:chloroform:water extracts of intact onion tissue are run on ion-exchange columns (Dowex 1) and eluted with increasing concentrations of acetic acid to fractionate flavor precursors, amino acids, and sugars, (0.1-M), glutamic acid (0.2 M), and peptides (1.0 and 2.0 M). Thin layer electrophoresis is used to remove interfering amino acids from 0.1 and 1.0-M fractions. The flavor precursors and two peptide fractions are each separated on silica gel TLC, in solvent I (methyl ethyl ketone:pyridine:water:glacial acetic acid; 8:15:15:2 v/v) and then solvent II (n-propanol:water:propylacetate:acetic acid:pyridine; 120:60:20:4:1 v/v) in the same direction.

Compounds are visualized by ninhydrin and measured by scanning densitometry. The use of the internal standards S-butyl cysteine sulfoxide, S-methyl glutathione, and γ-glutamyl glutamic acid (for fractions 0.1 1.0, and 2.0 M) ensures that losses of endogenous compounds are accounted for and that the method is quantitative.

C. MEASUREMENT OF ALLIINASE
Measurement of alliinase has usually been made by determining the substrate-dependent rate of production of pyruvate. Pyruvate production can be monitored either by taking samples over a time course and derivatizing the pyruvate with 2,4-dinitrophenylhydrazine, and reading the concentration of the derivative spectrophotometrically, or continuously by coupling pyruvate production with NADH oxidation using an excess of the enzyme lactate dehydrogenase. NADH can be monitored directly by either spectrophotometry or fluorometry. In either case it is necessary to extract the enzyme into a suitable buffer solution and remove the endogenous substrate and pyruvate by dialysis or gel chromatography.

D. MEASUREMENT OF γ-GLUTAMYL TRANSPEPTIDASE
γ-Glutamyl transpeptidase is usually assayed by following the hydrolysis of the synthetic substrate γ-glutamyl-p-nitroanilide. The assay for onion enzyme has been described by Schwimmer and Austin.[38]

E. NONSPECIFIC METHODS OF MEASUREMENT

An early method of flavor estimation was developed by Platenius.[110] Onions were steam distilled, and the volatile sulfide in the distillate oxidized to sulfate in bromine water. Excess bromine was driven off by heating; the sulfate was precipitated with barium chloride and determined by weight. Total volatile sulfur has also been used in subsequent method.[111]

Saguy et al. have used the chemical oxygen demand (COD) method to estimate the volatile aroma compounds in onions.[99] A distillate of the product is oxidized by potassium dichromate, and the volatile compounds are measured spectrophotometrically (650 nm). The correlation between COD values and taste and odor thresholds was good for fresh onions (r = 0.97). For dried onions the COD values correlated with taste scores, but not with odor thresholds, as off-odor developed during storage.

All of the above methods are physicochemical involving a similar response from compounds with like structures, whereas what our taste sensations perceive is a complex mixture of sweetness, pungency, bitterness, and volatiles. No one method can measure so many chemically diverse compounds. Conversely, not all the compounds measured by any one method are directly perceived as flavor. To circumvent this problem many methods have used correlations between taste perceptions and amounts of the compound being measured, e.g., pyruvate and lacrimatory effect; COD, and odor threshold.

There are advantages and disadvantages to each of the available methods and in selecting a method the following factors need to be considered:

1. Is a quantitative, semiquantitative, or comparitive method required? Quantitative methods need calibration by standards which may have to be chemically synthesized or purified from natural sources. This is time consuming and often expensive.
2. What is the end use, e.g., an approximate screening method for plant selections, a comparison of food products, or a detailed biochemical study? Different methods are applicable to different uses.
3. What kind of plant material is available? It is easier to estimate flavor in fresh material than in dehydrated material.

Although the above methods are wide ranging, alternative new methods are still feasible. The use of high performance liquid chromatography, particularly for flavor precursors and peptides, has not been reported in the literature and could be promising.

VII. DISTRIBUTION OF FLAVOR

A. LOCATION WITHIN THE PLANT

The components of flavor, i.e., flavor precursors, alliinase, and peptides, were first isolated from the onion and garlic bulbs. Flavor precursors have also been reported in the leaf blades, base plate, and roots of onion, garlic, and leek.[112-114] Flavor precursors were absent from the seeds of onion,[114] although γ-glutamyl peptides were present.[25,32,34]

Alliinase has been isolated from onion and garlic bulbs, but was negligible in amount in seeds.[115]

The scape and inflorescence produce the characteristic taste when eaten, indicating the presence of both alliinase and flavor precursors, but flavor is negligible in the cortical pith of leaf blades and scape.

Flavor precursors were not uniformly distributed within the bulb scales and leaf blades.[112,113] In the onion bulb flavor precursor concentration was highest in the base plate and innermost bulb scales, progressively decreasing to the outermost scales which were dried and almost free from flavor. Young leaf blades were higher in flavor-precursor concentration than old leaf blades. A similar pattern was found in leeks. In garlic the storage leaf accounted for the bulk of the clove as well as its flavor precursor concentration.

In qualitative studies based on histochemical detection of sulfur compounds Becker and Schuphan showed a higher concentration of flavor precursors/peptides in the bundle sheaths of onion bulb scales than in the surrounding cortical cells.[116] This suggests either the transport or the synthesis of flavor precursors within the bundle sheaths.

B. LOCATION WITHIN THE CELL

Allium flavor occurs only when the cells are disrupted and the compartmentalization of the cell is destroyed. It is generally considered that the alliinase and the flavor precursors are separate *in vivo* and only react on rupture or wounding of the cells. In order to determine the cellular localization of the enzyme and substrate, Lancaster and Collin prepared protoplasts from inner scales of onion bulbs and using the appropriate osmoticum lysed the protoplasts to produce intact vacuoles and a cytoplasmic fraction. The enzyme alliinase was shown to occur in the vacuole and the flavor precursors in the cytoplasm.[117]

An ultrastructural comparison of onion bulb cells (containing flavor) and onion callus cells (lacking flavor) showed the presence of distinctive vesicles in the cytoplasm of onion bulb cells but not in callus cells. Shoots differentiated from callus did contain vesicles and flavor. Turnbull et al. suggested that the flavor precursors may be stored in the vesicles.[118]

VIII. FACTORS DETERMINING FLAVOR

It is well known that not every onion tastes the same; differences exist between varieties and within varieties grown under different conditions, and at different growth stages. In this section we willl discuss the main factors influencing flavor in onions, mainly in terms of their influence on flavor precursor levels and alliinase levels.

A. GENETIC FACTORS

The genotype of an onion determines its potential for flavor production, although that potential may be modified by the environment. Flavor levels (using several different methods) have been determined in over 50 varieties of onions (Table 9).

Because of the ease of measurement pyruvate levels are the most preferred method for flavor determination. There is a tenfold range in flavor. Some of the strongest onions are those which have been bred for high dry matter and processing, e.g., Dehyso. Some of the mildest onions are Japanese varieties and the Early Grano-type varieties.

Although flavor has been determined in a large number of onion varieties, there have been no published studies on the inheritance of the flavor precursors in onions.

The flavor precursor *S*-methyl cysteine sulfoxide also occurs in forage rate (*Brassica napus* L.). Studies of the inheritance of *S*-methyl cysteine sulfoxide showed that both additive and dominance components were significant, but the greater mean squares of the additive gene effect indicated its greater importance.[93] Narrow sense heritability for methyl cysteine sulfoxide was high (50.8). *S*-Methyl cysteine sulfoxide content was not correlated with yield.

The inheritance of the flavor precursors in onions is an area which would repay further research.

B. ONTOGENETIC FACTORS

The flavor precursors and alliinase which give rise to flavor are an integral part of the *Allium* plants' metabolism. They are synthesized and degraded, and net measured; levels are the difference between these two processes. One might expect that the net amount of flavor precursors and alliinase would vary during the plant's life cycle, and this is in fact so.

1. Germinating Seeds

Seeds contain negligible amounts of flavor precursors and alliinase, but considerable

TABLE 9
A Comparison of Flavor Strength in Onion Varieties

Variety	Pyruvate (μmol/g fresh wt)	Thiopropanal S-oxide (μmol/g dry wt)	Volatile sulfur (ppm)	Flavor precursors (μmol/g fresh wt)	Ref.
Rijnsburger	8—18, 8.5	33.1			96,98,119
Express yellow OX	3, 5.8	35.1			96,98,119
Extra early Kaizuka	3				119
Imai early yellow	2				119
Buffalo	4—9				119
Granex 33 Hybrid	7—8				119
Keepwell	11				119
Senshyu semi-globe yellow	4—8, 6.4	37.3			96,98,119
Ebenezer	12		156		119,120
Giant Zittau	3				119
Australian brown	10				119
Excellent	11				119
Hyduro	10—15				119
Hygro	8—17				119
Revro	9				119
Solidor	11—12				119
Mammoth red	9—20				119
Brown beauty	11—12				119
Downings yellow globe	10				119
Early yellow globe	10		94		119,120
Espagnol	9				119
Granada	11				119
Topaz	12				119
Vela	10				119
Yellow sweet Spanish	2—11				119
Ailsa Craig	11—16				119
Brunswick	10				119
Southport red globe	9—18				119
Red torpedo	9				119
White Portugal	13—15				119

TABLE 9 (continued)
A Comparison of Flavor Strength in Onion Varieties

Variety	Pyruvate (μmol/g fresh wt)	Thiopropanal S-oxide (μmol/g dry wt)	Volatile sulfur (ppm)	Flavor precursors (μmol/g fresh wt)	Ref.
Perfecto blanco	7				119
Sunburn	9	32.9			96,119
Miracle	10.4	39.0			96,119
Keep well	5.6	43.3			96,119
Tropic ace	9.3	32.7			96,119
Amber express	7.4	37.3			96,119
Dragon eye	9.0	46.6			96,119
Pukekohe longkeeper				8—21	121
Dehyso				21.7	121
Sapporo yellow				8.7	121
Early longkeeper				6—27	121
Early Grano			68		120
Yellow Bermuda			72		120
Crystal wax			97		120
White Creole			129		120
Red Creole			155		120
Brigham Yellow Globe			117		120
Yellow Globe Danvers			124		120
Utah Sweet Spanish			98		120
Mountain Danvers			126		120
Red Wethersfield			123		120
Californian Red				3—7	157
Storage Red				11	157
Southport White Globe				10	157

amounts of γ-glutamyl peptides. Freeman found that alliinase activity was detectable within a few days of the onset of germination, increased rapidly as germination proceeded, and reached a maximum value after 15 to 20 d.[115] This is the only published information on ontogenetic effects on alliinase. However, now that alliinase has been characterized and can be extracted quantitatively, one might hope for more studies on the levels of activity of alliinase in onion plants. Although absent from seeds, flavor precursors accumulated rapidly as seeds germinated and the cotyledon emerged.[114] Levels of S-propenyl cysteine sulfoxide were very high in the hypocotyl. Flavor precursor accumulation was initially dependent on resources from the seed, probably from the metabolism of γ-glutamyl peptides, but as the cotyledon leaf developed, accumulation was dependent on photosynthetic products. S-Methyl cysteine sulfoxide accumulation was higher in dark grown seeds.

2. Seedling to Ripe Bulb

Flavor precursors are thought to be synthesized in the leaf blades and transported to the scales of the bulb where they are stored.[113] Younger leaf blades are a more productive source of flavor precursor than older leaf blades. Thus, flavor precursors are laid down during vegetative growth. Flavor precursor content in the scales (and, thus, collectively the bulb) increased during bulbing, and then gradually decreased as the bulb matured and ripened.[123] This was the pattern found also by Platenius and Knott,[124] measuring volatile sulfur, Saghir et al.[125] using headspace analysis of volatiles, and Malkki et al.[126] measuring pyruvate.

Lancaster et al.[123] found a small end of season increase in flavor precursor content of bulbs, as the leaf blades died off. Nitrogen compounds were found to be transported from dying leaf blades to the bulb at the end of the season,[127] and the increased flavor precursor content could be derived from this supply.

The advice given by Platenius and Knott in 1935 "that in studying pungency in different varieties of onion or in determining ecological factors influencing pungency it is important to take into consideration the stage of maturity" remains relevant.[124]

3. Stored Bulbs

Although the dry bulb onion is a nongrowing, resting stage it is still metabolically active. Respiration rates of Rijnsburger onions stored at 7.5°C increased from 5.1 to 14.0 mg CO_2 $kg^{-1}h^{-1}$ during the 263 d of storage, when sprouting occurred.[128] What are the changes in flavor during storage and subsequent sprouting? Platenius and Knott measured volatile sulfur content of three varieties of stored onions and found a 25, 26, and 44% increase during storage.[124] Freeman and Whenham measured pyruvate, thiopropanal S-oxide, and headspace volatiles of freeze-dried samples of stored onions and found a gradual (twofold) increase during storage followed by a sharp drop after 240 d storage.[128] They interpreted these changes in terms of an increasing incidence of sprouting during storage. During sprouting transpeptidase levels rise and γ-glutamyl peptides are metabolized to free flavor precursors;[38] these are then available to alliinase, and the result is increased flavor. The sharp drop in flavor at the end of the storage period may be attributable to the metabolism and translocation of the flavor precursors themselves for nutrients for developing shoots.[128]

Research into ontogenetic changes in flavor has been confined to stages up to the sprouting of stored bulbs. The changes in flavor during the rest of the onion's life cycle from sprouted bulb to flower and seedhead formation remain to be described.

C. ENVIRONMENTAL FACTORS

It is generally observed that the same onion cultivar at the same developmental stage varies in flavor strength with different seasons. Bedford commented that seasonal variations in pungency levels occurred in the 33 onion varieties she tested.[119] As early as 1944 Platenius commented that "Italian Red Onions imported from Italy were consistently milder than bulbs of the same variety grown in New York".[120] Despite this general awareness of the

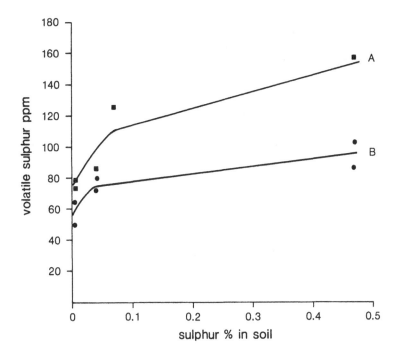

FIGURE 3. Effect of soil type (% sulfur) on the volatile sulfur content of two varieties of onions. (A) Yellow Bermuda (■—■); (B) Yellow Globe Danvers (●—●).

effect of environment on flavor strength, it has been very difficult to generate precise, predictive relationships between environment and flavor. The three environmental factors which have been investigated are temperature, water supply, and sulfur nutrition.

1. Temperature

The bulbing response in onions is in part determined by temperature. Plants grown at higher temperatures bulb more rapidly than those grown at lower temperatures. Thus, in any study on the effect of temperature on flavor it is difficult to separate a temperature effect per se from an ontogenetic effect. Platenius grew onions at 10 to 15, 15 to 21, and 21 to 27°C, with a resulting threefold increase in volatile sulfur at the higher temperature compared to the lower temperature.[120] Although the onions grown at 21 to 27°C were more developmentally advanced, and thus had a higher flavor content, the striking differences obtained from the treatment would suggest that the effect of temperature on flavor warrants further study.

2. Water Supply

The water content of the soil in which they are grown has a marked effect on the flavor of vegetables. An abundance of water produces lush, often flavorless vegetables whereas those grown with restricted water supply are often smaller and less attractive in appearance but richer in flavor. This effect of water regime has been documented for onions by Plantenius and by Freeman and Mossadeghi.[120,129] Platenius grew Early Grano, Ebenezer, and Utah Sweet Spanish onions under natural rainfall (240 mm) and irrigation (430 mm). Under the dryer conditions the onions were stronger (as measured by volatile sulfur content), but smaller, although the changes in each characteristic were not comparable. The onions were reduced 30, 51, and 47% in weight (for the three varieties, respectively), but increase in flavor was only 18, 3, and 6%, respectively. A similar increase in pungency under dry conditions was found for Yellow Globe Danvers onions grown in tubs in a glasshouse, although under those conditions both the irrigated and nonirrigated onions were very small.

Freeman and Mossadeghi also found an increase in flavor strength (as measured by headspace analysis, pyruvate, and sensory evaluation) for nonirrigated field-grown onions. In these experiments increase in flavor was correlated with a reduction in yield.

3. Sulfur Nutrition

Experimentation to find the relationship between flavor strength and soil sulfur supply has been approached in different ways.

A comparison of flavor strength of onions grown in different soil types, with a constant amount of fertilizer[120] — Two varieties of onions were grown in drums of peat, sandy loam, and sandy soil with a fertilizer application (NPK 5-10-5) of 184 kg/ha. There was a marked difference in the flavor strength of the bulbs depending on the soil type in which they were grown (Figure 3). The most pungent onions were harvested from the peat (which has a higher sulfur content), the mildest from the sandy soil.

A comparison of the flavor strength of onions grown in sulfur-deficient sand, in pots, and watered with nutrient solutions of differing sulfur content — Freeman and Mossadeghi found that at none or very low sulfate supply (0 to 0.05 meq l^{-1}) spring onion plants grew poorly and had low to nonexistent flavor strength.[130] Plants supplied with some sulfate (0.1 meq l^{-1}) grew well, but had a sulfur content and a flavor strength comparable to the deficient plants. Increasing levels of sulfate (1 to 3 meq l^{-1}) did not increase the size of the plants, but did increase their flavor strength to a level normally expected. It would be of interest to know how sulfate levels between 0.1 and 1.0 meq l^{-1} affected growth and flavor. Freeman and Mossadeghi found comparable results with garlic and wild onion (*A. vineale*).[131]

A comparison of the flavor strength of onions grown in one soil type, with increasing supply of sulfur fertilizer[132,133] — Application of sulfur at 1.8 to 5.5 kg/ha to a Texas fine sandy loam produced a 10% increase in bulb size and a small (but not significant) increase in flavor strength compared to nonfertilized soil.[132]
Application of large doses of sulfur (up to 73 kg/ha) did not significantly increase bulb size, but did increase flavor strength 2.5-fold as measured by volatile sulfur content. However, the results obtained from this sort of experiment depended on the sulfur content of the soil itself. In both of the above experiments the onions grown in the soild without fertilizer were of a good size and within the range of normal flavor strength. Field experiments on onions at three sites in the U.K. failed to demonstrate any growth or flavor intensity responses to heavy applications of calcium sulfate.[134]

The experiments outlined above indicate that sulfur is required for both growth and flavor of onions; a deficiency of sulfur reduces growth and flavor. Available sulfur is used preferentially for growth, and when growth requirements have been met, sulfur becomes available for incorporation into the biosynthetic pathway leading to flavor precursors.

In order to produce very strongly flavored onions, substantial amounts of sulfur (>18 kg/ha) need to be applied. Thus, it should theoretically be possible to supply a minimum level of sulfur which ensures good growth, but mild flavor.

IX. BIOSYNTHESIS OF FLAVOR PRECURSORS

The biosynthesis of the flavor precursors and other nonprotein sulfur amino acids involves the interaction of the carbon, nitrogen, and sulfur pathways within the plant. Granroth and Virtanen[52] and the group headed by Suzuki[51,135,136] initiated studies on the biosynthesis of the *S*-alk(en)yl cysteine sulfoxides in onion and garlic in the early 1970s. More recent work on the uptake and reduction of sulfate in *Lemna* and *Chlorella* and the biosynthesis of glutathione in tobacco cells throws light on the sulfur biosynthetic pathway in onions.[137-139] In addition current work in our laboratory elucidates the interrelationship between flavor precursors and their peptides.[140]

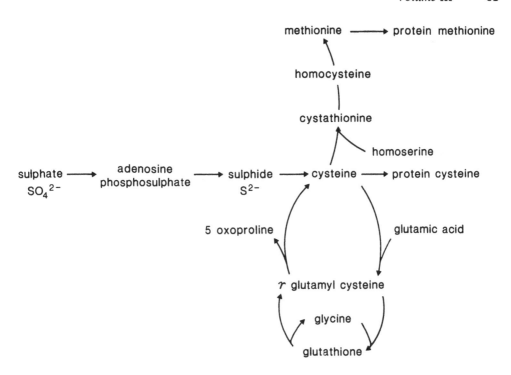

FIGURE 4.   Summary of reduction and assimilation of sulfur in plants.

The current state of knowledge about the biosynthesis of flavor precursors will be discussed in more detail below.

A. UPTAKE, REDUCTION, AND ASSIMILATION OF SULFATE

Sulfur is taken up from the soil by the roots as sulfate (SO_4^{2-}). Most of the sulfate is transported in the xylem to leaf tissue, reduced to sulfide, and assimilated into cysteine in light-dependent reactions. However, some reduction of sulfate and assimilation into cysteine can also take place in the roots. Cysteine is the principal starting metabolite for the synthesis of other sulfur-containing metabolites. From cysteine are formed: protein cysteine, methionine, protein methionine, and glutathione via their intermediates, as shown in Figure 4. The reader is referred to Anderson[137] and Giovanelli et al.[138] for further details on sulfur reduction and assimilation.

Glutathione may itself be recycled to cysteine.[139] In most plants 90% of sulfur is in the form of cysteine and methionine, which are themselves incorporated into proteins. In alliums most of the sulfur is in the form of nonprotein amino acid derivatives.

B. BIOSYNTHESIS OF FLAVOR COMPOUNDS

The earliest biosynthetic studies of alliums were by Suzuki et al.[53,135,136] who studied the incorporation of labeled sulfur into sulfur compounds of garlic. They found:

1. Labelled sulfate taken up by roots was incorporated into cysteine, methionine, *S*-allyl cysteine sulfoxide, *S*-methyl cysteine sulfoxide, and six sulfur peptides, among which were *S*-2-carboxy propyl glutathione, and γ-glutamyl *S*-methyl cysteine sulfoxide and γ-glutamyl methyl cysteine.[135]
2. Radioactivity from uniformly labeled valine was incorporated into *S*-2-carboxy propyl glutathione and *S*-2-carboxy propyl cysteine, probably via methacrylic acid which is an intermediate catabolite of valine.[53]

3. ^{35}S-labeled methionine fed to plants was incorporated into *S*-methyl cysteine and *S*-methyl cysteine sulfoxide, suggesting a thio-methyl transfer mechanism.[136]

Granroth carried out parallel studies on the biosynthesis of flavor precursors in onions, using roots, bulbs, and sprouted leaves.[54] He confirmed the incorporation of radioactivity from valine via methacrylic acid into *S*-2-carboxy propyl glutathione and *S*-2-carboxy propyl cysteine and eventually into *S*-propenyl cysteine sulfoxide.

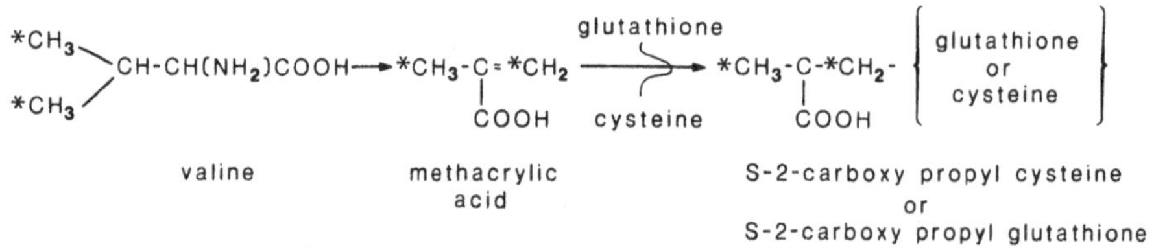

$$
\begin{array}{c}
*CH_3 \\
\quad\quad CH\text{-}CH(NH_2)COOH \longrightarrow *CH_3\text{-}C\text{=}*CH_2 \xrightarrow[\text{cysteine}]{\text{glutathione}} *CH_3\text{-}C\text{-}*CH_2\text{-} \left\{ \begin{array}{c} \text{glutathione} \\ \text{or} \\ \text{cysteine} \end{array} \right\} \\
*CH_3 \\
\end{array} \tag{6}
$$

valine   methacrylic   S-2-carboxy propyl cysteine
                  acid   or
                                       S-2-carboxy propyl glutathione

However, studies on the incorporation of labeled serine led him to suggest that *S*-propyl cysteine sulfoxide was derived from thioalkyl conjugation with serine.

$$ \text{serine} \xrightarrow{\text{propyl-}S} S\text{-propyl cysteine} \rightarrow S\text{-propyl cysteine sulfoxide} \tag{7} $$

S-Methyl cysteine sulfoxide was thought to be formed from both thioalkylation of serine and methylation of cysteine.

$$
\begin{array}{c}
\text{CH}_3\text{-}S \\
\text{serine} \longrightarrow \\
\quad\quad\quad\quad\quad S\text{-methyl cysteine} \rightarrow S\text{-methyl cysteine sulfoxide} \\
\text{cysteine} \longrightarrow \\
\text{CH}_3
\end{array} \tag{8}
$$

He also found that uptake of labeled sulfur compounds resulted in many labeled sulfur peptides as well as free flavor precursors. These early studies gave rise to two important questions:

1.   What was the relationship between peptides and free flavor precursors? Were the γ-glutamyl peptides biosynthetic intermediates or metabolites of the flavor precursors?
2.   Did the structurally homoglogous flavor precursors have a common pathway?

We have used pulse chase experiments with radioactively labeled sulfate to investigate the sequence of appearance of sulfur compounds and thus to determine the biosynthetic relationship between free flavor precursors and peptides.[140]

Onion seedlings were deprived of sulfate for 1 d. The roots were then given a 15-min exposure to high specific-activity sulfate (^{35}S) solution, rinsed, and supplied with unlabeled nutrient solution for the duration of the experiment. Seedling leaf blades were harvested at 6 h, 1, 2, 3, and 7 d after the labeled sulfate pulse. The flavor precursors and peptides were extracted from the leaf blades and purified and separated on ion exchange chromatography by elution with increasing concentrations of acetic acid.

The 0.1 *M* acetic acid fraction contained free flavor precursors and cysteine and methionine, the latter two amino acids being subsequently removed by electrophoresis. 1.0-

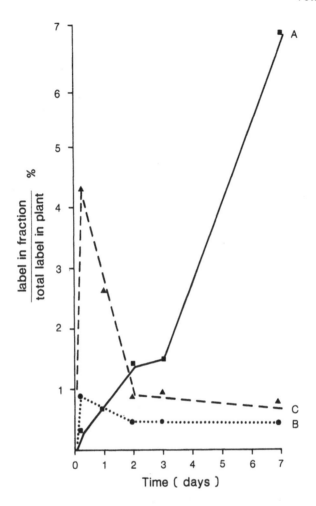

FIGURE 5. Time course study of appearance of labeled sulfate in 0.1 *M* (A), 1.0 *M* (B), and 2.0 *M* (C) fractions following a 15-min pulse of ³⁵S-labeled sulfate at time t = 0. Radioactivity in each fraction is expressed as a percentage of total leaf radioactivity.

and 2.0-*M* acetic acid fractions contained a range of sulfur peptides, γ-glutamyl *S*-propenyl cysteine sulfoxide, and *S*-methyl glutathione being known to elute in 1.0*M* acetic acid and glutathione, *S*-2-carboxy propyl glutathione, γ-glutamyl *S*-methyl cysteine and γ-glutamyl methionine eluting in 2.0 *M* acetic acid.

Radioactivity in each of these fractions was determined by liquid scintillation counting and expressed as a percentage of the total activity in the leaf blades. Figure 5 shows the amount of radioactivity in each of the fractions for the duration of the experiment. The 1.0 and 2.0-*M* acetic acid fractions showed maximum radioactivity 6 h after the pulse of labeled sulfate, dropping by the second day to a level of about 1% which was maintained thereafter. The 0.1-*M* fraction, containing only free flavor precursors, was very low at 6 h but gradually increased in radioactivity over 7 d. These results indicated that labeled sulfur was first incorporated into γ-glutamyl peptides and later into free flavor precursors, suggesting that the γ-glutamyl peptides are intermediates on the pathway to free flavor precursors.

We have also compared the sequence of appearance of sulfur compounds in alliums with different flavor precursor profiles, and the results are summarized here. Garlic, onion, and *A. albopilosum* have a predominance of *S*-allyl, *S*-propenyl, and *S*-methyl cysteine sulfoxides, respectively. Feeding the S sulfate to *A. albopilosum* in particular means that the appearance of sulfur compounds leading to only *S*-methyl cysteine sulfoxide can be determined.

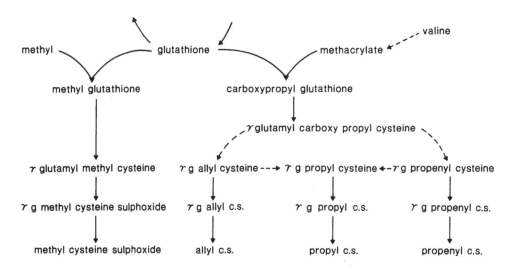

FIGURE 6.   Proposed biosynthetic pathway to flavor precursors. c.s. = cysteine sulfoxide; γg = γ-glutamyl.

A 15-min pulse of high specific-activity sulfate was fed in the transpiration stream to leaf blades of onion, garlic, and *A. albopilosum*. The leaf blades were harvested after 1- and 24-h incubation with nonlabeled nutrient solution. The sulfur peptides and free flavor precursors were extracted, purified, separated by TLC, and measured for radioactivity. *A. albopilosum* leaf blades after 1 h contained six labeled sulfur peptides, three of which cochromatographed with glutathione, S-2-carboxy propyl glutathione, and γ-glutamyl S-methyl cysteine. No label was detectable in S-methyl cysteine sulfoxide until after 24-h incubation. The results indicate that peptides are intermediate in the formation of S-methyl cysteine sulfoxide.

Garlic and onion leaf blades had similar patterns of labeled peptides; there were, in addition to the six labeled sulfur peptides found in *A. albopilosum*, a further six labeled peptides after 1 h. Two of these were identified as γ-glutamyl S-propenyl cysteine sulfoxide and γ-glutamyl methionine. It was not possible to separate S-allyl, S-propenyl, and S-propyl γ-glutamyl cysteines and γ-glutamyl sulfoxides.

Free flavor precursors were not detected until after 24-h incubation.

A biosynthetic pathway from sulfate to flavor precursors needs to account for:

1.   Known pathway of sulfate assimilation and reduction and the glutathione cycle
2.   The labeling of peptides before free flavor precursors
3.   The labeling patterns in *A. albopilosum* and the similar labeling of peptides in garlic and onions

The pathway outlined in Figure 6 meets these criteria. Glutathione is the starting compound for all sulfoxides. Methylation of glutathione produces the tripeptide methyl glutathione, the glycine moiety, which is cleaved to produce γ-glutamyl S-methyl cysteine. Oxidation of the sulfur produces γ-glutamyl S-methyl cysteine sulfoxide. The final step is the hydrolysis of the γ-glutamyl residue to produce the free S-methyl cysteine sulfoxide.

Methacrylic acid, derived from valine, reacts with glutathione to produce S-2-carboxypropyl glutathione. Cleavage of the glycine moiety produces γ-glutamyl S-2-carboxypropyl cysteine. The remaining three flavor precursors are derived from this compound. It is suggested that decarboxylation of the alkyl chain produces either allyl or propenyl residues. Oxidation of the allyl residue to propyl (as would happen in garlic) or of the propenyl residue to propyl (in onion) is suggested as a possible reaction pathway. The final steps in the pathway are the oxidation of sulfur in γ-glutamyl S-alk(en)yl cysteine to produce the sulfoxide, and the hydrolysis of the γ-glutamyl residue to produce the free *S*-alk(en)yl cysteine sulfoxides.

All of the intermediates in the pathway with the exception of γ-glutamyl *S*-allyl cysteine sulfoxide, have been reported to occur in either onion or garlic or chives (see Table 2).

In the biosynthetic pathway outlined in Figure 6 the enzyme γ-glutamyl transpeptidase has an important function in hydrolyzing the γ-glutamyl moiety to produce the free flavor precursors. The pH optimum of the enzyme in the hydrolytic mode is 6.5, a pH which is physiologically sound. When the enzyme was not expressed, one would expect the peptides to accumulate, which would explain the situation found in the stored bulb of peptides but no transpeptidase. During sprouting the enzyme is synthesized, and one would expect the final step in the pathway to be activated and peptides to be metabolized to free flavor precursors.

C. BIOSYNTHESIS IN UNDIFFERENTIATED CELLS

Undifferentiated cells of alliums grown in tissue culture produce minimal flavor when they are cut or eaten. The enzyme alliinase has been shown to occur in onion callus and in garlic callus although at slightly lower levels in the latter compared to intact plants.[141-143] An analysis for flavor precursors showed that only *S*-methyl cysteine sulfoxide was present and at lower concentrations than in the intact plant.[144] Thus, the lack of flavor in callus was due to low levels of flavor precursors. However, when callus was redifferentiated into roots and shoots, flavor precursor synthesis occurred and flavor was detectable.[141,143,145] Selby et al. fed radioactive-labeled intermediates into onion callus to determine which parts of the biosynthetic pathway were active or blocked. They found that ^{14}C-cysteine was incorporated into *S*-methyl cysteine sulfoxide, as would be expected. The feeding of ^{14}C-valine and methacrylic acid did not result in the production of *S*-propyl and propenyl cysteine sulfoxides, but feeding [14]C *S*-2-carboxypropyl cysteine did result in limited flavor precursor synthesis.[146] Callus cultures containing *S*-2-carboxypropyl cysteine in the agar had the aroma of onions when crushed. They also found that *S*-propyl cysteine, fed to callus, was oxidized to *S*-propyl cysteine sulfoxide, demonstrating oxidase activity in the callus. Selby et al. concluded that in callus tissue of onions the step just prior to the formation of the carboxypropyl derivative is blocked, but that enzymes for subsequent steps are present.[146]

The biosynthetic pathway to flavor precursors, in Figure 6, proceeds via γ-glutamyl intermediates. Selby et al., however, found that the free cysteine derivatives, *S*-2-carboxypropl cysteine and *S*-propyl cysteine, were active in callus, indicating that the enzymes in the pathway act on the alkyl chain end of the intermediates.

Undifferentiated cells are an important tool in the elucidation of the biosynthetic pathway. The induction of the enzyme which catalyzes the metabolism of *S*-2-carboxypropyl glutathione may be the key to full flavor expression in onion and garlic callus.

X. POSSIBLE ROLE OF FLAVOR PRECURSORS IN ALLIUMS

The distinctive flavors of alliums have established the plants as an essential part of most of the cuisine of the world. The function of the flavor precursors in the plants themselves still remains a subject of speculation. Plant secondary products, i.e., those compounds not directly involved in the primary plant functions of growth, respiration, photosynthesis, used to be thought of as waste product, of no value to the plant. As knowledge of the interaction between plants and other organisms and of the metabolic turnover of many secondary products has grown, ideas about the role of plant secondary products have developed, too.

It is generally considered that plant secondary products function as chemical defense mechanisms or as storage and/or transport compounds. For a detailed consideration of these roles in plants the reader is referred to the chapter by Bell.[147]

A. CHEMICAL DEFENSE

For a compound to have a role in chemical defense against insects or pathogens one would expect it to confer on the plant a resistance to a range of pests and pathogens. One might also expect to find that in the course of evolution some organisms had developed biochemical modifications which enabled them to breach this general resistance and become serious pests. The literature on alliums contains some well-documented examples of insects and fungi which have evolved very host-specific, specialized interactions with alliums.

First, the onion maggot (*Hylema antiqua* Meig.) is a serious pest of onions. The fly is attracted to the plant by the volatile sulfur compounds produced by decaying or damaged onions.[148,149] The flavor precursors themselves play an important role in oviposition. Cultivars susceptible to *H. antiqua* contain higher levels of flavor precursors than resistant ones, suggesting an obligate relationship between the fly and onions.[150]

Second, the leek moth (*Acrolepiopsis assectella*) is attracted to the alliums, particularly leek, by the labile "propanethio sulfenic acid." The mining larvae of this pest develop only on *Allium* plants, their feeding being stimulated by the flavor precursors in the cells. Young pupae of this moth are parasitized by the wasp *Diadromus pulchellus,* the locomotor activity and kinesis of the wasp being increased by the volatile sulfur compounds from leek leaves damaged by moth larvae.[151]

The other main insect to cause damage to onions is the onion thrips (*Thrips tabaci* Lind.). However, there are, in general, few insect pests of alliums.

The antimicrobial effects of the sulfur volatiles of onion and garlic are part of folklore and scientific literature. It was during an investigation of the antimicrobial effect of garlic that *S*-allyl cysteine sulfoxide was first characterized by Stoll and Seebeck.[6,7] Extracts of onion and garlic are known to have an antimicrobial effect on many plant pathogens. The reader is referred to the review by Fenwick and Hanley for more details.[152] Despite the general antimicrobial effect of garlic and onion there are some economically significant diseases of alliums, e.g., white rot, neck rot, soft rot, downy mildew. White rot (*Sclerotium cepivorum*) has evolved a very host-specific, specialized interaction with alliums. It is exclusively virulent on alliums, and the sclerotia only germinated in the soil when alliums or their extracts were present.[153]

B. STORAGE AND/OR TRANSPORT

The flavor precursors are part of a metabolic network of sulfur compounds, integrated into the plants' whole metabolism. γ-Glutamyl peptides of the flavor precursors occur in large amounts in dormant bulbs and in seeds, both of which are traditionally regarded as "storage organs". In addition, the peptides contain a nitrogen:sulfur ratio of 2:1. Thus, it has generally been considered that peptides have a function as storage of nitrogen and sulfur for use in sprouting or germination.[128]

Recent work on the biosynthesis of flavor precursors, which has shown that the peptides are intermediates in the pathway (see Section IX) raises the possibility of an alternative role for the peptides. Kasai and Larson have documented evidence of a role for peptides in the transport of amino acids across membranes.[19] A role in the transport of reduced sulfur around the plant has been suggested for the tripeptide glutathione.[139]

The possibility that γ-glutamyl peptides are transported around the cell or the plant before being hydrolyzed to the free-flavor precursors cannot be discounted.

XI. CONCLUSIONS

Onions are grown, processed, and consumed mainly for their distinctive and attractive flavor. Although the reaction of enzyme on sulfoxide flavor precursors to produce flavor is chemically complex, it has been well researched and documented. Advances in the methods of measurement of flavor have been made, but alternative methods are still possible.

However, the factors which regulate the levels of enzyme and flavor precursors are less well researched and impact greatly on the agricultural production of alliums. Knowledge of the genetic basis for inheritance of flavor is almost negligible, despite the large numbers of onion cultivars developed and the years of onion breeding which have occurred. Progress has been made on the biosynthetic regulation of the flavor precursors, but not of alliinase. Current understanding of environmental and management influences on levels of flavor precursors and alliinase is sparse. However, greater knowledge of these influences would enable alliums to be cultivated and handled to produce the desired flavor intensity and quality.

REFERENCES

1. **Moore, H. E.,** The cultivated *Alliums, Baileya,* 2, 103, 1954.
2. **Virtanen, A. I. and Spare, C. G.,** Isolation of the precursor of the lachrymatory factor in onion *(Allium cepa), Suom. Kemistil. B,* 34, 72, 1961.
3. **Moisio, T., Space, C. G., and Virtanen, A. I.,** Mass spectral studies of the chemical nature of the lachrymatory factor formed enzymically from *S*-(1-propenyl)-cysteine sulphoxide isolated from onion *(Allium cepa), Suom. Kemistil., B,* 35, 29, 1962.
4. **Whitaker, J. R.,** Development of flavour, odour and pungency in onion and garlic, *Adv. Food Res.,* 22, 73, 1976.
5. **Fenwick, G. R. and Hanley, A. B.,** The Genus *Allium.* I and II, *CRC Crit. Rev. Food Sci. Nutr.,* 22, 199, 1985.
6. **Stoll, A. and Seebeck, E.,** Alliin, the pure mother substance of garlic oil, *Experientia,* 3, 114, 1947.
7. **Stoll, A. and Seebeck, E.,** *Allium* compounds. I. Alliin, the true mother compound of garlic oil, *Helv. Chim. Acta,* 31, 189, 1948.
8. **Virtanen, A. I. and Matikkala, E. J.,** Isolation of *S*-methyl cysteine sulphoxide and *S-n*-propyl cysteine sulphoxide from onion *(Allium cepa)* and the antibiotic activity of crushed onion, *Acta Chem. Scand.,* 13, 1898, 1959.
9. **Spare, C. G. and Virtanen, A. I.,** On the lacrimatory factor in onion *(Allium cepa)* vapours and its precursor, *Acta Chem. Scand.,* 17, 641, 1963.
10. **Carson, J. F., Lundin, R. E., and Lukes, T. M.,** The configuration of (+) *S*-(1-propenyl)-L-cysteine-*S*-oxide from *Allium cepa, J. Org. Chem.,* 31, 1634, 1966.
11. **Carson, J. F. and Boggs, L. E.,** The synthesis and base-catalysed cyclization of (+)- and (−)-*cis-S*(1-propeny l)-L-cysteine sulphoxides, *J. Org. Chem.,* 31, 2862, 1966.
12. **Freeman, G. G. and Whenham, R. J.,** A survey of volatile components of some *Allium* species in terms of *S*-alk(en)yl-L-cysteine sulphoxides present as flavour precursors, *J. Sci. Food Agric.,* 26, 1869, 1975.
13. **Saghir, A. R., Mann, L. K., Bernhard, R. A., and Jacobsen, J. V.,** Determination of aliphatic mono and disulphides in *Allium* by Gas Chromatography and their distribution in the common food species, *J. Am. Soc. Hortic. Sci.,* 84, 386, 1963.
14. **Hashimoto, S., Miyazawa, M., and Kameoka, H.,** Volatile flavour components of *Allium grayi* Regal, *J. Sci. Food. Agric.,* 35, 353, 1984.
15. **Bernhard, R. A.,** Chemotaxonomy: distribution studies of sulphur compounds in *Allium, Phytochemistry,* 9, 2019, 1970.
16. **Fujiwara, M., Yoshimura, M., Tsuno, S., and Murakami, F.,** "Allithiamine" a newly found derivative of vitamin B, on the Allicin homologues in the vegetables, *J. Biochem.,* 45, 141, 1958.
17. **Hegnauer, R.,** *Chemotaxonomie der Pflanzen,* Vol. 2, Birkhauser Verlag, Basel, 1963.
18. **Tsuno, S.,** The nutritional value of *Allium* plants. XVII. Formation of *S*-2-(*n*-(2-methyl-4-amino-5-pyrimidyl methyl formamido)-5-hydroxy-2-pentene-3yl) allyl disulphide with *Ipheion uniflorum, Bitamin,* 14, 665, 1958.
19. **Kasai, T. and Larson, P. O.,** Chemistry and biochemistry of γ-glutamyl derivatives from plants including mushrooms (Basidiomycetes), in *Progress in the Chemistry of Natural Products,* Vol. 39, Herz, W., Grisebach, H., and Kirby, G. W., Eds., Springer Verlag, Vienna, 1980, 173.
20. **Virtanen, A. I. and Matikkala, E. J.,** New γ-glutamyl peptides in onion *(Allium cepa).* I. γ-Glutamyl phenylalanine and γ-glutamyl 1-*S*-(B-carboxy, B-methylethyl)-cysteinyl glycine, *Suom. Kemistil., B,* 33, 83, 1960.

21. **Matikkala, E. J. and Virtanen, A. I.,** Isolation of γ-L-glutamyl arginine and γ-glutamyl-S-(2-carboxy-n-propyl)-L-cysteine from *Allium cepa* (onion), *Suom. Kemistil. B*, 43, 435, 1970.

22. **Suzuki, T., Sugii, M., and Kakimoto, T.,** New γ-glutamyl peptides in garlic, *Chem. Pharm. Bull.*, 9, 77, 1961.

23. **Suzuki, T., Sugii, M., and Kakimoto, T.,** γ-L-Glutamyl-S-allyl-L-cysteine, a new γ-glutamyl peptide in garlic, *Chem. Pharm. Bull.*, 10, 345, 1962.

24. **Sugii, M., Suzuki, T., Nagasawa, S., and Kawashima, K.,** Isolation of γ-L-glutamyl-S-allylmercapto-L-cysteine and S-allyl mercapto-L-cysteine from garlic, *Chem. Pharm. Bull.*, 12, 1114, 1964.

25. **Matikkala, E. J. and Virtanen, A. I.,** A new γ-glutamyl peptide, γ-L-glutamyl-S-(prop-1-enyl)-L-cysteine in the seeds of chives *(Allium schoenoprasum)*, *Acta Chem. Scand.*, 16, 2461, 1962.

26. **Virtanen, A. I. and Matikkala, E. J.,** Structure of γ-glutamyl peptide 4 isolated from onion *(Allium cepa)*. γ-L-glutamyl-S-propenyl cysteine sulphoxide, *Suom. Kemistil.*, *B*, 34, 84, 1961.

27. **Virtanen, A. I. and Matikkala, E. J.,** Proofs of the presence of γ-L-glutamyl-S-(1-propenyl)cysteine sulphoxide and cycloalliin as original compounds in onion *(Allium cepa)*, *Suom. Kemistil.*, *B*, 34, 114, 1961.

28. **Virtanen, A. I.,** Antimikrobielle and antithyroide Stoffe in einigen Nahrungspflanzen, *Qual. Plant. Mater. Veg.*, 18, 8, 1969.

29. **Virtanen, A. I. and Mattila, I.,** γ-L-glutamyl-S-allyl-L-cysteine in garlic *(Allium sativum)*, *Suom. Kemistil.*, *B*, 34, 44, 1961.

30. **Virtanen, A. I., Hatanaka, M., and Berlin, M.,** γ-L-Glutamyl-S-n-propyl cystein in Knoblauch *(Allium sativum)*, *Suom. Kemistil.*, *B*, 35, 52, 1962.

31. **Matikkala, E. J. and Virtanen, A. I.,** New γ-glutamyl peptides isolated from the seeds of chives: N, N'-bis-(γ-glutamyl)-cysteine, *N*,*N*'-bis-(γ-glutamyl)-3-3'-C2-methylethylene-1,2-dithiol-dialanine, γ-glutamyl-S-propyl cysteine, *Acta Chem. Scand.*, 17, 1799, 1963.

32. **Matikkala, E. J. and Virtanen, A. I.,** Synthesis of 3,3'-C2-methlethylene-1,2-dithio) dialanine, an amino acid found as γ-glutamylpeptide in the seeds of chive *(Allium schoenoprasum)*, *Acta Chem. Scand.*, 18, 2009, 1964.

33. **Matikkala, E. J. and Virtanen, A. I.,** A new type of γ-glutamyl tripeptide, γ-glutamyl-S-(prop-lenyl) cysteinyl-S-(prop-1-enyl) cysteine sulphoxide, *Suom. Kemistil.*, *B*, 39, 201, 1966.

34. **Matikkala, E. J. and Virtanen, A. I.,** γ-Glutamyl peptidase (glutaminase) in germinating seeds of chives, *(Allium schoenoprasum)*, *Acta Chem. Scand.*, 19, 1258, 1965.

35. **Matikkala, E. J. and Virtanen, A. I.,** γ-Glutamyl peptidase in sprouting onion bulbs, *Acta Chem. Scand.*, 19, 1261, 1965.

36. **Nissen, P.,** Uptake mechanisms: inorganic and organic, *Annu. Rev. Plant Physiol.*, 25, 53, 1974.

37. **Poole, R. J.,** Transport in cells of storage tissue, in *Encyclopaedia of Plant Physiology*, (new series), Vol. 2A, Luttig, V. and Pitman, M. G., Eds., Springer, Berlin, 1976, 229.

38. **Schwimmer, S. and Austin, S. J.,** γ-Glutamyl transpeptidase of sprouted onion, *J. Food Sci.*, 36, 807, 1971.

39. **Penninckx, M. J. and Jaspers, C. J.,** Molecular and kinetic properties of purified γ-glutamyl transpeptidase from yeast *(Saccharomyces cerevisiae)*, *Phytochemistry*, 24, 1913, 1985.

40. **Kean, E. A. and Hare, E. R.,** γ-Glutamyl transpeptidase of the Ackee plant, *Phytochemistry*, 19, 199, 1980.

41. **Steinkamp, R. and Rennenberg, H.,** γ-Glutamyl transpeptidase in tobacco suspension cultures: catalytic properties and subcellular localisation, *Physiol. Plant.*, 61, 251, 1984.

42. **Kasai, T., Ohmiya, A., and Sakamura, S.,** γ-Glutamyl transpeptidases in the metabolism of γ-glutamyl peptides in plants, *Phytochemistry*, 21, 1233, 1982.

43. **Thompson, J. F., Turner, D. H., and Gering, R. K.,** γ-Glutamyl transpeptidase in plants, *Phytochemistry*, 3, 33, 1964.

44. **Goore, M. V. and Thompson, J. F.,** γ-Glutamyl transpeptidase from kidney bean fruit. I. Purification and mechanism of action, *Biochim. Biophys. Acta*, 132, 15, 1967.

45. **McIntyre T. and Curthoys, N.,** Comparison of the hydrolytic and transfer activities of rat renal γ-glutamyl transpeptidase, *J. Biol. Chem.*, 254, 6499, 1979.

46. **Tate, S. S. and Meister, A.,** Interaction of γ-glutamyl transpeptidase with amino acids, dipeptides and derivatives and analogs of glutathione, *J. Biol. Chem.*, 249, 7593, 1974.

47. **Jaspers, C. J., Gigot, D., and Penninckx, M. J.,** Pathways of glutathione degradation in the yeast *Saccharomyces cerevisiae*, *Phytochemistry*, 24, 703, 1985.

48. **Schwimmer, S. and Austin, S. J.,** Enhancement of pyruvic acid release and flavour in dehydrated *Allium* powders by γ-glutamyl transpeptidases, *J. Food Sci.*, 36, 1081, 1971.

49. **Schwimmer, S.,** Flavour enhancement of *Allium* products. U.S. Patent 3,725,085.

50. **Matikkala, E. J. and Virtanen, A. I.,** On the quantitative determination of the amino acids and γ-glutamyl peptides of onion, *Acta Chem. Scand.*, 21, 2891, 1967.

51. **Sugii, M., Suzuki, T., and Nagasawa, S.,** Isolation of (−)S-propenyl-L-cysteine from garlic, *Chem. Pharm. Bull.*, 11, 548, 1963.

52. **Granroth, B. and Virtanen, A. I.,** S-(2-Carboxypropyl) cysteine and its sulphoxide as precursors in the biosynthesis of cycloalliinin, *Acta Chem. Scand.,* 21, 1654, 1967.
53. **Suzuki, T., Sugii, M., and Kakimoto, T.,** Metabolic incorporation of L-valine-C14 into S-2-carboxypropylglutathione and S-2-carboxypropyl cysteine in garlic, *Chem. Pharm. Bull.,* 10, 328, 1962.
54. **Granroth, B.,** Biosynthesis and decomposition of cysteine derivatives in onion and other *Allium* species, *Ann. Acad. Sci. Fenn. Ser. A,* 154, 1970.
55. **Virtanen, A. I. and Matikkala, E. J.,** A new sulphur containing amino acid in the onion, *Suom. Kemistil.,* B, 29, 134, 1956.
56. **Virtanen, A. I. and Spare, C. G.,** Isolation of the precursor of the lacrimatory factor in onion, *(Allium cepa), Suom. Kemistil. B,* 34, 72, 1961.
57. **Ettala, T. and Virtanen, A. I.,** Labelling of sulphur-containing amino acids and γ-glutamyl peptides after injection of labelled sulphate into onion, *Acta Chem. Scand.,* 16, 2061, 1962.
58. **Virtanen, A. I. and Matikkala, E. J.,** Proofs of the presence of γ-L-glutamyl-S-(1-propenyl)-L-cysteine sulphoxide and cycloalliinin as original compounds in onion, *(Allium cepa), Suom. Kemistil. B,* 34, 114, 1961.
59. **Muller, A. I. and Virtanen, A. I.,** On the biosynthesis of cycloalliinin, *Acta Chem. Scand.,* 19, 2257, 1965.
60. **Stoll, A. and Seebeck, E.,** Über den Enzymatischern abbau der Alliins und die Eigenshaften der Alliinase, *Helv. Chim. Acta,* 22, 197, 1949.
61. **Tsuno, S.,** Alliinase in *Allium* plants, *Bitamin,* 14, 659, 1958.
62. **Jacobsen, J. V., Yamaguchi, M., Mann, L. K., Howard, F. D., and Bernhard, R. A.,** An alkylcysteine sulfoxide lyase in *Tulbaghia violacea* and its relation to other alliinase-like enzymes, *Phytochemistry,* 7, 1099, 1968.
63. **De Lima, D. C.,** Isoenzymes cleaving S-cysteine derivatives from cauliflower, M. S. thesis, University of California, Davis, 1974.
64. **Schwimmer, S. and Kjaer, A.,** Purification & specificity of the C-S-lyase of *Albizzia lophanta, Biochem. Biophys. Acta,* 42, 316, 1960.
65. **Nomura, J., Nishizuka, Y., and Hayaishi, O.,** S-Alkylcysteinase; enzymatic cleavage of S-methyl cysteine and its sulfoxide, *J. Biol. Chem.,* 238, 1441, 1963.
66. **Schwimmer, S. and Guadagni, D. G.,** Kinetics of the enzymatic development of pyruvic acid and odour in frozen onions treated with cysteine C-S lyase, *J. Food Sci.,* 33, 193, 1968.
67. **Kazaryan, R. A. and Goryachenkova, E. V.,** Alliinase: purification and characterization, *Biokhimiya,* 43, 1905, 1978.
68. **Tobkin, H. E.,** Onion Alliinase: Purification and Characterization of the Homogenous Enzyme, M.S. thesis, University of California, Davis, 1979.
69. **Tobkin, H. E. and Mazelis, M.,** Alliin lyase: preparation and characterization of the homogeneous enzyme from onion bulbs, *Arch. Biochem. Biophys.,* 193, 150, 1979.
70. **Ogasawdra, P. and Mazelis, M.,** reported in **Tobkin, H. E. and Mazelis, M.,** Alliin lyase: preparation and characterisation of the homogeneous enzyme from onion bulbs, *Arch. Biochem. Biophys.,* 193, 150, 1979.
71. **Schwimmer, S., Ryan, C. A., and Wong, F.,** Specificity of L-cysteine sulfoxide lyase and partially competitive inhibition by S-alkyl-L-cysteines, *J. Biol. Chem.,* 239, 777, 1964.
72. **Brosin, J. M.,** A purification and partial characterisation of alliinase from *Allium cepa,* M.S. thesis, University of California, Davis, 1969.
73. **Schwimmer, S.,** Enzymatic conversion of trans-(+)-S-1-propenyl-L-cystein-S-oxide to the bitter and odour-bearing components of onion, *Phytochemistry,* 7, 401, 1968.
74. **Kupiecki, F. P. and Virtanen, A. I.,** Cleavage of alkyl cysteine sulphoxides by an enzyme in onion, *(Allium cepa), Acta Chem. Scand.,* 14, 1913, 1969.
75. **Mazelis, M. and Crews, L.,** Purification of the alliin lyase of garlic, *Allium sativum* L., *Biochem. J.,* 108, 725, 1968.
76. **Kazaryan, R. A., Kocherginskaya, S. A., and Goryachenkova, E. V.,** Reaction of alliinase with some inhibitors, *Bioorg. Khim.,* 5, 1691, 1979.
77. **Wilkins, W. F.,** Ph.D. thesis, Cornell University, Ithaca, NY, 1961.
78. **Brodnitz, M. H. and Pascale, J. V.,** Thiopropanal S-oxide: a lacrimatory factor in onions, *J. Agric. Food Chem.,* 19, 269, 1971.
79. **Block, E., Revelle, L. K., and Bazzi, A. A.,** The lacrimatory factor of the onion: an NMR study, *Tetrahedron Lett.,* 21, 1277, 1980.
80. **Block, E., Penn, R. E., and Revelle, L. K.,** Structure and origin of the onion lacrimatory factory. A microwave study, *J. Am. Chem. Soc.,* 101, 2200, 1979.
81. **Boelens, M., deValois, P. J., Wobben, H. J., and van der Gen, A.,** Volatile flavour compounds from onion, *J. Agric. Food Chem.,* 19, 984, 1971.
82. **Abraham, K. O., Shankaranarayana, M. L., Raghavan, B., and Natarajan, L. P.,** *Alliums* — varieties, chemistry and analysis, *Lebensm. Wiss. Technol.,* 9, 193, 1976.

83. **Freeman, G. G. and Whenham. R. J.,** Synthetis *S*-alk(en)yl-L-cysteine sulphoxides-allinase fission products: simulation of flavour components of *Allium* species, *Phytochemistry*, 15, 521, 1976.

84. **Freeman, G. G. and Whenham, R. J.,** Thiopropanal *S*-oxide, alk(en)yl thiosulphinates and thiosulphonates: simulation of flavour components of *Allium* species, *Phytochemistry*, 15, 187, 1976.

85. **Brodnitz, M. H., Pascale, J. V., and Van der Slice, L. V.,** Flavour components of garlic extract, *J. Agric. Food Chem.*, 19, 273, 1971.

86. **Schwimmer, S. and Friedman, M.,** Genesis of volatile sulphur-containing food flavours, *Flavour Ind.*, 3, 137, 1972.

87. **Brodnitz, M. H., Pollock, C. L., and Vallon, P. P.,** Flavour components of onion oil, *J. Agric. Food Chem.*, 17, 760, 1969.

88. **Galetto, W. G. and Hoffman, P. G.,** Synthesis and flavour evaluation of some alkylthiophenes. Volatile components of onion, *J. Agric. Food Chem.*, 24, 852, 1970.

89. **Yamanishi, T. and Orioka, K.,** Chemical studies on the change in flavour and taste of onions by boiling, *J. Home Econ.*, 6, 45, 1955.

90. **Schwimmer, S.,** Development of a bitter substance in onion juice, *Food Tech.*, 21, 292, 1967.

91. **Shannon, S., Yamaguchi, M., and Howard, F. D.,** Precursors involved in the formation of pink pigments in onion purees, *J. Agric. Food Chem.*, 15, 423, 1967.

92. **Schwimmer, S.,** Characterisation of *S*-propenyl-L-cysteine sulphoxide as the principal endogenous substrate of L-cysteine sulphoxide lyase in onion, *Arch. Biochem. Biophys.*, 130, 312, 1969.

93. **Cavallito, C. J., Buck, J. S., and Suter, C. M.,** Allicine, the antibacterial principle of *Allium sativum*. II. Determination of the chemical structure, *J. Am. Chem. Soc.*, 66, 1952, 1944.

94. **Lukes, T. M.,** Thin-layer chromatography of cysteine derivatives of onion flavour compounds and the lacrimatory factor, *J. Food Sci.*, 662, 1971.

95. **Tewari, G. M. and Bandyopadhyay, C.,** Quantitative evaluation of lacrimatory factor in onion by thin-layer chromotography, *J. Agric. Food Chem.*, 23, 645, 1975.

96. **Freeman, G. G. and Whenham, R. J.,** A rapid spectrophotometric method of determination of thiopropanal *S*-oxide (lacrimator) in onion (*Allium cepa*) and its significance in flavour studies, *J. Sci. Food Agric.*, 26, 1529, 1975.

97. **Schwimmer, S. and Weston, W. J.,** Enzymatic development of pyruvic acid in onion as a measure of pungency, *J. Agric. Food Chem.*, 9, 301, 1961.

98. **Schwimmer, S. and Guadagni, D. G.,** Relation between olfactory threshold concentration and pyruvic acid content of onion juice, *J. Food Sci.*, 27, 94, 1962.

99. **Saguy, M., Mannheim, C. H., and Peleg, Y.,** Estimation of volatile onion aroma and flavour compounds, *J. Food Technol.*, 5, 165, 1970.

100. **Mazza, G., LeMaguer, M., and Hadziyev, D.,** Headspace sampling procedures for onion (*Allium cepa* L.) aroma assessment, *Can. Inst. Food Sci. Technol. J.*, 13, 87, 1980.

101. **Grenby, T. H. and Young, L.,** Biochemical studies of toxic agents. XII. The biosynthesis of *n*-propyl-mercapturic acid from *n*-propyl halides, *Biochem. J.*, 75, 28, 1960.

102. **Freeman, G. G. and Whenham, R. J.,** The use of synthetic (+)-*S*-1-propyl-L-cysteine sulphoxide and of alliinase preparations in studies of flavour changes resulting from processing of onion (*Allium cepa* L.), *J. Sci. Food Agric.*, 26, 1333, 1975.

103. **Stoll, von A. and Seebeck, E.,** Über die Spezifitat der Alliinase und die Synthese inmehrer dem Alliin verwandter Verbindungen, *Helv. Chim. Acta.*, 32, 866, 1949.

104. **Lancaster, J. E. and Kelly, K. E.,** Quantitative analysis of the *S*-alk(en)yl-L-cysteine sulphoxides in onion (*Allium cepa* L.), *J. Sci. Food Agric.*, 34, 1229, 1983.

105. **Barnsley, E. A., Thomson, A. E. R., and Young, L.,** Biochemical studies of toxic agents. XV. The biosynthesis of ethyl mercapturic acid sulphoxide, *Biochem. J.*, 90, 588, 1964.

106. **Nishimura, H., Mizuguchi, A., and Mizutani, J.,** Stereo selective synthesis of *S*-(*trans*-prop-1-enyl)-L-cysteine sulphoxide, *Tetrahedon Lett.*, 37, 3201, 1975.

107. **Granroth, B.,** Separation of *Allium* sulphur amino acids and peptides by thin layer electrophoresis and thin layer chromatography, *Acta Chem. Scand.*, 22, 3333, 1968.

108. **Bieleski, R. L. and Turner, N. A.,** Separation and estimation of amino acids in crude plant extracts by thin layer electrophoresis and chromatography, *Anal. Biochem.*, 17, 278, 1966.

109. **Shaw, M. L., Lancaster, J. E., and Lane, G.,** Quantitative analysis of the major γ-glutamyl peptides in onion (*Allium cepa* L.), *J. Sci. Food Agric.*, in press.

110. **Platenius, H.,** A method for estimating the volatile sulphur content and pungency of onions, *J. Agric. Res.*, 51, 847, 1935.

111. **Farber, L.,** The chemical evaluation of the pungency of onion and garlic by the content of volatile reducing substances, *Food Technol. (Chicago)*, 11, 621, 1957.

112. **Freeman, G. G.,** Distribution at flavour components in onion (*Allium cepa* L.), leek (*Allium porrum*) and garlic (*Allium sativum*), *J. Sci. Food Agric.*, 26, 471, 1975.

113. **Lancaster, J. E., McCallion, B. J., and Shaw, M. L.,** The dynamics of the flavour precursors the S-alk(en)yl-1-cysteine sulphoxides, during leaf blade and scale development in the onion, *(Allium cepa), Physiol. Plant.,* 66, 293, 1986.

114. **McCallion, B. J. and Lancaster, J. E.,** Changes in the content and distribution, in different organs, of the flavour precursors, the S-alk-(en)yl-1-cysteine sulphoxides, during seedling development of onions *(Allium cepa)* grown under light and dark regimes, *Physiol. Plant.,* 62, 370, 1984.

115. **Freeman, G. G.,** Factors affecting flavour during growth, storage and processing of vegetables, in *Progress in Flavour Research,* Land, D. J. and Nursten, H. E., Eds., Applied Science Publishers, Barking, U.K., 1979, chap. 20.

116. **Becker, A. and Schuphan, W.,** Ein Beitrag zur Biogenese und Biochemie antimikrobielle Wirkender antherischer ole der Kuchenzweibel *(Allium cepa* L.), *Qual. Plant.,* 25, 107, 1975.

117. **Lancaster, J. E. and Collin, H. A.,** Presence of alliinase in isolated vacuoles and of alkyl cysteine sulphoxides in the cytoplasm of bulbs of onion *(Allium cepa),* Plant Sci. Lett., 22, 169, 1981.

118. **Turnbull, A., Galpin, I.J., Smith, J. L., and Collin, H. A.,** Comparison of the onion plant *(Allium cepa)* and onion tissue culture. IV. Effect of shoot and root morphogenesis on flavour precursor synthesis in onion tissue culture, *New Phytol.,* 98, 257, 1981.

119. **Bedford, L. V.,** Dry matter and pungency tests on British grown onions, *J. Natl. Inst. Agric. Bot. (G.B.),* 16, 58, 1984.

120. **Platenius, H.,** Factors affecting onion pungency, *J. Agric. Res.,* 62, 371, 1944.

121. **Lancaster, J. E., Reay, P. F., Mann, J. D., Bennett, W. D., and Sedcole, J. R.,** Quality in New Zealand grown onion bulbs—a survey of chemical and physical characteristics: flavour precursors, dry matter, sugars, colour and hardness, *N. Z. J. Agric. Res.,* 16, 279, 1988.

122. **Paul, N. K., Johnston, T. D., and Eagles, C. F.,** Inheritance of S-methyl-L-cysteine sulphoxide and thiocyanate contents in forage rape *(Brassica napus* L.), *Theor. Appl. Genet.,* 72, 706, 1986.

123. **Lancaster, J. E., McCallion, B. J., and Shaw, M. L.,** The levels of S-alk(en)yl-L-cysteine sulphoxides during the growth of the onion *(Allium cepa* L.), *J. Sci. Food Agric.,* 35, 415, 1984.

124. **Platenius, H. and Knott, J. E.,** The pungency of the onion bulb as influenced by the stage of development of the plant, *Proc. Am. Soc. Hortic. Sci.,* 33, 481, 1935.

125. **Saghir, A. R., Mann, L. K., and Yamaguchi, M.,** Composition of volatiles in *Allium* as related to habitat, stage of growth and plant part, *Plant Physiol.,* 35, 681, 1965.

126. **Malkki, Y., Nikkila, O. E., and Aalto, M.,** The composition and aroma of onions and influencing factors, *J. Sci. Agric. Soc. Finl.,* 50, 103, 1978.

127. **Nilsson, T.,** The influence of the time of harvest on the chemical composition of onions, *Swed. J. Agric. Res.,* 10, 77, 1980.

128. **Freeman, G. G. and Whenham, R. J.,** Effect of overwinter storage at three temperatures on the flavour intensity of dry bulb onions, *J. Sci. Food Agric.,* 27, 37, 1976.

129. **Freeman, G. G. and Mossadeghi, N.,** Studies on the relationship between water regime and flavour strength in watercress *(Rorippa nasturtium-aquaticum* [L.] Hayek), cabbage *(Crassica oleracea capitata)* and onion *(Allium cepa), J. Hortic. Sci.,* 48, 365, 1973.

130. **Freeman, G. G. and Mossadeghi, N.,** Effect of sulphate nutrition on flavour components on onion *(Allium cepa), J. Sci. Food Agric.,* 21, 610, 1970.

131. **Freeman, G. G. and Mossadeghi, N.,** Influence of sulphate nutrition on the flavour components of garlic *(Allium sativum)* and wild onion *(A. vineale), J. Sci. Food Agric.,* 22, 330, 1971.

132. **Paterson, D. R.,** Sulphur fertilization effects on onion yield and pungency, *Tex. Agric. Exp. Stn. Prog. Rep.,* 3551, June 1979.

133. **Kumar, K. and Sahay, R. K.,** Effect of sulphur fertilization on the pungency of onion, *Curr. Sci.,* 23, 368, 1954.

134. **Freeman, G. G. and Whenham, R. J.,** Nature and origin of volatile flavour components of onions and related species, *Int. Flavours Food Additives,* 7, 222, 1976.

135. **Suzuki, T., Sugii, M., and Kakimoto, T.,** New γ-glutamyl peptides in garlic, *Chem. Pharm. Bull.,* 9, 77, 1961.

136. **Sugii, M., Nagasawa, S., and Suzuki, T.,** Biosynthesis of S-methyl-L-cysteine and S-methyl-L-cysteine sulphoxide from methionine in garlic, *Chem. Pharm. Bull.,* 11, 135, 1963.

137. **Anderson, J. W.,** Assimilation of inorganic sulfate into cysteine, in *The Biochemistry of Plants,* Vol. 5, Stumpf, P. K. and Conn, E. E., Eds., Academic Press, New York, 1980, chap. 5.

138. **Giovanelli, J., Mudd, H. S., and Datko, A. H.,** Sulphur amino acids in plants, in *The Biochemistry of Plants,* Vol. 5, Stumpf, P. K. and Conn, E. E., Eds., Academic Press, New York, 1980, chap. 12.

139. **Rennenberg, H.,** Glutathione metabolism and possible biological roles in higher plants, *Phytochemistry,* 21, 2771, 1982.

140. **Lancaster, J. E., Shaw, M. L., and Lane, G.,** γ-Glutamyl peptides in the biosynthesis of S-alk(en)yl cysteine sulphoxides in *Allium,* Phytochemistry, 28, 455, 1989.

141. **Freeman, G. G., Whenham, R. J., Mackenzie, I. A., and Davey, M. R.,** Flavour components in tissue cultures of onion *(Allium cepa* L.), *Plant Sci. Lett.,* 3, 121, 1974.

142. **Selby C. and Collin, H. A.,** Clonal variation in growth and flavour production in tissue cultures of *Allium cepa* L., *Ann. Bot. (London),* 40, 911, 1976.

143. **Malpathak, N. P. and David, S. B.,** Flavour formation in tissue cultures of garlic (*Allium sativum* L.), *Plant Cell Rep.,* 5, 446, 1986.

144. **Selby, C., Galpin, I. J., and Collin, H. A.,** Comparison of the onion plant (*Allium cepa*) and onion tissue culture. I. Alliinase activity and flavour precursor compounds, *New Phytol.,* 83, 351, 1979.

145. **Turnbull, A., Galpin, I. J., Smith, J. L., and Collin, H. A.,** Comparison of the onion plant (*Allium cepa*) and onion tissue culture. IV. Effect of shoot and root morphogenesis on flavour precursor synthesis in onion tissue culture, *New Phytol.,* 87, 257, 1981.

146. **Selby, C., Turnbull, A., and Collin, H. A.,** Comparison of the onion plant (*Allium cepa*) and onion tissue culture. II. Stimulation of flavour precursor synthesis in onion tissue cultures, *New Phytol.,* 84, 307, 1980.

147. **Bell, E. A.,** The physiological role(s) of secondary (natural) products, in *The Biochemistry of Plants,* Vol. 7, Stumpf, P. K. and Conn, E. E., Eds., Academic Press, New York, 1908, chap. 1.

148. **Ishikawa, Y., Ikeshoji, T., and Matsumoto, Y.,** A propyl thio moiety essential to the oviposition attractant and stimulant of the onion fly (*Hylemya antiqua*), *Appl. Entomol. Zool.,* 13, 115, 1978.

149. **Dindonis, L. L. and Miller, J. R.,** Host finding responses of onion fly (*Hylemya antiqua*) and seed corn flies (*Hylemya platura*) to healthy and decomposing onions (*Allium cepa*) and several synthetic constituents of onion, *Environ. Entomol.,* 9, 467, 1980.

150. **Ikeshoji, T.,** *S*-Propenylcysteine sulphoxide in exudates of onion roots and its possible decompartmental-isation in root cells by bacteria into attractant of the onion maggot, *Hylemya antiqua.* (Diptera: Antho-myiidae), *Appl. Entomol. Zool.,* 19, 159, 1984.

151. **Thibout, Auger, J., and LeComte, C.,** Host plant chemicals responsbile for attraction and oviposition in *Acrolepiopsis assectella,* Proc. 5th Int. Symp. Insect. - Plant Relationships, Pudoc, Wageningen, Visser, J. H. and Minks, A. K., Eds., 107.

152. **Fenwick, G. R. and Hanley, A. B.,** The genus *Allium.* III, CRC Crit. Rev. Food Sci. Nutr., Vol. 22, 1986.

153. **Esler, G. and Coley-Smith, J. R.,** Flavour and odour characteristics of species of *Allium* in relation to their capacity to stimulate germination of sclerotia of *Sclerotium cepivorum, Plant Pathol.,* 32, 13, 1983.

154. **Boland, M. J.,** unpublished.

155. **Boland, M. J. and Kennedy,** in preparation.

156. **Lancaster, J. E.,** unpublished results.

157. **Shaw, M.,** personal communication.

Chapter 4

PROCESSING OF ALLIUMS; USE IN FOOD MANUFACTURE

G. R. Fenwick and A. B. Hanley

TABLE OF CONTENTS

I. INTRODUCTION

Alliums have been grown for many thousands of years for their therapeutic and pro-phylactic properties, their religious significance, and their flavor and taste.[1] The Chinese, Sumerians, Indians, and Ancient Egyptians are all known to have consumed onions and garlic over 4000 years ago, and it is probable that processes were developed very early on to reduce the lachrimatory consequences of the former. Among others, Hippocrates (430 B.C.) and Theophrastus (322 B.C.) have described the consumption of onions and garlic in the Mediterranean areas, and later Apiccus (230 A.D.) listed a number of recipes in which onion was used as a seasoning. Hippocrates, the Elder Pliny (79 A.D.), and, above all, Dioscorides (1st Century A.D.) have referred to a plethora of medical uses and described numerous potions for both internal and external application.

By the early Middle Ages onions had become widespread throughout Europe, being especially important as a source of food for the peasants at the end of the winter when alternatives were scarce. Onions were by no means only a poor man's food, being included in a handwritten cookery book "The fame of Cury", completed by the chef to King Richard II of England and also being mentioned in what was probably the first printed cookbook, that of the Renaissance librarian to the Vatican.

Today alliums are used for their flavor, aroma, and taste, being prepared domestically or forming raw materials for a variety of food manufacturing processes (dehydration, freez-ing, canning, and pickling). Onions were among the earliest vegetables to be processed, canned, dried, and frozen. While the sun drying of onions and other alliums has been practiced in the tropics for many centuries, it is of variable effectiveness. Onions were among the vegetables included in early dehydration experiments, dating back to the late 18th Century. In retrospect the initiation of a major program of onion and garlic dehydration in California in 1923 may be seen as highly significant.[2] By 1941, dehydrated onion production was established in the civilian market, the products being widely used, especially in the manufacture of other processed foods.

It was, however, World War II which gave added impetus to the development of commercial food-drying techniques. The widespread distribution of the Allied Forces, the requirement for light-weight, easily handled products, for emergency long-term food storage, and the limitation in the numbers of ships available for transportation of provisions all led to a rapid expansion of dehydrated food production facilities. This has continued still further in the past 40 years to meet the demands of the food industry and the consumer.

The present chapter will concentrate on factors affecting processing quality and will contrast the natures of domestic cooking and industrial processing and their effects on end product composition and quality. While emphasis is placed mainly upon onions and garlic, consideration is also given, where appropriate, to other alliums.

II. PRODUCTION

According to the 1987 yearbook of the Food and Agriculture Organization of the United Nations,[3] total production of dry bulb onions in that year exceeded 25 million tonnes, representing an increase of almost a third over the last decade. The main production was in Asia (49%), Europe, including the U.S.S.R. (27%), and the Americas (17%). Developing countries accounted for 58% of the total.

Global garlic production in 1987 was almost 2.7 million t, almost two thirds of which was grown in Asia, especially China and Turkey.[3] Annual sales of onion spices and sea-sonings in Japan for 1984 and 1985 were 3000 and 2850 t, respectively.[4] Over three quarters of the global production was from developing countries. The increase in global production of 34% since 1974 to 1976 is mainly a result of the growth of the Asian crop. In 1984 and

1985 Japanese sales of garlic spices and seasonings were 2200 and 2100 t, respectively. In the latter year this total was made up of retail (190 t), catering (360 t), and, mainly, bulk (1550 t) sectors.[4]

Data on other alliums are not as readily available. Both leeks and chives are widely cultivated; production of both is mainly centered in Europe. The production of leeks within the European Economic Community (EEC) currently exceeds 500,000 t, with more than one half this amount being grown in France. Production in the U.K. was 45,000 t in 1985, the crop having a value of 18 million British pounds. In 1985 approximately 1100 t of chives were grown in Denmark, about 95% of which were freeze-dried. This represents almost a third of the world production, other processing countries including Peru and the U.S.[5] Three quarters of the EEC production of shallots (34,000 t) are grown in France with smaller amounts, approximately 5000 t, from the Netherlands.

III. γ-IRRADIATION

According to data cited by Salunke and Desai[6] postharvest spoilage losses of onions are 16 to 35%, with even higher figures being possible in tropical countries. Over the last decade shrinkage and loss of U.S.-grown onions has averaged 10.2%, with a high point of 12.6% in 1977.[3] Losses in Yellow Globe, Grano-Granex, and Spanish-type onions were reported to be 0.2 to 3.7, 3.0 to 4.2, and 6.4 to 8.0%, respectively, over a 3-year period,[7] according to data for the metropolitan New York area. Curing and storage will be dealt with in detail elsewhere, but a consideration of γ-irradiation is deemed pertinent, because of present interest in this technique by industry and the concerns of consumers over its introduction. However, since the topic has been critically reviewed by Thomas[8] (as part of a comprehensive coverage of radiation preservation of plant foods) and considered by Urbain,[9] it will only briefly be discussed here.

The use of γ-irradiation for the inhibition of sprouting in onions and garlic was described over 30 years ago by Dallyn et al.,[10] who reported that doses of 37 to 74 Gray (1 Gray [Gy] = 100 rad) were effective in preventing sprouting in yellow and white onions. In the intervening period many studies have been conducted, (References 1 and 8) on the application of γ-irradiation to sprout inhibition. There are considerable differences in the varieties, sizes, and agronomic histories of the onions and garlic which have been examined, and this, in no little part, explains the variations in the effectiveness of sprout control, optimal irradiation dose, and its timing. As Thomas[8] has indicated, most workers have found a dose of 20 to 120 Gray to be effective, carried out soon after harvesting, i.e., when onions are dormant.

It is obviously important that γ-irradiation should not adversely affect the composition (and thereby flavors and processing characteristics) of onions and garlic. Bandyopadhyay et al.[11] have observed no adverse effects of γ-irradiation (60 to 500 Gray) on skin color or color stability during storage, and other authors have arrived at similar conclusions. There is evidence that γ-irradiation causes darkening of inner buds or growth center.[8] This occurs, apparently, irrespective of onion type, level and time of irradiation, and subsequent storage, although the extent of darkening may be affected by these factors. Nevertheless, it appears that the darkening is rarely such as to exclude use of the product for subsequent dehydration. Firmness and texture of onion scales was not adversely affected by γ-irradiation.[12,13] Most workers have found γ-irradiation at levels considered to be economically viable, to have little effect on composition,[1,8] although Cuo et al.[14] have recently found 100 to 500 Gray γ-irradiation to significantly reduce vitamin C content.

There have been numerous studies examining the relationship between γ-irradiation and flavor in garlic and, especially, onion.[1,8] In summary it seems that such treatment may well lead to a reduction in the development of flavor, odor, and lachrimatory principle, probably caused by damage to the alliinase enzyme, but subsequent storage restores them to levels equivalent to those of nonirradiated bulbs. It is likely, therefore, that these findings are of scientific, rather than practical or economic, significance.

A number of workers have examined the toxic effects, if any, which follow the consumption of γ-irradiated onions.[1,8] No dose-related effects have been found in rats, dogs, or pigs, and the joint FAO/IAEA/WHO Expert Committee on Wholesomeness of Irradiated Food[15] has concluded that γ-irradiation of any food commodity with <10 kGray causes no toxic hazard and that hence no further toxicological testing should be necessary. Countries which have given agreement for the irradiation of bulb crops and their sale for human consumption include Belgium, Bulgaria, Canada, Czechoslovakia, France, Hungary, Brazil, the Netherlands, S. Africa, Spain, Thailand, and the U.S.S.R. Further details are given in recent reviews.[1,8] Consumer acceptablity of irradiated food products is, however, likely to remain the major stumbling block to their introduction.

IV. PROCESSING

It is appropriate to consider the advantages and disadvantages of raw, as compared to processed, flavor ingredients in the food industry. Raw alliums possess the ideal "balanced" flavor profile and provide desirable textural and water-retaining properties when incorporated into food products. On the other hand, the bulk of the raw material is water and thus lacks flavor and aroma. Moreover, handling and processing will inevitably result in the surface of the vegetable carrying a heavy load of bacteria.[16] This has been tragically borne out recently in the U.S. where sautéed onions in hot sandwiches were responsible for a large outbreak of botulism. Twenty-eight people were hospitalized, one not surviving the poisoning.[17] Since constancy of supply is an important factor for use in food processing, the fresh product has to be stored or preserved, for example, by peeling and freezing.

Reduction in bulk, by dehydration or flavor extraction, results in lower transportation and storage costs and reduced seasonal fluctuations in cost, quality, and availability. However, processing may introduce undesirable appearance changes and will cause modification of the natural "balanced" aroma and flavor. Heath[18] has listed flavor profiles of fresh onion and garlic, their dehydrated powders and oils which clearly illustrate this fact. In many cases modern automated food-processing plants are unable to handle fresh alliums, and a range of alternative processed onion and garlic products have been developed to meet the particular needs of individual market sectors.

This has led to the introduction of sophisticated processing technology and matching of product characteristic to market specification. Among processed ingredients currently available are dehydrated products, essential oils, oleoresins, or products employing a solid carrier. Depending upon the nature of the raw material and the type and conditions of processing, the end product may possess specific flavor profiles, e.g., of Spanish, Egyptian, or English onion, cooked, roasted, fried, or sautéed onion. The commercial advantages of natural and nature-identical flavorings are increasingly important, given the current attitudes of the consumer toward artificial food ingredients. Other technologies are concerned with processing the vegetable for edible purposes, e.g., canning, freezing, pickling. Increasingly, the therapeutic properties of alliums have been recognized in the processing of onion and garlic capsules, tablets, and even in the development of deodorized "health" products.

A. ONION AND GARLIC OILS

Onion oil is obtained by the distillation of minced onions which have been allowed to stand for a period of hours prior to distillation. The oil, a dark amber liquid, is obtained in 0.002 to 0.03% yield depending upon source and processing conditions. One gram of onion oil is considered to possess the flavoring strength of 4.4 kg fresh onions or 500 g onion powder.[19] The oil is free from microbiological contamination, and its use in food manufacture

TABLE 1
Uses and Concentrations
(ppm) of Onion and Garlic
Oil in Various Foodstuffs[20]

	Onion	Garlic
Baked products	1.9	6.0
Beverages	0.5	0.01—0.3
Condiments	2.2	16
Chewing gum	—	12
Confectionery	0.5	2.9
Ice Cream	0.5	40
Pickles	10	—
Meat	16	—

does not introduce unwanted color or moisture into the final product. However, the very concentration of the onion aroma makes it difficult to handle, and, in some processes, the flavor was subsequently readily lost. Moreover, because of the removal of volatiles during distillation the flavor is incomplete (often possessing a "boiled" character) and variable. According to data cited by Fenaroli[20] onion oil has been reported to be used in nonalcoholic beverages, ice cream and ices, confectionery, baked goods, condiments, meats, and pickles (Table 1).

Volatile oils comprise 0.1 to 0.25% of the fresh weight of garlic, thus being present in levels approaching tenfold that of onions. Garlic oil, recovered by steam distillation of freshly ground cloves, is a reddish-brown overpowering liquid. One gram is equivalent in flavoring terms to 900 g fresh garlic or 200 g dehydrated powder. The pungency of the product makes it difficult to use directly, and it is commonly diluted in vegetable oil or encapsulated. Garlic oil, and other products, have been discussed recently by Raghavan et al.[21] Fenaroli[20] lists garlic oil being used in beverages, ice cream and ices (at a rather surprising 40 ppm), confectionery, baked goods, chewing gum, and condiments.

B. ONION AND GARLIC JUICE

Onion juice is obtained by repeated expression of onion tissue, flash heating (140 to 160°C), then cooling to 40°C. The product is then carefully evaporated to 72 to 75% solids to facilitate preservation. The juice contains both flavor and aroma-active compounds and their precursors, as well as sugars. Thus, concentration will result in nonenzymic, Maillard reactions between sugars and amino acids and associated darkening of the product. During processing the aromatic components may be removed so that the product has a low flavor profile (which may, however, be increased by suitable blending with onion oil) and is very viscous. Moreover, its dark color makes it impossible to use in light-colored food products. The dark brown juice may be mixed with a support, such as propylene glycol, lecithin, or glucose to yield an oleoresin having a flavor intensity approximately ten times that of onion powder or a 100-fold that of fresh onions.[19]

Recently a liquid-liquid extraction plant has been opened in the U.K., having the capacity to process 5 to 10,000 t annually. After inspection, washing, cleaning, and pulping, the semisolid mass is mixed with a food-grade solvent which extracts the characteristic volatiles of fresh onion. The slurry is then pumped to special column extractors where the onion aromatics are subjected to continuous extraction into the solvent, which is then separated from the aqueous, sugar-rich phase. The volatile solvent fraction is evaporated to produce a concentrated onion extract and the solvent returned to the process. The aqueous phase is subsequently processed to a clear viscous syrup. The syrup and the volatile extracts are blended and homogenized to produce a standardized emulsion-liquid extract possessing 35 to 100 times the flavor strength of fresh onions. Garlic oleoresin is a dark viscous liquid, having 12 times the flavor of dehydrated garlic or 50 times that of fresh garlic cloves. The disadvantages of its use are as stated above for onion.

C. SOLID FLAVORINGS

In some instances, product formulation may necessitate the use of solid material. One such product, onion and garlic salt, is described below. In such circumstances the essential oils or oleoresins may be "fixed" by dispersal onto an appropriate support, e.g., dextrose, sugar, or salt, and blended until a uniform dispersion is produced. Flour- or rusk-based supports may also be used, but these will increase the extent of microbial contamination of the final product. To ensure a free-flowing product, an allowed additive such as calcium stearate may be added.[19] Another technique which has found extensive use is that of microencapsulation;[22] this describes the coating of the flavor-active components with a continuous film of polymer, the particle size of the capsules thus produced being between 0.1 and 4000 μm. Not only does this process facilitate ease of handling and standardization, but it also provides protection against oxidative and light-induced chemical changes.

A number of polymers may be used in microencapsulation, and these, and the means of processing, enable specific characteristics (e.g., heat resistance, pressure rupturing, slow release) to be achieved. For use in the food industry the polymers must be edible and nontoxic, and among the most commonly used are modified starches, gelatin, and gums. It is obviously also necessary that the polymer selected should not react with the internal flavoring compounds or the external components of the final food product. Encapsulated products are relatively expensive, possessing by weight only one fifth of the essential oil or oleoresin used in their formulation. Encapsulated garlic products are generally based upon gum acacia, gum Arabic, or modified starch. The exact flavor strength depends upon the conditions of manufacture but is generally two to four times that of garlic powder.

Onion and garlic pastes are formulated from suitable flavors and viscous edible bases. Thus, a recent patent for the former comprises onion powder (19%), onion juice (32%), roast onion oil (7%), salt (9%), seasoning (1%), lactic acid (1%), and glycerin.[23]

D. DEHYDRATED ONION AND GARLIC

A schematic representation of the process for onion and garlic dehydration[24] is shown in Figure 1. Onions for dehydration are generally white or yellow varieties, bred for strong pungency and high solids content. The latter may be determined chemically[25] or by non-invasive physicochemical methods, e.g., near infrared reflectance (NIR) spectroscopy.[26] According to Pruthi[27] the onions should ideally possess white flesh, full-globe to tall-globe shape to facilitate trimming, be of medium diameter (5 to 6 cm), and be able to be stored for long periods. Important cultivars for dehydration include White Creole, Southport White Globe, and Grano. While cv. Grano has been extensively used for processing in regions where labor and energy are cheap, it possesses low dry-matter content, yellow skin, and lacks the globe shape suited to modern dehydrating practices. Although yellow and red varieties have been used in commercial dehydration, they are usually inferior to white onion varieties. A frequently cited problem of yellow onions is their inferior taste and flavor, generally ascribed to the concentration of bitter quercetin in the flesh.

The raw bulbs should be free from rot, damage, or disease. After thorough washing the root bases and tops are removed and the outer, bacterially contaminated layers removed by lye (alkali) treatment or flaming. Following washing (with high-pressure water jets) the peeled onions are sliced or chopped. Dehydration occurs in a series of hot-air tunnels, the onion slices being placed on stainless-steel conveyors. Drying is achieved by forced hot air with the total process divided into three stages, drying at 75, 65, and 55 to 60°C, the conditions of dehydration becoming milder as the moisture content falls. Failure to adhere

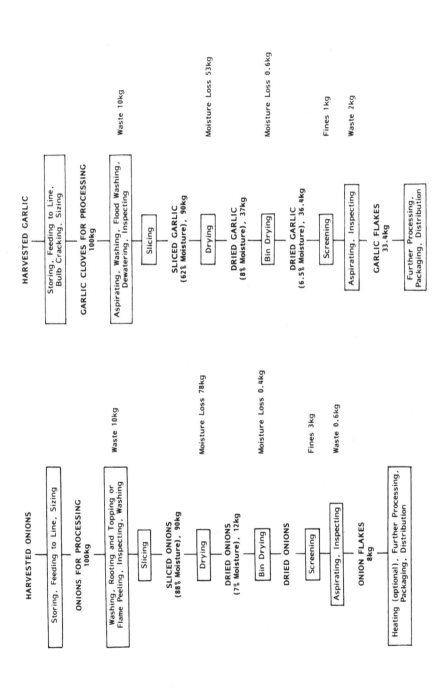

FIGURE 1. A schematic representation of the process for onion and garlic dehydration. (From Van Arsdel, B. S., Copley, M. J., and Morgan, A. I., Eds., *Food Dehydration: Practices and Applications*, Vol. 2, 2nd ed., AVI Publishing, Westport, CT, 1973.

to this stepwise process may lead to undesirable chemical reactions occurring within the product and subsequent darkening. The process is completed by placing the dehydrated onion pieces in bins where the final moisture content (~4%) is achieved via the circulation of warm air currents. A number of these aspects are described in detail by Hanson.[29] Problems, such as scorching and agglomeration, have been overcome by the use of fluidized bed dehydration techniques whereby contact between onion pieces is minimized.[30]

Bacterial contamination is not removed by these processes, and so additional sterilization treatments are required. It is necessary, however, to critically consider the high total bacterial and spore counts exhibited by such products. The levels set by many bodies for the total count in dehydrated onion and garlic is relatively meaningless since it is only certain kinds of organisms which cause concern. In particular it is far more important that a limit for coliforms (especially *Escherichia coli*) be set and that pathogens and *Salmonella* be completely absent. Pruthi[27] has discussed the microbiology of dehydrated onion in some detail. Exposure to ethylene oxide has been widely used and, if properly conducted, reduces bacterial counts very effectively without adversely affecting flavor. There are, however, concerns over the use of this chemical, and other workers have suggested the application of brining[31] or of hydrogen peroxide.[32] γ-Irradiation (<20 kGray) has also been used successfully, but there is still consumer and legislative resistance to its general use.[1] Farkas and El-Nawawy[33] reported that a dose of 5 to 6 kGray was necessary to achieve 99% reduction of microbial contamination, while Silberstein et al.[34] investigated the effect of γ-irradiation energies of 4 to 15 kGray on the bacterial contents of dehydrated onion. A dose of 9 kGray was effective in treating all qualities of onion powder. Taste and chemical analyses failed to detect any adverse effects following the γ-irradiation of onion powder with low-energy doses (4 to 40 kGray), either before or after storage. At much higher doses (270 kGray) a "cooked" onion flavor became apparent. If necessary and permissible, anticaking agents (e.g., calcium stearate) may be added. To maintain quality the product is packed in cans, pouches, or foil-laminated drums. The product may be used as such, converted into powder, granules, or flakes, or used in the formulation of other products. Toasted onion products, having a characteristic flavor of sautéed onion, may be manufactured by additional heating (e.g, 70 to 175°C for 30 min).

Farrell[19] lists 12 different types of dehydrated onion products: powdered, granulated, coarse, ground, minced, chopped, chopped special, large chopped, large chopped special, random chopped, sliced, and diced, each of which may be produced in toasted and untoasted form, the former having a somewhat lower moisture content (~3 vs. ~4%, respectively). The approximate composition of onion and garlic powders is shown in Table 2. The British Standard[35] specification for dehydrated onion recognizes four categories: powder, grits, flakes, and slices or rings. According to the International Standard[36] moisture contents for powder and grits should be <6%, with others being <8%. Total ash, acid insoluble ash, and extraneous material should be <5.5, <0.5, and <0.5, respectively. Generally 8 to 10 kg of raw onions furnish 1 kg of dehydrated product. Dehydrated onion is used widely in sauces (catsups), soups, mayonnaise, salad dressing, sweet pickles, and dog food. In 1973 Van Arsdel et al.[24] reported that tomato catsup and chili sauce contained about 1 and 4% fresh onion equivalent by weight and that almost 500 t of dehydrated onion were used in these products annually in the U.S., with about half this amount being for comminuted meat products.

When onion appearance and texture are not required, onion powder is utilized. While this may be formed by the grinding of onion pieces, a stronger flavored product is produced by spray drying. For this onions are washed, freed of debris, rinsed, and blended to a puree (optimum particle size <0.3 mm). Dextrose (30 to 40%) is added and the mixture spray dried at <68°C. The product may be further dried (4 min at 65 to 68°C), possesses excellent keeping and sensory qualities, and has a low bacterial contamination as indicated by a low bacterial plate count.

TABLE 2
Approximate Composition
(100 g^{-1}) of Onion and
Garlic Powders[19]

	Onion	Garlic
Water	5.0 g	6.5 g
Energy	347 kcal	332 kcal
Protein	10.1 g	16.8 g
Fat	1.1 g	0.8 g
Carbohydrate	80.7 g	72.2 g
Fiber	5.7 g	1.9 g
Ash	3.2 g	3.3 g
Ca	363 mg	80 mg
Fe	3 mg	3 mg
Mg	122 mg	58 mg
P	340 mg	417 mg
K	943 mg	1.1 g
Na	54 mg	26 mg
Zn	2 mg	3 mg
Vitamins	15 mg[a]	—

[a] Ascorbic acid.

Farrell[19] describes dehydrated garlic as being available in six forms, powdered, minced, coarse, granulated, chopped, and diced, depending upon increasing particle size. Moisture content should not exceed 6.75%, with hot-water insolubles <12.5%. It is generally considered that 5 kg of fresh garlic cloves yield 1 kg of dehydrated product. On the basis of chemical, microbial, and sensory studies, Pruthi et al.[37] concluded that 60°C was the maximum temperature for dehydrating garlic, and this has been confirmed by recent work.[38] A variety of techniques amenable to underdeveloped countries for the dehydration of garlic have been examined, and hot-air drying was considered to be the most economical.[39] Storage of garlic powder was best achieved in cans, the use of sealed polyethylene bags being considered unsatisfactory.[40] Little additional benefit accrued from sealing under nitrogen. Calcium stearate (<2%) may be used to prevent caking.[19] Garlic powder has sometimes a "boiled" characteristic absent in the profile of the fresh clove. "Odorless" garlic powders for use as health supplements may be produced by inactivating alliinase,[27] by contacting garlic with fumaric acid/fumarates,[41] or by physically separating garlic powder and alliinase.[42] Freeze-drying of the garlic/cyclodextrin (70:30) mixture has also been claimed to produce an odorless product.[43] According to Van Arsdel et al.[24] a rather large quantity of dehydrated garlic is used in canned dog food. In addition, it finds extensive use in processed luncheon meats and sausage, salad dressing, sauce, and soup mixes.

Both dehydrated onion and garlic are prone to discoloration. Darkening is assocated with nonenzymic, Maillard reactions[44] and may be prevented by reducing processing temperatures. The use of sulfur dioxide has been suggested but significantly reduced garlic pungency and biological activity.[45] A green discoloration[46] associated with the maceration of garlic tissue may be minimized if dormant tissue is processed, and this has been recently reexamined by Lukes.[47] Much more significant, however, is the development of a pink discoloration during onion processing.[48,49]

The problem, which results from interaction of alliinase-derived products and amino acids to yield an intermediate which can further react with carbonyl components to form the colored species, has been minimized or prevented by adding whitening agents or antioxidants. A patent granted to the Beatrice Food Corporation[50] involves the use of cysteine

(0.05 to 0.3%, by weight), followed by heat treatment (71 to 81°C for 5 to 240 s). An antioxidant such as ascorbic acid or citric acid (up to 0.3% by weight) may be added, the pungency being increased by inclusion of propionaldehyde (0.02 to 0.1%, by weight).

The dehydration exhaust emissions from onion and garlic processing plants contain highly odiferous disulfides at levels sufficient to infringe environmental pollution regulations. Odors and vapors may be controlled by scrubbers (containing alkali) and sprays which will dissolve ~90% of the products. Further removal may be achieved through catalytic incineration. It is possible to add oxidizing agents (permanganate, chlorine) to the scrubber/spray systems. McGowan et al.[51] have examined the feasibility of controlling such pollution by using ozone to oxidize the pungent sulfur compounds; the authors concluded that while the latter are controlled (60 to 90% of the pungent gaseous components being destroyed) deodorization was not feasible in less than 5 s.

E. OTHER DEHYDRATED PRODUCTS

Dehydrated chives are used as an accompaniment and garnish. In the last decade flash freezing has been developed as a method for retaining the delicate aroma and flavor of the fresh plant. The cleaned, sorted, and trimmed chives are cut into small pieces (3 mm), flash frozen, and then freeze-dried. The product is packed in nitrogen-flushed polyethylene bags and stored under cool (21°C), dry conditions. Commercial specification dictates a water content of 1 to 3%, pathogen-free, standard plate count <20,000 g^{-1}, coliform count <20 g^{-1}, 1 g of product being equivalent to 12 g of fresh chives. Reconstitution with water is not usually necessary; rather, the moisture of the host product is generally sufficient to quickly reconstitute the dehydrated chives. According to Poulsen and Nielsen[52] fast-frozen, low-pressure, dried products were lighter in color than those frozen slowly and dehydrated at 1.5 mmHg pressure.

According to Van Arsdel et al.[24] dehydrated leeks are used in soups such as vichyssoise, cream of leek, and in dry soup blends generally. The perishable nature of the fresh vegetable means that rapid processing is desirable, although if storage is necessary, conditions of 0°C, RH 90 to 95% are recommended. After washing and removal of the roots, the leek is diced, sprayed with bisulfite/sulfite mixture (final concentration 500 to 1000 ppm in the dried product). The dehydration is carried out in a similar manner to that of onions.

Green onions are grown for dehydration purposes in California. The onions are harvested by hand when the bulbs are 6 to 9 mm in diameter and thereafter processed similarly to leeks. The dehydrated product is available in flake or minced form. Dry green onions are used as seasonings, in salads, and in cottage cheese. They can be used to replace chives or shallots at a significantly reduced cost.

F. ONION AND GARLIC SALT

These comprise free-flowing, uniformly blended dry mixtures of noniodized salt, the appropriate *Allium* powder, and calcium stearate, the approximate compositions being shown in Table 3. According to U.S. military specification,[19] onion salt should contain not more than 2.5% moisture, not more than 73% salt, less than 25% moisture-free onion powder, and 1 to 2% calcium stearate. The final product should be so constituted that not less than 90% by weight should pass through a U.S. Standard No. 2 sieve. Accordingly, one tablespoon is considered equivalent to a medium-sized onion in flavoring value.

The specification for garlic salt, the composition of which is shown in Table 3, is broadly similar except that the salt content should no be more than 81% and the moisture-free white garlic powder should be between 18 to 19%. Garlic salt has much wider culinary potential than the powder, and one tablespoon is considered to be equivalent to a clove of fresh garlic.

For obvious reasons neither product should be used by persons who must guard against high sodium intakes. However, for the same individuals and others who are health conscious,

TABLE 3
Approximate Composition (100 g^{-1}) of Onion and Garlic Salt[19]

	Onion salt	Garlic salt
Water	1.3 g	1.4 g
Energy	87 kcal	63 kcal
Protein	2.5 g	3.2 g
Fat	0.3 g	0.1 g
Carbohydrate	20.2 g	13.8 g
Fiber	1.4 g	0.4 g
Ash	75.6 g	81.5 g
Ca	280 mg	220 mg
Fe	1 mg	1 mg
Mg	31 mg	11 mg
P	85 mg	79 mg
K	239 mg	212 mg
Na	29.1 mg	31.4 g
Zn	1 mg	1 mg
Vitamins	4 mg[a]	—

[a] Ascorbic acid.

the American Heart Association has recommended as a salt substitute[53] one part each of basil, black pepper, garlic powder, mace, marjoram, onion powder, parsley, sage, savory, and thyme to half a part of cayenne pepper.

G. PICKLED ONIONS

Pickled onions are popular in the U.K., comprising approximately 20% of the total pickle market (£92 million in 1984).[54] Onions are a component of other pickled products, including piccalilli (9% total U.K. market) and sweet pickle (36%). In addition to commercial production, the pickling of onions is still carried out in the home. Two main types of onions are used for pickling, the traditional "brown" onions (typically between 28 to 45 mm) and the smaller, silverskin (pearl or cocktail) onions (10 to 28 mm).

Varieties of brown onion have been developed for pickling; the main requirements, in addition to size, are shape and color. Ideally, as in "New Brown Pickling", the onion should be spherical and white or light brown in color. McGeary[55] has recently described the effects of plant density on physical and composition characteristics of pickling onions. Peeling is carried out mechanically by flaming, and thereafter hand finished. The peeled onions are subjected to a rapid brine, typically 24 to 96 h fermentation in 10% brine, lactic acid being added to control fermentation.[56] Fermentation is effected by organisms converting sugars to lactic acid, carbon dioxide, and traces of acetic acid and alcohol. After such brine fermentation the onions are washed, placed in jars, and steeped in vinegar, usually spirit vinegar, with caramel being added for coloring purposes. Pasteurization (80° in the center of the pack for 10 min) is carried out before the packs are stored. The white color of the onions, and the sparkle of the liquor, may be retained by addition of bisulfite. Occasionally yellow spots are seen in the pickled bulbs. This is a result of flavonoid-initiated chemical reaction and may be overcome by selecting flavonoid-free varieties.

Silverskin onions are fermented in 10 to 15% brine, containing 0.5% lactic acid. Thereafter the brine is diluted to 5%, packed into appropriate containers, and further diluted with vinegar to give a final brine concentration of 2%. Pasteurization is usually not employed because of adverse effects on texture and flavor. Stroup et al.[57] have examined the factors affecting acid equilibrium developed in acidified, canned pearl onions. The time required for the centers of the onions to achieve an equilibrium pH of 4.6 (at which growth of *Clostridium botulinum* is inhibited) depended upon acid type and initial acid concentration. The authors concluded that the observed rate of acidification (pH <4.6 in 7 d) was sufficient to prevent growth of this pathogen. Pickled onions have, not unexpectedly, a characteristic sour taste[58] and lack fresh onion character. Consumption in the U.K. is greatest in the winter months and averages 10.5 g per person per week. Rol and Gersons[59] have found that after fermentation, silverskin onions are able to be stored in brine for up to 2 years without significant deterioration, providing that the onion brine was fortified with lactic, rather than acetic, acid after fermentation.

H. OTHERS

Canned and bottled onions are available in the U.S. These are used domestically and by the catering industry. While canning is not as common in Europe, it is noteworthy that onions were among the earliest vegetables subjected to canning experimentation. Thorne[60] cites 300 lb of onions and 2511 lb pickled onions among the 26 ton of canned food taken by Ross on his expedition to Antarctica in 1839, the first major expedition in which canned provisions were a predominant part of the food supplies. Frozen onion products, especially onion rings, have become a significant product in the U.S. The rings are coated in batter and frozen directly or following frying in oil. The former are used mainly by caterers, the latter are used both industrially and domestically. In this context the finding[61] that onions rings had a greater effect on the corn oil used for frying than, for example, potato chips is of interest.

Originally the rings were machine produced but, to reduce wastage and improve product formulation, extrusion processes have been developed. In these the food material is forced to flow, under a variety of mixing, heating, and shearing conditions, through a nozzle which is designed to form and/or puff dry the ingredients. Rings of any desired diameter and thickness can be obtained, alginate being added to bind the rings and "stick on" the coatings, usually based on breadcrumbs.[29] The rings are marketed in bags or tubes.

According to Pederson,[62] the Chinese were the first to preserve vegetables by fermentation processes. Locally grown onions were among the fermented vegetable mixture given as rations to the builders of the Great Wall of China by Emperor Ch'in Shih Huang Ti (3rd Century B.C.). Today onions and garlic (and occasionally leek) are included in a variety of fermented food products, including kimchi (Korea), paw tsay (China), Malaysian and Egyptian pickles, enjera (enferfer, Ethiopia), ragi (Indonesia), soysauce (Indonesia), and chuanao (Thailand).[63]

The addition of onion and garlic to meat products and meat stews in the Middle Ages assisted, but probably only to a small degree, in the control of microbial spoilage given the broad range of antifungal and antibacterial activities of their sulfur volatiles.[64] Italian salamis, Genoa, Milano, Siciliano, and Cappicola are heavily flavored with garlic before drying, and Hungarian and German sausages also contain this flavoring. Fenwick and Hanley[64] have described cases of dermatitis and asthma relating to the handling of dehydrated garlic in sausage manufacture.

El-Khateib et al.[65] have recently concluded that the addition of onion and garlic (each 5 g 100 g^{-1}) to the Egyptian meat preparation "kofta" improves the microbiological qualities by decreasing the multiplication of *Salmonella* organisms. Similar effects have been noted by other workers in other meat-based products, and a patent has been filed citing the effectiveness of onion, garlic, and cinnamaldehyde mixtures in food as inhibitors of microbial growth and spoilage.[66] However, as DeWit and colleagues[67] have clearly warned, the use of these alliums to prevent the formation of food spoilage bacteria should not be encouraged because they are not effective against all types of *C. botulinum*.

The use of raw onion in sausage products may cause problems due to the introduction of excessive moisture. Thus, Baumgarten[68] has critically examined the trade literature on onion sausage in the light of suggestions that 10% extraneous water (due to 10 to 15% raw onions) be allowed. According to the code of good commercial practice a maximum of 5% onion is permitted for onion sausage.

V. EFFECTS OF STORAGE AND PROCESSING ON CHEMICAL COMPOSITION AND FLAVOR

The proximate composition of fresh and processed alliums is shown in Table 3. The composition of individual cultivars of *Allium* species varies considerably and processing, like the conditions of growth, harvesting, and storage, has a significant effect on overall product composition. Various studies[25] have revealed the storage of onions to have only a small effect on dry-weight content, although the relative proportions of reducing and non-reducing sugars vary greatly. The effect of onion storage on the color of onion flakes stored for 15 months has also been examined. Flakes made from onions stored at higher temperatures were lighter in color, a finding attributed to the lower reducing-sugar content of bulbs stored under these conditions.[71] According to Darbyshire, who examined bulbs of cv. Golden Brown Lockyer stored at 4, 15, 25, or 37° for 12 weeks, sucrose, glucose, and fructose contents increased from outer to inner leaf bases with absolute concentration increasing with rising storage temperature.[72] A rapid increase in fructose content was noted after 8 weeks in the bulbs stored at low temperatures, and it was considered that this was associated with fructan hydrolysis.[73] Fructose content has been suggested as a chemical index of onion storage, but additional work on the effect of storage conditions and of preharvest chemical treatments is necessary before the reliability of the method can be assessed.[74]

The keeping quality of dehydrated kibbled onions stored for 39 weeks under a range of conditions and temperatures has been examined.[75] Storage in cans was found to be best, with little nonenzymic browning occurring at 15°C. The effect of storage conditions on dried half-rings and powder prepared from three onion varieties has been examined by de Aguirre et al.[76] The pyruvate content of the products decreased by 50 to 75% over 180 d storage at 23°C (RH 65%). Sensory quality was better for samples stored under N_2 in cans than for those in polyethylene bags within paper sacks. Hamed and Foda[77] have examined the effect of various drying processes on the chemical composition of Egyptian onions and concluded that freeze-drying gave the best results, both in reducing undesirable color changes and in maintaining desirable chemical composition. Later chemical and sensory studies arrived at the same conclusion.[78] However, because of the cost associated with this process, it seems unlikely to find application outside of the preparation of premium-grade products. The effect of cooking and canning on onion composition has been examined. The latter study revealed that the processed product was a better source of Cl, Ca, Cu, Fe, Ni, Na, and Sn than the fresh vegetable.[79]

The effect of γ-irradiation on onion and garlic bulbs has been much studied. While high levels of γ-irradiation have produced direct or indirect evidence (e.g., change in sensory characteristics) of alteration in chemical composition, most studies have failed to identify any adverse effects at levels considered to be realistic of commercial practice.[1,8] An adverse effect on ascorbate levels has, however, been commented upon by various workers.[15]

The storage of onions has been found to be associated with an increase in flavor strength. Although in sensory studies on stored cv. Rijnsberger onions (average 6°C for 40 to 224 d) Freeman and Whenham[80] were unable to identify any differences associated with storage, chemical analysis of the same products showed higher levels of flavor precursors in the bulbs stored for longer periods. Further studies revealed that while these precursors increased up to 190 to 200 d of storage, thereafter a leveling off or significant reduction occurred, although even in the latter case, final levels remained above those of the fresh vegetable.[81]

It has been suggested that these reductions in flavor-precursors late in storage are a consequence of greatly increased rates of respiration and associated weight loss, under which conditions flavor-precursors are utilized as sources of nitrogen- and sulfur-containing nutrients necessary for growth and increased metabolic activity.[81]

Many authors, including Pruthi,[27] Bernard,[82] Boelens et al.,[83] and Freeman and Whenham[80,81] have demonstrated the processing of onion and garlic to have an adverse effect on the quality and intensity of flavor. Thus, dehydration has been associated with losses of >90% of the flavor intensity (which, if occurring industrially, would obviously represent a potential environmental hazard of major proportions). Chemical and sensory studies have revealed large losses in flavor intensity to also result from pickling (with or without pasteurization), boiling, frying, freezing, and canning. The underlying cause(s) of such flavor loss varies with the nature of the process.[78,84] Thus, boiling was observed to have little effect on flavor-precursor content but led, not unexpectedly, to inactivation of alliinase; in contrast, pickling reduced both precursor levels and inactivated the flavor-releasing enzyme. Reduction of flavor intensity as a result of endogenous alliinase inactivation clearly suggests the possiblity of flavor regeneration following addition of an exogenous source of this enzyme. This has been demonstrated by Schwimmer,[85] and, in agreement with the above, the characteristic flavor of onion was reproduced when alliinase was added to autoclaved fresh onions. In contrast similar addition to bottled cocktail onions was without effect while treatment of canned onions resulted in the formation of a sauerkraut-like odor. The absence of lachrimatory character from these products suggests that the precursor *trans*(+)-*S*-(1-propenyl)-L-cysteine sulfoxide may be particularly susceptible to thermal processing, possible yielding the flavor-inactive cycloalliin.[25]

As has been mentioned above, reduction in flavor intensity during the commercial dehydration of onion and garlic represents both a potential environmental problem and a significant economic loss. Mazza and Le Maguer[86] have examined the origins of such losses and concluded that they occur prior to the realization of a critical moisture content, whereupon the flavor may be envisaged as being sealed within the product. Accelerated drying would thus be expected to benefit retention of flavor volatiles. A significant enhancement of flavor intensity has also been claimed[87] to follow the rehydration of fried onions with cysteine solution rather than water, but more recent work[88] indicates that the effect is time dependent and selective, increasing the concentration of certain volatiles only, thus not producing the desired "balanced" note.

Problems of undesirable bitterness which may occur in onion juice and macerates have been attributed,[89] with little hard evidence, to interaction between alliinase and *trans*-(+)-*S*-(1-propenyl)-L-cysteine sulfoxide and may be overcome by initial acidification to pH 3.9 followed by adjusting to pH 5.5 to 6.0.

Differences, both qualitative and quantitative, are observed in the headspace volatiles above cut, chopped, or macerated onion, extracted and steam-distilled oils. The amount of the volatiles is related to the degree of tissue disruption, and the composition will be largely dependent upon the period between such disruption and subsequent sampling.[25] Freshly chopped onions contained lower amounts of mono-, tri-, and higher sulfides and thiophenes than did cooked onion or heated onion oils, a finding attributed to the thermal decomposition of the initially formed alk(en)yl disulfides. Analysis of the headspace volatiles above rehydrated, dried onions revealed only a fraction (2%) of the volatiles occurring above freshly cut onions.[82] Qualitative differences were also noted, these being responsible for the difference in overall character from fresh onion.

Various compounds have been suggested as imparting characteristic flavor to raw and processed onions. Methyl propyl disulfide, methyl propyl trisulfide, and dipropyl disulfide were reported to be the major flavor components of onion oil,[90] while the flavor of fresh, raw onion has been attributed to alk(en)yl thiosulfonates.[91] The sweetness of cooked onions is thought to be due to propanethiol[92] while the flavor (and that of steam-distilled onion oil) was due primarily to propyl- and 1-propenyl- di- and trisulfides.[83] As might be expected, the higher temperatures of frying result in an increase in secondary and tertiary reactions, with dimethylthiophenes being important in producing the characteristic "fried" flavor notes.[83] Ledl[93] has conducted an investigation into the volatiles resulting from frying onion in butter. More than 120 compounds were identified with the majority containing sulfur. A number of furans were also observed which were considered to have arisen during heating from the sugars present in the onion. The major volatile components of roasted onion (140°C) were, according to Kimura et al.,[94] 2,4-dimethylthiophene, methyl propyl trisulfide, and *cis*- and *trans*-propyl(1-propenyl) trisulfide.

Wu and co-workers[95-97] have carried out a number of investigations into the effect of cooking practices on the chemical composition of shallot volatiles; approximately 0.03, 0.012, and 0.005% (volume/wet weight basis) volatile oils were obtained from raw, baked, and deep-fried shallots, respectively.[95] Cooking led to an increase of dimethylthiophenes and unsaturated trisulfides while saturated trisulfides and unsaturated disulfides decreased. Baked shallot possessed larger proportions of oxygen-containing compounds (which would not be expected to contribute significantly to the flavor) while frying led to an enhancement of dimethylthiophenes, saturated alkyl disulfides, and thiols.[96] An investigation of a shallot-flavored rice product revealed baking or deep-fat frying to increase contents of thiophenes, saturated sulfides and trisulfides, unsaturated sulfides, and pyrazines.[97]

The effect of packing on the quality and volatile aroma constituents of Chinese chives (*A. tuberosum*) has revealed an enhancement of the latter when a thin vinyl coating was used.[98] Seal packaging and precooling has been shown to improve the keeping quality of this vegetable.[99] Later studies have examined the effect of a vinyl covering on the growth, quality, and chemical composition of Chinese chives. Covering reduced the sugar and chlorophyll contents.[100]

VI. UPGRADING WASTE PRODUCTS OF *ALLIUM* PROCESSING

Waste products from *Allium* processing plants have been examined with a view to their upgrading. Thus, onion juice concentrate prepared from such waste has been shown to be a viable source of yeast-derived alcohol, maximum yields of 5.6% being achieved when the concentrate was utilized as the sole culture medium (200 g/l).[101] The skins remaining from the processing on both onion and garlic have been shown to be a good source of pectin.[102] From the standpoint of chemical composition the waste products of *Allium* processing may be seen as sources of biologically active principles, but it is not thought that these are economical. Thus, the pigments of red and yellow onions may have relevance as natural colorants; in former times cyanidin glycosides from yellow onions were used as fabric dyes and for coloring Easter eggs in Germany.[25] Akaranta and Odozi[103] have furthermore suggested that components of the polyphenolic fraction obtained from red onion skins may have use as a natural antioxidant, while the skins themselves have been demonstrated to remove mercury from aqueous systems and thus may have value in cleaning up contaminated wastewater or aqueous environments.[104]

The possibility that onion wastewater may, after inoculation with suitable microorganisms (e.g., *Klebsiella* species), yield a product having potential as an attractant for pests of the onion crop has been discussed.[64] Once again the economics would appear to be against this idea. Since Okada and Fujiwara[105] have recently been granted a patent for the use of garlic extract as a stabilizer of vitamin B$_1$ in red fish meat, it may be that suitable by-products of the *Allium* processing industry can be obtained for this purpose.

A number of reports of illness and losses in stock and domestic animals resulting from acute anemia brought about by the disulfides contained therein have been reported following

TABLE 4
Proximate Composition (%) of Raw and Processed Alliums

	Moisture	Protein	Fat	Carbohydrate	Ash	Energy	Ref.
Onion							
Raw	92.8	0.9	tr	5.2		23	69
Boiled	96.6	0.6	tr	2.7		13	69
Fried	42.0	1.8	33.3	10.1		345	69
Dehydrated	4.0	8.7	1.3	82.1	3.9	350	70
Powder	6.2	23.2	0.5		3.3		27
Sauce	81.5	2.9	6.4	7.9		99	69
Garlic							
Raw	61.3	6.2	0.2	30.8	1.5	137	70
Powder	6.1	3.0	6.2		3.5		27
Salt	10.0	8.8	5.6		82.2		27
Leek							
Raw	86.0	1.9	tr	6.0		31	69
Boiled	90.8	1.8	tr	4.6		24	69

the consumption of alliums.[64] This clearly indicates that onion and garlic waste should not be used to supplement normal livestock rations.

ACKNOWLEDGMENT

The patience and persistence of Mrs. D. Davies in typing this chapter is gratefully acknowledged.

REFERENCES

1. **Fenwick, G. R. and Hanley, A. B.,** The genus *Allium.* I, *CRC Crit. Rev. Food Sci. Nutr.,* 22, 19, 1985.
2. **Pardieck, J. B.,** My four decades in dehydration, *Act. Rep. QM Food Container Inst.,* 12(2), 142, 1960.
3. *Food and Agriculture Organisation 1987 Production Yearbook,* Vol. 41, Food and Agriculture Organization, Rome, 1989.
4. *Wines, Spirits, Provisions Mon. Stat.,* 28, 81, 1986.
5. **Poulsen, N.,** private communication.
6. **Salunke, D. K. and Desai, B. B.,** Onions and garlic, in *Postharvest Biotechnology of Vegetables,* Vol. 2, CRC Press, Boca Raton, FL, 1984, 23.
7. **Ceponis, M. J. and Butterfield, J. E.,** Dry onion losses at the retail and consumer levels in metropolitan New York, *Hortic. Sci.,* 16, 531, 1981.
8. **Thomas, P.,** Radiation preservation of foods of plant origin. II. Onions and other bulb crops, *CRC Crit. Rev. Food Sci. Nutr.,* 21, 95, 1984.
9. **Urbain, W. M.,** *Food Irradiation,* Academic Press, Orlando, FL, 1986.
10. **Dallyn, S. L., Sawyer, R. L., and Sparrow, A. H.,** Extending onion storage life by gamma irradiation, *Nucleonics,* 13, 48, 1955.
11. **Bandyopadhyay, C., Tewari, G. M., and Sreenivasan, A.,** Studies on some chemical aspects of gamma irradiated onions, in *Radiation Preservation of Food,* International Atomic Energy Agency, Vienna, 1973, 11.
12. **Gruenwald, T.,** Studies on sprout inhibition of onions by irradiation in the Federal Republic of Germany, in *Food Preservation by Irradiation,* Vol. 1., International Atomic Energy Agency, Vienna, 1978, 123.

13. **Kalman, B.,** Larger scale irradiation and marketing of onions, *Food Irradiat.*, 6, 16, 1982.
14. **Cuo, A.-X., Wang, G. Z., and Wang, Y.,** Biochemical effects of irradiation on potato, onion and garlic in storage. I. Changes of major nutrients during storage, *Yuan Tzu Neng Nung, Yeh Ying Yung*, 1, 16, 1981.
15. *Wholesomeness of Irradiated Food*, Tech. Rep. Ser. 659, World Health Organization, Geneva, 1981.
16. **Heath, H. B.,** The microbiology of onion products, *Food*, 5, 22, 1983.
17. **Solomon, H. M. and Kautter, D. A.,** Growth and toxin production by *Clostridium botulinum* in sauteed onions, *J. Food Prot.*, 49, 618, 1986.
18. **Heath, H. B.,** *Flavour Technology: Profiles, Products, Applications.* AVI Publishing Co., Westport, CT, 1978.
19. **Farrell, K. T.,** *Spices, Condiments and Seasonings*, AVI Publishing, Westport, CT, 1985.
20. **Furia, T. E. and Bellanca, N., Eds.,** *Fenaroli's Handbook of Flavour Ingredients*, 2nd ed., CRC Press, Boca Raton, FL, 1975.
21. **Raghavan, B., Abraham, K. O., and Shankaranarayana, M. L.,** Chemistry of garlic and garlic products, *J. Sci. Ind. Res.*, 42, 401, 1983.
22. **Blenford, D.,** Microencapsulation, fully protected, *Food*, July, 43, 1986.
23. **Kominato, J. and Yanagida, K.,** Manufacture of onion pastes; Japanese Patent, 6,185,162,1986.
24. **Van Arsdel, B. S., Copley, M. J., and Morgan, A. I., Eds.,** *Food Dehydration: Practices and Applications*, Vol. 2, 2nd ed., AVI Publishing, Westport, CT, 1973.
25. **Fenwick, G. R. and Hanley, A. B.,** The genus *Allium*. II, *CRC Crit. Rev. Food Sci. Nutr.*, 22, 273, 1985.
26. **Birth, G. S., Dull, G. G., Renfroe, W. T., and Kays, S. J.,** Non-destructive spectrophotometric determination of dry matter in onions, *J. Am. Soc. Hortic. Sci.*, 110, 297, 1985.
27. **Pruthi, J. S.,** Spices and condiments, chemistry, microbiology and technology, *Adv. Food Res. Supp.*, 4, 198, 1980.
28. **Jones, H. A. and Mann, L. K.,** *Onions and Their Allies*, Interscience, New York, 1963, 70.
29. **Hanson, L. P.,** *Commercial Processing of Vegetables*, Noyes Data Corp., Park Ridge, NJ, 1975.
30. **Williams-Gardner, A.,** *Industrial Drying*, Gulf Publishing, Houston, 1977, 172.
31. **Firstenberg, G., Mannheim, C. H., and Cohen, A.,** Microbial quality of dehydrated onions, *J. Food Sci.*, 39, 685, 1974.
32. **Gilmore, J. and Scarlett, J. A.,** U.S. patent, 3,728,134,1973.
33. **Farkas, J. and El-Nawawy, A. S.,** Effect of gamma irradiation on the viable cell count and some other quality characteristics of dried onions, *Acta Aliment.*, 2, 437, 1973.
34. **Silberstein, O., Galetto, W., and Henzi, W.,** Irradiation of onion powder: effect on microbiology, *J. Food Sci.*, 44, 975, 1979.
35. Dehydrated Onion, British Standard Organisation, B.S. 6205, British Standards Institute, London, 1981.
36. Dehydrated Onion - Specification, International Standard ISO 5559, International Organization for Standardization, Geneva, 1981.
37. **Pruthi, J. S., Singh, L. J., and Lal, G.,** Determination of the critical temperature of dehydration of garlic, *Food Sci.*, 8, 436, 1959.
38. **Raghavan, B., Abraham, K. O., and Shankaranarayana, M. L.,** Flavour losses during dehydration of garlic and onion, *PAFAI J.*, 11, 1986.
39. **Pruthi, J. S., Singh, L. J., Kalbag, S. S., and Lal, G.,** Effect of different methods of dehydration on the quality of garlic powder, *Food Sci.*, 8, 444, 1959.
40. **Singh, L. J., Pruthi, J. S., Sankaran, A. N., Indiramma, K., and Lal, G.,** Effect of type of packaging and storage temperature on flavour and colour of garlic powder, *Food Sci.*, 8, 457, 1959.
41. **Kikkoman, K.K.,** Odourless garlic product, Japanese Patent, 5703,341,1982.
42. **Hess, H., Mehn, S., and Schoenmann, H.,** Garlic-containing oral preparation, German Patent 3541,304,1987.
43. **Kawashima, Z.,** Manufacture of odourless garlic powder, Japanese Patent, 6191,128,1986.
44. **Eriksson, C.,** *Progress in Food and Nutrition Science. Maillard Reactions in Food*, Pergamon Press, Oxford, 1981.
45. **Pruthi, J. S., Singh, L. J., and Lal, G.,** Some technological aspects of dehydration of garlic—a study of some factors affecting the quality of garlic powder during dehydration, *Food Sci.*, 8, 441, 1959.
46. **Joslyn, M. A. and Sano, T.,** The formation and decomposition of green pigment in crushed garlic tissue, *Food Res.*, 21, 170, 1956.
47. **Lukes, T. M.,** Factors governing the greening of garlic puree, *J. Food Sci.*, 51, 1577, 1986.
48. **Shannon, S., Yamaguchi, M., and Howard, F. D.,** Reactions involved in formation of a pink pigment in onion purees, *J. Agric. Food Chem.*, 5, 417, 1967.
49. **Shannon, S., Yamaguchi, M., and Howard, F. D.,** Precursors involved in the formation of pink pigments in onion puree, *J. Agric. Food Chem.*, 15, 423, 1967.
50. **Li, K.H., Bundus, R. H., and Noznick, P. P.,** Prevention of pink colour in white onions, U.S. Patent, 3,352,691,1967.

51. **McGowan, C. L., Bethea, R. M., and Tock, R. W.**, Feasibility of controlling onion and garlic dehydration odours with ozone, *Trans. Am. Soc. Agric. Eng.*, 22, 899, 1979.

52. **Poulsen, K. P. and Nielsen, P.**, Freeze drying of chives and parsley-optimisation attempts, *Bull. Inst. Int. Froid*, 59, 1118, 1979; Abstr. Cl-77.

53. American Heart Association, *Nat. Foods Merchantizer*, 5, 20, 1983.

54. **Hilliam, M. A.**, *Food Market Updates, No. 12*, British Food Manufacturers Research Association, Leatherhead, 1985.

55. **McGeary, D. J.**, The effects of plant density on the shape, size, uniformity, soluble solids content and yield of onions suitable for pickling, *J. Hortic. Sci.*, 60, 83, 1985.

56. **Anderson, K. G.**, Pickles and sauces, in *Food Industries Manual*, 21st edn., Ranken, M. D., Ed., Leonard Hill, Glasgow, 1984, 286.

57. **Stroup, W. H., Dickerson, R. W., Jr., and Johnson, M. R.**, Acid equilibrium development in mushrooms, pearl onions and cherry peppers, *J. Food Protect.*, 48, 590, 1985.

58. **Gummery, C. S.**, A review of commercial onion products, *Food Trade Rev.*, 452, 1977.

59. **Rol, W. and Gersons, L.**, The addition of lactic acid during the brining of silverskin onions, *Food Process. Mark.*, 33, 13, 1965.

60. **Thorne, S.**, *The History of Food Preservation*, Parthenon Publishing Kirby Lonsdale, 1986.

61. **Abdel-Aal, M. H. and Karara, H. A.**, Changes in corn oil during deep fat frying of foods, *Lebensm. Wiss. Technol.*, 19, 323, 1986.

62. **Pederson, C. S.**, *Microbiology of Food Fermentations*, 2nd ed., AVI Publishing, Westport, CT, 1979.

63. **Steinkraus, K. H., Ed.**, *Handbook of Indigenous Fermented Foods*, Marcel Dekker, New York, 1983.

64. **Fenwick, G. R. and Hanley, A. B.** The genus *Allium*. III, *CRC Crit. Rev. Food Sci. Nutr.*, 23, 1, 1985.

65. **El-Khateib, T., Schmidt, U., and Leistner, L.**, Wirkung von Knoblauch auf Salmonellen in ägyptischer Kofta, *Fleischwirtschaft*, 66, 1973, 1986.

66. **Sanick, I. H.**, German Patent, 2,423,076,1974.

67. **De Wit, J. C., Notermans, S., Gorin, N., and Kampelmacher, E. H.**, Effect of garlic oil or onion oil on toxin production by *Clostridium botulinum* in meat slurry, *J. Food Prot.*, 42, 222, 1979.

68. **Baumgarten, H. J.**, Zwiebelwurst Ermittlung der Verkehrsauffassung, *Fleischwirtschaft*, 65, 798, 1985.

69. **Widdowson, E. M. and McCance, R. A.**, *The Composition of Foods*, 3rd ed., Spec. Rep. Ser. No. 297, Her Majesty's Stationery Office, London, 1960.

70. **Watt, B. K. and Merrill, A. L.**, Composition of Foods; Raw, Processed, Prepared, Agricultural Handbook, No. 8, U.S. Department of Agriculture, Washington, D.C., 1963.

71. **Yamaguchi, M., Pratt, H. K., and Morris, L. L.**, Effect of storage temperature on keeping quality and composition of onion bulbs and on subsequent darkening of dehydrated flakes, *Proc. Am. Soc. Hortic. Sci.*, 69, 421, 1957.

72. **Darbyshire, B.**, Changes in the carbohydrate content of onion bulbs stored for various times at different temperatures, *J. Hort. Sci.*, 53, 195, 1978.

73. **Edelman, J. and Jefford, T. G.**, The mechanism of fructosan metabolism in higher plants as exemplified in *Helianthus tuberosum*, *New Phytol.*, 67, 517, 1963.

74. **Rutherford, R. P. and Whittle, R.**, Methods of predicting long term storage of onions, *J. Hortic. Sci.*, 59, 537, 1984.

75. **Peleg, Y., Mannheim, C. H., and Berk, Z.**, Changes in quality of dehydrated, kibbled onions during storage, *J. Food Sci.*, 35, 513, 1970.

76. **de Aguirre, A. M., Travaglini, D. A., Baldini, V. L. S., Silveira, E. T. F., and de Campos, S. D.**, Deterioration of quality of dried onions during storage, *Bol. Inst. Tecnol. Aliment. Sao Paulo*, 22, 91, 1985.

77. **Hamed, M. G. E. and Foda, Y. H.**, Freeze drying of onions, *Z. Lebensm. Unters. Forsch.*, 130, 220, 1966.

78. **Freeman, G. G. and Whenham, R. J.**, Changes in onion (*Allium cepa* L.) flavour components resulting from some postharvest processes, *J. Sci. Food Agric.*, 25, 499, 1974.

79. **Lopez, A. and Williams, H. L.**, Essential elements in fresh and canned onions, *J. Food Sci.*, 44, 887, 1979.

80. **Freeman, G. G. and Whenham, R. J.**, Flavour changes in dry bulb onions during overwinter storage at ambient temperature, *J. Sci. Food Agric.*, 25, 517, 1974.

81. **Freeman, G. G. and Whenham, R. J.**, Effect of overwinter storage at three temperatures on the flavour intensity of dry bulb onions, *J. Sci. Food Agric.*, 27, 37, 1976.

82. **Bernhard, R. A.**, Comparative distribution of volatile aliphatic disulphides derived from fresh and dehydrated onions, *J. Food Sci.*, 33, 298, 1968.

83. **Boelens, M., deValois, P. J., Wobben, H.J., and van der Gern, A.**, Volatile flavour compounds from onions, *J. Agric. Food Chem.*, 19, 984, 1971.

84. **Freeman, G. G. and Whenham, R. J.**, The use of synthetic ($\pm$)-S-1-propyl-L-cysteine sulphoxide and of alliinase preparations in studies of flavour changes resulting from processing of onion (*Allium cepa*) L., *J. Sci. Food Agric.*, 26, 1333, 1975.

85. **Schwimmer, S.,** Alteration of the flavour of processed vegetables by enzyme preparations, *J. Food Sci.,* 28, 460, 1962.
86. **Mazza, G. and Le Maguer, M.,** Volatiles retention during the dehydration of onion (*Allium cepa* L), *Lebensm, Wiss. Technol.,* 12, 333, 1979.
87. **Schwimmer, S. and Guadagni, D. G.,** Cysteine-induced odour intensification in onions and other foods, *J. Food Sci.,* 32, 405, 1967.
88. **Mazza, G., Le Maguer, M., and Hadziev, D.,** The effect of cysteine on volatile development in dehydrated onions (*Allium cepa* L.) *Can. Inst. Food Sci. Technol. J.,* 12, 43, 1979.
89. **Schwimmer, S.,** Enzymatic conversion of trans(+)-S-(1-propenyl)-L-cysteine-S-oxide to the bitter and odour bearing components of onion, *Phytochemistry,* 7, 401, 1968.
90. **Galetto, N. G. and Bednarczyk, A. A.,** Relative flavour concentration of individual volatile components of the oil of onion, *(Allium cepa), J. Food Sci.,* 40, 1165, 1975.
91. **Brodnitz, M. H., Pascale, J. V., and Vock, M. H.,** Propyl propene thiosulphonate and novel flavouring compositions containing it and processes for preparing them, British Patent, 1,313,813,1973.
92. **Yamanishi, T. and Orioka, A.,** Chemical studies on the change of flavour and taste of onion by boiling, *J. Home Econ.,* 6, 45, 1955.
93. **Ledl, F.,** Untersuchung des Rostzwiebelsaromas, *Z. Lebensm. Unters. Forsch.,* 157, 229, 1975.
94. **Kimura, K., Nishimura, H., Kimura, I., Iwata, I., and Mizutani, J.,** Changes in flavour components of onion by roasting, *Nippon Eiyo Shokuryo Gakkaishi,* 37, 343, 1984.
95. **Wu, C. M., Chou, C. C., Chen, M. H., and Wu, C. M.,** Volatile flavour components from shallots, *J. Food Sci.,* 47, 606, 1982.
96. **Wu, C. M. and Wu, L.-P.,** Flavour chemistry of *Allium cepa* L. var. *aggregatum* and its application in foods, *Chem. Abst.,* 98, 12443q, 1983.
97. **Wu, J. L. and Wu, C. M.,** Flavour chemistry of shallot and its application in canned rice rood, *Chem. Abst.,* 8, 177797q, 1983.
98. **Saito, S. and Takama, F.,** Effect of vinyl film covering on the growth, quality and chemical composition of vegetables. IV. Effect on the quality and volatile aroma constituents of the Chinese chive, *Allium tuberosum* Rottler et Springel, *J. Agric. Sci. (Tokyo),* 21, 177, 1976.
99. **Ishii, K. and Okubo, M.,** The keeping quality of Chinese chive (*Allium tuberosum* Rottler) by low temperature and seal-packaging with poly-ethylene bag, *J. Jpn. Soc. Hortic. Sci.,* 53, 87, 1984.
100. **Saito, S. and Takahashi, Y.,** Effect of vinyl covering on the growth, quality and chemical composition of vegetables. V. Effect on growth, sugars and chlorophyll contents of the Chinese chive (*Allium tuberosum* Rottler et Sprengel), *J. Agric. Sci. (Tokyo),* 29, 122, 1984.
101. **Abdel-Fatteh, A. R., Abou-Zeid, A.-Z. A., and Farid, M. A.,** Production of ethyl alcohol by *Saccharomyces cerevisiae* including utilization of onion juice, *Agric. Wastes,* 9, 101, 1984.
102. **Alexander, M. M. and Sulebele, G. A.,** Pectic substances in onion and garlic skins, *J. Sci. Food Agric.,* 24, 611, 1973.
103. **Akaranta, O. and Odozi, T. O.,** Antioxidant properties of red onion skin (*Allium cepa*) tannin extracts, *Agric. Wastes,* 18, 299, 1986.
104. **Asai, S., Konishi, Y., and Tomisaki, H.,** Separation of mercury from aqueous mercuric chloride solutions by onion skins, *Sep. Sci. Technol.,* 21, 809, 1986.
105. **Okada, A. and Fujiwara, K.,** Stable vitamin B$_1$ additive for fish feeds, Japanese Patent, 6,236,153,1987.

Chapter 5

THERAPEUTIC AND MEDICINAL VALUES OF ONIONS AND GARLIC

K. Thomas Augusti

TABLE OF CONTENTS

I. THERAPEUTIC AND MEDICINAL VALUES OF ONIONS
(*ALLIUM CEPA* L.)

A. Traditional Uses of Onions

Onions are widely used in all parts of the world as a flavoring vegetable in various types of food. According to traditional medical literature,[1] they are a source of many vitamins and are useful in fever, dropsy, cattarrh, and chronic bronchitis, mixed with common salt, onions are a domestic remedy in colic and scurvy. Roasted or otherwise they are applied as a poultice to indolent boils, bruises, wounds, etc., to relieve hot sensations and applied to the navel for dysentery and fever. Raw onion has a antiseptic value throughout the alimentary canal that is more effective than when roasted or cooked. Eaten raw it is also a diuretic and emmenagogue. Warm juice is dropped into the ear to relieve ear ache, applied hot to the soles of feet in convulsive disorders. It is sniffed in epistaxis. It is applied to eyes in dimness of vision and locally to allay the irritation of insect bites and scorpion stings, and of skin diseases. It is given as an antidote in tobacco poisoning. Mixed with mustard oil in equal proportions it is a good application for rheumatic pains and other inflammatory swellings. Onions eaten with unrefined sugar stimulate the growth of children. Oil contained in the bulbs is a stimulant, diuretic, and expectorant. Mixed with vinegar, onions are useful in cases of sore throat. Cooked with vinegar they are given in jaundice, splenic enlargement, and dyspepsia. In malarial fevers they are eaten twice a day with remarkable relief. Roasted onions mixed with cumin, sugar candy, and butter oil are a demulcent of great benefit in piles.[1] Fresh onion juice is moderately bactericidal. The essential oil (0.05% of bulbs) contains a heart stimulant, increases pulse volume and frequency of systolic pressure and coronary flow, and stimulates the intestinal smooth musculature and the uterus. It promotes bile production and reduces blood sugar.[2] Extracts of onion when given orally to diabetic dogs[3] or injected into rabbits[4] showed an action similar to that of insulin. In addition to the hypoglycemic principle a hyperglycemic substance is also present.

B. Chemistry

Fresh onion contains about 86.8% moisture, 11.6% carbohydrate including 6 to 9% soluble sugars, 1.2%protein, 0.1% fat, 0.2 to 0.5% calcium, 0.05% phosphorus, and traces of Fe, Al, Cu, Zn, Mn, and I. Vitamins A, B_1, B_2, C, nicotinic acid, and pantothenic acid are also present in onion, their quantities being 25 IU, 64 mg, 79 mg, 11 mg, 0.77 mg, and 1.42 mg per 100 g onion, respectively.[5,6] Onion contains an acrid volatile oil (0.05%) with a pungent smell. The oil is rich in sulfur and contains a variety of aliphatic disulfides including allyl or propenyl propyl disulfide, dipropyl disulfide, methyl propyl disulfide, and their trisulfides. However, the chief component is propenyl propyl disulfide, an isomer of allyl propyl disulfide.[7,8]

The precursors of onion oil are the cysteine sulfoxide derivatives of methyl, allyl, propenyl, and poropyl groups, viz.:

$R-S-CH_2-CH-(NH_2)COOH$ where R is any of the alkyl or allyl groups CH_3, C_3H_5, C_3H_7,
|
O

$(C_nH_{2n} - 1)$ and $(C_nH_{2n} + 1)$, etc. These amino acids are known as alliins. Onion contains an enzyme called alliinase which is released on crushing the vegetable and which converts alliins to disulfide oxides. These oxides are allicin-type compounds.[8-11]

$$2R-S-CH_2CH(NH_2)COOH \xrightarrow{\text{alliinase}} R_1-S-S-R_2 + 2NH_3 + 2CH_3CO-COOH \quad (1)$$
$$\text{O} \qquad\qquad\qquad\qquad\qquad \text{O (allicin)}$$

If R = CH_3, C_3H_5, C_3H_7, etc., then the allicin will be a mixed type as methyl propyl, allyl propyl, propenyl propyl, etc.

methyl cysteine sulfoxide + propyl cysteine sulfoxide →

methyl propyl sulfide sulfoxide + ammonia + pyruvic acid   (2)

These allicin-type compounds may rearrange to disulfides and thiosulfonates as follows:

$$2R_1-\underset{\underset{O}{|}}{S}-S-R_2 \rightarrow R_1-S-S-R_2 + R_1-\underset{\underset{O}{|}}{\overset{\overset{O}{}}{S}}-S-R_2 \qquad (3)$$

C. Antibacterial Property

Fresh onion juice is moderately bactericidal because of the action of allicin-type compounds. Allicins and disulfides interact with −SH(thiol) compounds like cysteine and prevent their incorporation into proteins.

$$R_1-\underset{\underset{O}{|}}{S}--S--R_2 + 2R_3SH \rightarrow R_1-S-S--R_3 + R_2-S-S--R_3$$
$$+ H_2O$$

allicins   thiols   mixed disulfides   (4)

$$R_1-S-S--R_2 + R_3SH \rightarrow R_1-S-S--R_3 + R_2SH \qquad (5)$$

Such reactions[11-13] inhibit the growth of bacteria.

D. Blood Sugar-Lowering Effects

The hypoglycemic action of onion extract was reported for the first time by Collip[3,14,15] just after the discovery of insulin. He isolated a fraction from sprouting onions that lowered blood sugar when given orally and demonstrated its effects in fasting and depancreatized dogs. These findings were confirmed by Janot and Laurin,[4] Laurin,[16] Kreitmair,[17] and Brahmachari and Augusti.[18,19] The principle was active on injection also.[4] In addition to the hypoglycemic substance a fraction which raises blood sugar was also reported to be present in onions. Mathew and Augusti[20] separated these two fractions from ether-oil extracts from fresh onions. Petroleum ether-soluble fractions of the diethyl ether onion extract contained the hyperglycemic fraction. The insoluble component contained the hypoglycemic fraction. The hyperglycemic fraction A initially produced hyperglycemia followed by hypoglycemia at the third hour after the dose in fasting animals. Thy hypoglycemic fraction B at 250mg/ kg body weight produced only hypoglycemia during a 4-h period in fasting rabbits. Fraction A had 34% and fraction B had 97% activity of the same dose of the antidiabetic drug tolbutamide. Fraction B also effectively improved glucose tolerance as compared to tolbutamide in normal alloxan-induced diabetic rabbits.[21] Gas chromatographic analysis[8] of fraction A showed that it contained methan and propane thiols (CH_3SH, C_3H_7SH); dimethyl disulfide ($CH_3-S-S-CH_3$), an isomer of allyl propyl disulfide, viz. propenyl propyl disulfide ($C_3H_5S-S-C_3H_7$), dipropyl disulfide ($C_3H_7-S-S-C_3H_7$), methyl propyl disulfide ($CH_3-S-S-C_3H_7$), and related compounds, while fraction B contained dimethyl sulfide, methyl propenyl disulfide ($CH_3-S-S-CH=CH-CH_3$), and traces of allyl thiol ($CH_2=CH-CH_2-SH$). In both the fractions the disulfides were the major sulfur compounds, and most of them were unsaturated. The hyperglycemic action of fraction A may be attributed to its thiol compounds as they inactivate insulin, and both the 3-h-delayed hypoglycemic action of A and the steady hypoglycemic action of B may be attributed to their unsaturated disulfides and related compounds.[8] Paper chromatography on fraction B[22] showed the presence of *S*-methyl cysteine sulfoxide and *S*-propyl cysteine sulfoxide, the precursors of the disulfides and their oxides (allicins) in onions, All these sulfur compounds can remove thiols, which inactivate insulin, by oxidation or thiol disulfide exchange reactions.[13,23] (See Equations 4 and 5.)

$$R_1S-CH_2CH(NH_2)COOH \rightarrow R_1-S-CH_2-CH(NH_2)COOH +$$
$$\underset{O}{\mid}$$
$$+ R_2-SH \qquad\qquad R_2-S-S-R_2 + H_2O$$
$$\text{Thiol} \tag{6}$$

Here alkyl cysteine sulfoxide is converted to alkyl cysteine, and thiol is converted to a disulfide. In addition the sulfoxide amino acid can react with NADPH.[24]

$$R_1-S-CH_2CH(NH_2)COOH + NADPH + H^+ \rightarrow R-S-CH_2CH(NH_2)COOH \tag{7}$$
$$\underset{O}{\mid} \qquad\qquad\qquad\qquad\qquad + NADP^+ + H_2$$

Also oxidation of NADPH[23] may prevent reduction of insulin.

The steam-distilled volatile oil of onion is a mixture of disulfides, the chief component of which is an isomer of allyl propyl disulfide,[8] viz., propenyl propyl disulfide ($C_3H_5-S-S-C_3H_7$). A shift of the double bond from carbon one to carbon three makes these isomers: $CH_2=CH-CH_2$ allyl group; $CH_3-CH=CH$-propenyl group. These disulfides are formed by the oxidation of different thiols of both alkyl and allyl groups such as CH_3, C_3H_5, C_3H_7, etc. Onion oil demonstrated hypoglycemic action in normal mice,[8] alloxan diabetic rabbits,[21,25] and human beings.[26] In humans blood sugar reduction was accompanied by a small but significant increase in insulin level which effectively maintained a steady fatty acid level in fasting subjects. Sulfide bonds, unsaturation, and oxide function of the onion principles may be responsible for their biological action.[8,27] Recently, another antihyperglycemic agent present in onions has been identified, viz., diphenylamine.[28] This is the first nonsulfur compound of onion claimed to have hypoglycemic action.

E. Lipid-Lowering Effects

The lipid-lowering effects of onion and garlic extracts were first reported by Augusti and Mathew.[29] According to Sharma et al.[30] onion juice prevented the rise of serum cholesterol in rabbits fed a high cholesterol diet. Bordia et al.[31] showed that onion oil could prevent the accumulation of lipid in the aorta of cholesterol-fed rabbits. During the last decade the hypolipidemic effects of onion and garlic have been attributed to their sulfur-containing compounds. Sulfur-sulfur bonds or sulfur-monoxide bonds with unsaturated side chains are a common feature of the onion and garlic principles. Garlic products contain a higher proportion of unsaturated compounds. Aqueous extracts as well as oil of onion counteracted the lipogenic effects of sucrose, alcohol, and high-fat diets, in experimental animals, and the mechanism of action was attributed to the properties of sulfur compounds in the active fractions (see Equations 1 to 7). The disulfides in the oil of onions react with the −SH group enzymes and NADPH, which are necessary for both cholesterol and lipid synthesis and inactivate them by thiol disulfide reactions and oxidation, respectively, as follows:

$$R_1-S-S-R_2 + R_3-SH \rightarrow R_1S-S-R_3 + R_2-SH \tag{as in 5}$$

$$NADPH + H^+R_1\text{--}S\text{--}S\text{--}R_2 \rightarrow NADP^+ + R_1\text{--}SH + R_2\text{--}SH \tag{8}$$

$$\text{If } R_1 = C_3H_5, \quad \text{then} \quad C_3H_5SH + NADPH + H^+ \rightarrow C_3H_7\text{--}SH + NADP^+ \tag{9}$$

Sulfoxides present in onion principles may also oxidize NADPH and prevent lipogenesis (see Equation 7). By reduction, unsaturated allyl and propenyl groups (C_3H_5 –) become the propyl group (C_3H_7 –). The inactivation of oxidation of the thiol group enzymes and NADPH, needed for lipid synthesis by the onion sulfur compounds, may partly explain the lipid-lowering effects of onion and its products.[32] Furthermore, Itokawa et al.[33] found that the sulfur amino acids, S-methyl cysteine sulfoxide, present in both onion and cabbage, and S-allyl cysteine sulfoxide present in both onion and garlic, could counteract the hypercholesterolemic effects of high-fat cholesterol diets. These are the precursors of various alkyl disulfides and their oxides, and as all these sulfur compounds are hypolipidemic in action, sulfur-sulfur bonds, unsaturated side chains, and sulfur-oxide bonds may have a role in their biological activities.

Jain[34] reported that onion juice could not prevent cholesterol-fed atherosclerosis in rabbits, whereas garlic juice had a significant ameliorating effect. An explanation may be that the concentration of active principle in the juice of onions is very low as compared to that of garlic juice. Augusti reported in one of his papers[35] that regular use of onion by a diabetic patient lowered his insulin dosage considerably. Giving volatile onion disulfides significantly counteracted the lipogenic effects of diabetes in rats.[36]

In the Jain community of India different individuals in the same families have widely varying eating habits with regard to onion and garlic. The people who have onion and garlic consume 50 to 80 g onions and 7 to 10 g garlic daily. They have significantly lower body fats, and their serum levels of cholesterol, triglycerides, β-lipoprotein, and phospholipids are significantly lower than those who abstain from onion and garlic.[37] Singh and Kanaharaj[38] reported that incubation of cholesterol-enriched erthrocytes with onion extract brought back the cholesterol level in the former to normal. Tables 1 to 3 show the effects of feeding onion oils to rats on various components of fat metabolism.

F. Antiplatelet-Aggregation Effects

Mittal et al.[39] and Baghurst et al.[40] reported that the oral administration of onion decreased blood platelet aggregation in human and animal subjects. Makheja et al.[41-43] demonstrated that garlic and onion oil fractions containing allicin (C_3H_5–S–S–C_3H_5) or paraffinic poly-

$$R - (S)_n - R$$
$$\qquad\ |$$
$$\qquad\ O$$

sulfides (R – (S)$_n$ –R) effectively inhibited ADP, arachidonic acid, or collagen-induced platelet aggregation. These oily fractions of onion and garlic suppressed thromboxane (a breakdown product of arachidonic acid) synthesis completely and induced a redistribution in the products of the lipogenase pathway. Dimethyl trisulfide (CH_3–S–S–S–CH_3) was one of the most active polysulfides to act on both aggregation and arachidonic acid metabolism as measured by thromboxane B_2 synthesis.[43] According to Vatsala and Singh[44] oral administration of onion extract prevented the distortion of erythrocytes and their tendency to aggregate in the blood of cholesterol-fed rabbits.

G. Fibrinolytic Effects

Gupta et al.[45] showed that onion and its products enhanced fibrinolysis in human beings. This was later confirmed by others[46,47] who showed that fatty meals decrease the capacity of blood to lyse the fibrin clot. Addition of raw or fried onions along with a fatty meal not only prevented a decrease in fibrinolysis, but actually enhanced it. Augusti et al.[48] attributed the fibrinolytic activity of onion to sulfur compounds in onion, viz., disulfides in oil and a

TABLE 1

Effects ($\pm LSD_{0.05}$) of Onion and Garlic oil[a] at 100 mg/kg/d on Sucrose (73% of the Diet) fed Rats. Changes after 2 Months Treatment are shown. Values are Averages from 6 Rats.[32]

	Groups of rats			
Parameter studied	Normal	Sucrose diet fed	Sucrose diet & onion oil	Sucrose diet & garlic oil
Blood sugar (mg/100 ml)	87 ± 13	96 ± 15	77 ± 23	76 ± 9
Serum cholesterol (mg/100 ml)	84 ± 11	111 ± 8	79 ± 7	80 ± 5
Serum triglyceride glycerol (mg/100 ml)	20 ± 3	51 ± 4	27 ± 3	24 ± 2
Liver cholesterol (mg/g tissue)	7.4 ± 1	10.8 ± 0.6	6.1 ± 0.7	6.7 ± 0.8
Liver triglyceride glycerol (mg/g tissue)	3.0 ± 0.2	7.5 ± 0.5	3.0 ± 0.1	2.8 ± 0.1
Liver total lipids (mg/g tissue)	29.0 ± 5	43.0 ± 4.7	31.6 ± 3.2	30.4 ± 2.3

[a] Steam-distilled oil was prepared as described previously.[35]

TABLE 2

Effects ($\pm LSD_{0.05}$) of Onion and Garlic Oils at 100 mg/kg/d on Rats fed a High Fat Diet (18% Fat + 2% Cholesterol) for 1 month. Values are Averages of 6 Rats.[27,85]

	Groups of rats			
Parameter studied	Normal	High-fat diet fed	High-fat diet + onion oil	High-fat diet + garlic oil
Blood sugar (mg/100ml)	73.2 ± 9.9	70 ± 9	56 ± 5.2	60 ± 16
Serum cholesterol (mg/100ml)	69 ± 9	135 ± 15	100 ± 8	76 ± 8
Serum triglyceride glycerol (mg/100ml)	25 ± 3	44 ± 4	20 ± 4	16 ± 3
Liver cholesterol (mg/g tissue)	1.4 ± 0.12	3.5 ± 0.10	1.0 ± 0.05	1.3 ± 0.12
Liver triglyceride glycerol (mg/g tissue)	1.2 ± 0.1	2.2 ± 0.15	0.8 ± 0.09	0.6 ± 0.08
Liver total lipid (mg/g tissue)	31.2 ± 6	64 ± 1.7	32.0 ± 3.5	30.5 ± 8

TABLE 3

Effects of Garlic and Onion Oils at 100 mg/kg/d on Streptozotocin Diabetic Rats after 2 Months Feeding. Values are Averages of 5 Rats.[36,66]

	Groups of rats			
Parameter studied	Normal	Diabetic	Diabetic + onion oil	Diabetic + garlic oil
Blood sugar (mg/100 ml)	75 ± 10	300 ± 6	130 ± 15	125 ± 12
Serum cholesterol (mg/100ml)	80 ± 8	175 ± 10	90 ± 8	85 ± 7
Serum triglyceride glycerol (mg/100 ml)	22 ± 2	55 ± 12	27 ± 2	20 ± 1.5
Liver cholesterol (mg/g tissue)	2.0 ± 0.5	2.8 ± 0.6	1.6 ± 0.2	1.5 ± 0.3
Liver triglyceride glycerol (mg/g tissue)	1.3 ± 0.2	3.0 ± 0.5	1.5 ± 0.2	1.6 ± 0.1
Liver total lipids (mg/g tissue)	28.4 ± 2	44.4 ± 3	32.8 ± 2	30.0 ± 4

sulfur containing imino acid cycloalliin, both acting independently. While onion oil disulfides are obnoxious substances, cycloalliin is an odorless, acceptable compound for human consumption. However, the oil is more active than the imino acid. Two possible mechanisms were suggested. First, these substances may weaken blood clots by opening –S–S crosslinks between fibrin molecules through a disulfide exchange reaction and therefore render them more susceptible to natural fibrinolysis. Second, they may stimulate the release of plasminogen activator (a component of blood coagulation) from its sites of production in the walls of veins and elsewhere. These and perhaps other possibilities need to be explored.

Since an improved fibrinolytic activity may be a very important factor in the prevention and amelioration of coronary heart disease, some speculation about the prophylactic and therapeutic use of this common item of food is quite natural. The usefulness of onion for the prevention of heart disease is also indicated by the hypolipidemic effects reported above.

H. Other Effects

The inhibitory effects of an ethanolic extract of onions on allergic skin reactions in man as well as on allergen-induced bronchial asthma in man and guinea pigs have been shown.[49] Asthma-protective effects of isothiocyanates of onion, e.g., benzyl-isothiocyanate, were also described.[49]

II. THERAPEUTIC AND MEDICINAL VALUES OF GARLIC
(*ALLIUM SATIVUM L.*)

A. Traditional Uses of Garlic

Garlic is used as a spicy food item in all parts of the world, having significance as a folk remedy for many ailments. In the past two decades there have been a number of scientific reports supporting its therapeutic uses. Garlic has been described as a hot stimulant, carminative, antirheumatic, and alterative. Garlic oil is said to be a powerful antiseptic. It is used as vermifuge for expelling roundworms,[1] and has long been recommended for cure of a number of ailments,[50,51] viz., wounds, foul ulcers, pneumonia, bronchitis, atopic dyspepsia, and gastrointestinal disorders. Insecticidal, antibacterial, antifungal, antitumor, hypoglycemic, hypolipidemic, antiatherosclerotic, fibrinolytic, and antiplatelet-aggregation effects of garlic have been reported. Garlic is more important than onion in these respects.

B. Chemistry

Garlic contains about 62.8% water, 6.3% protein, 0.1% fat, 29% carbohydrate including 3.9% sucrose, traces of Ca, Fe, Zn, and phosphate salts, and small amounts of vitamins like thiamin, riboflavin, niacin, and ascorbic acid.[5,6] Garlic can yield 0.06 to 0.1% essential oil. The oil is made up of mainly diallyl disulfide and small amounts of allylpropyl disulfide.[1,2,5] This oil is formed from the decomposition products of an alliin (see Section I.B) called S-allyl cysteine sulfoxide (SACS). Garlic cloves contain an enzyme alliinase, which is released when they are crushed. Alliinase acts on SACS and produces allicin, ammonia, and pyruvic acid:

$$2C_3H_5\text{–}\underset{\underset{O}{|}}{S}\text{–}CH_2\text{–}CH(NH_2)COOH \xrightarrow{\hspace{1cm}} \overset{\text{allicin}}{C_3H_5\text{–}\underset{\underset{O}{|}}{S}\text{–}S\text{–}C_3H_5} + 2NH_3 +$$

alliin   alliinase $2CH_3CO \cdot COOH$

(10)

Later allicin rearranges to diallyl disulfide (DADS) and a thiosulfonate as follows:[9,10]

$$2C_3H_5\text{–}\underset{\underset{O}{|}}{S}\text{–}S\text{–}C_3H_5 \rightarrow C_3H_5\text{–}S\text{–}S\text{–}C_3H_5 + C_3H_5S\text{–}\underset{\underset{(O)_2}{|}}{S}\text{–}C_3H_5$$

allicin   DADS; thiosulfonate of DADS   (11)

DADS may further rearrange to tri- and polysulfides. Traces of methyl and propyl cysteine sulfoxides present in garlic may give rise to methyl and propyl disulfides and certain polysulfides [R– $(S)_n$ –R], e.g., methyl allyl trisulfide.[52]

C. Antimicrobial Action

Garlic has been reported to possess a broad spectrum antimicrobial property, inhibiting growth of a wide variety of fungi and bacteria including enterobacteria *in vitro*. The crude extract of garlic was found to be quite effective against Gram-negative as well as Gram-positive bacteria *in vitro*. It inhibited the growth of some bacterial cultures which were resistant to commonly used antibiotics.[52-55] Lilianna et al.[56] reported that bacteria resistant to eight main antibiotics were strongly suppressed by garlic extract. *In vivo* oral administration of the extract considerably reduced the count of viable Gram-negative bacteria present in the intestinal tract of poultry. Garlic was found to be bactericidal as well as bacteriostatic against enterotoxigenic *Escherichia coli* strains; 10% dilution of aqueous extract of garlic completely inhibited the growth of these bacteria. This suggests that garlic can be used as a prophylactic agent against enterotoxigenic *E. coli*-induced diarrhea.[55] Such effects of garlic may be due to its inhibition of protein synthesis.[57] In addition, reports showed that garlic extract exhibited promising antibacterial activity against several clinical strains of *Staphylococcus, Escherichia*, and *Pseudomonas*.[58] Although the mechanism of antibacterial action of garlic has not been clearly worked out, Willis[59] opined the allicin, that main active principle of garlic affects the activity of important metabolic enzymes, especially those having −SH groups.[60]

D. Blood Sugar-Lowering Effects

Laland and Havrevold[61] first reported on the hypoglycemic effect of garlic. They extracted from garlic an ether-soluble, steam-volatile, alkaloidal substance (probably a sulfur amino acid) which, when mixed with the disulfides found in garlic and injected into dogs and rabbits, showed a hypoglycemic action. The hypoglemic effect of garlic principle was later confirmed by others.[62-66] Zaman et al.[67] and Mahata et al.[68] administered garlic oil and raw garlic, respectively, to human volunteers and found considerable, but not statistically significant, reductions in blood sugar after 1 month, and that garlic oil[67] reduced normal blood sugar significantly after 2 months of regular use.

The hypoglycemic effect of garlic may be due to increased glycogenesis in the liver and better utilization of glucose in the peripheral tissues.[67] The blood sugar-lowering effect of garlic was ascribed to allicin and related disulfide-containing compounds. SH-group compounds are antagonistic to the action of insulin. Since allicin and the related disulfides can remove the thiols from a system by oxidation and thiol disulfide reactions (Equations 4 and 5), garlic and its disulfide principles may spare some insulin from inactivation by thiols.[63,66] This hypothesis is supported by the work of Chang and Jonson,[65] who found that garlic supplementation of a hypercholesterolemic diet given to rats decreased blood sugar and increased their liver glycogen and serum insulin level significantly as compared to control values. Work by Rasiah[69] also illustrated the beneficial effects of garlic oil in experimental diabetes. Oral administration of garlic oil at 100 mg/kg/d for 1 month to streptozotocin (an antibiotic which destroys the β cells of the pancreas) diabetic rats significantly reduced their blood sugar and lipids and enhanced the action of low doses of insulin (see Table 4). All these findings support the claim that garlic is beneficial to diabetics.

E. Hypolipidemic Effects

Temple[70] demonstrated that polysulfides resembling those in garlic have hypocholesterolemic action, and Augusti and Mathew[29] first reported that an aqueous extract of garlic has hypolipidemic action in normal rats. Later work showed that allicin also causes a fall in serum and tissue cholesterol and triglycerides. Rats fed a regular laboratory diet for 2 months were compared with those fed similarly, but with a daily supplement with allicin (100 mg/kg).[71] The total serum lipids, phospholipids, triglycerides, and cholesterol were significantly lower in the allicin-fed group as against the control group. Itokawa et al.[33]

TABLE 4
Effects of 1-Month Treatment with Insulin and Garlic Oil on Certain Parameters of Streptozotocin Diabetic Rats. Values are Averages of 5 Rats.[69]

	Groups of rats (dose units/kg/d)			
Parameter studied	I Control (no treatment)	II Insulin (10u/kg/d)	III Insulin (5u/kg/d)	IV Insulin (5µ/kg/d) + garlic oil (100 mg/kg/d)
Percentage weight increase or loss	−5 ± 1	+27 ± 2	+12 ± 1	+23 ± 3
Blood sugar (mg%)	280 ± 9.1	198 ± 5.8	134 ± 3.2	109 ± 38.9
Serum triglyceride (mg%)	380 ± 6	168 ± 2.5	332 ± 4.1	224 ± 4
Serum cholesterol (mg%)	450 ± 10	157 ± 3	303 ± 5.9	301 ± 11
Liver triglyceride (mg/g tissue)	40 ± 2	19 ± 1.1	28 ± 2.2	20 ± 1.7
Liver cholesterol (mg/g tissue)	15 ± 1	8 ± 0.3	11 ± 0.6	7 ± 0.5
Liver total lipids (mg/g tissue)	100 ± 3	64 ± 1.9	75 ± 3.8	69 ± 1.9
Kidney triglyceride (mg/g tissue)	25 ± 2	11 ± 0.7	13 ± 1	11 ± 3
Kidney cholesterol (mg/g tissue)	16 ± 1	7 ± 0.3	10 ± 1	8 ± 0.3

reported that oral administration of SACS, the precursor of allicin, markedly depressed the increase of plasma cholesterol in rats fed a hypercholesterolemic diet. These experiments show that the hypolipidemic action of garlic is due to the action of its sulfur-containing compounds and oils.

Bordia et al.[31] compared the effect of the essential oils of garlic and onion equivalent to 1 g raw cloves per kilogram body weight per day with the antihyperlipidemic agent clofibrate in its usual clinical dose of 33 mg/kg/d. Rabbits were fed a hypercholesterolemic diet. At the end of 3 months the scrum cholesterol in the control group was 572 mg/100 ml while the groups supplemented with clofibrate, onion, and garlic had significantly lower leves of 527, 488, and 458 mg/100 ml, respectively. The effect of garlic was significantly higher than that of clofibrate. Mahanta et al.[68] observed a significant reduction in serum cholesterol level when 50 g of raw garlic was fed daily to ten normal human volunteers for 1 month. However, Zaman et al.[67] did not find a significant hypocholesterolemic effect of garlic extract (32 mg per person per day) in 1- and 2-month feeding experiments on humans. They stated that the absence of the effect may have been due to too low a dose of garlic extract in their experiments. In other studies Bordia and Verma[72,73] fed rabbits with a hypercholesterolemic diet for 3 months. They were divided into two groups. One group was kept as control and the other group was given the regular diet supplemented with garlic extract equivalent to 2 g raw garlic per kilogram body weight per day. The group supplemented with garlic showed significantly greater decreases in low density lipoprotein (LDL) plus very low density lipoprotein (VLDL) and significant increases in high density lipoprotein (HDL) at 6,9, and 12 months. In another report an increased HDL level was demonstrated in rats fed freeze-dried garlic powder making up 2% of an otherwise atherogenic diet.[50] Jain[34] compared the effect of garlic and onion in rabbits fed a diet containing cholesterol at 0.5 g/d. The group supplemented with garlic juice equivalent to 25 g of raw garlic per day showed approximately 20% of the cholesterol level of the control group after 16 weeks. Onion juice supplementation as above failed to show such a significant effect on hypercholesterolemia. Using garlic oil, Jain and Konar[74] further showed that the cholesterol-lowering effect was dose related. Rabbits were fed a diet with 2 g/d of cholesterol for 16 weeks. Administration of garlic oil at the dose of 0.25, 0.5, and 1 g/d resulted in 6.2, 21, and 30% reductions in the serum cholesterol, respectively. The suggested mechanism of action of garlic was to increase the excretion of bile acid and sterol in the feces.

In the experiments of Chang and Johnson[65] significant hypolipidemic effects were found in rats fed with a 1% cholesterol diet containing 5% garlic and which were either intraperitoneally injected with ^{14}C-acetate or fed with ^{14}C-sucrose. Feeding of garlic (5 g/100 g diet) resulted in decreases of radioactivity by 56% in serum total lipid, 27% in serum total cholesterol, and 65% in serum triglyceride and free fatty acid. Kritchevsky[75] fed rabbits a 2 g cholesterol per 100 g diet plus 0.25 or 0.50% garlic oil. In test animals the serum cholesterol level was about 10% lower than that of the control group. In a later experiment Kritchevsky's group[76] observed an increased excretion of feces in rats fed garlic oil. Chi et al.[77] observed an increased excretion of acidic and neutral steroids by rats fed with lyophilized garlic powder along with a 1% hypercholesterolemic diet. In addition to this, garlic decreased plasma cholesterol, triglycerides, VLDL, and LDL cholesterol and increased HDL cholesterol to a considerable extent.

An epidemiological study comparing three populations with different dietary habits reported by Sainani et al.[37] was mentioned under onion. The group that consumed these vegetables liberally showed the lowest values of serum cholesterol and triglycerides. It may be deduced that the differences in cholesterol and triglycerides were indeed associated with the ingestion of onion and garlic. In Augusti's study[78] five patients with high serum cholesterol (275 to 350 mg/100 ml) were given fresh garlic juice at 0.5 ml/kg/d for 2 months. A drop of serum cholesterol occurred in all the patients varying from 21 to 28% with a mean drop of 28.5%. This decrease of cholesterol did not persist but returned to pretreatment levels within 2 months after garlic treatment stopped. Bordia and Bansal[79] reported that supplementation of a fatty diet with fresh garlic juice made from 50 g of garlic prevented a rise of serum cholesterol and fibrinogen in humans.

The effect of garlic on lipoproteins was studied in 62 patients with coronary heart disease.[80] Of these 31 randomly chosen patients were placed on essential oil of garlic at 0.25 mg/kg body weight for 10 months, while the others served as the control. Serum lipoproteins were checked bimonthly. In the control group the values remained constant throughout. The treatment group, however, had steady decreases of LDL + VLDL accompanied by a progressive rise of HDL. Lowering of LDL has also been reported in healthy individuals ingesting raw garlic.[81] Thus, the consumption of garlic appears to reduce serum lipids perhaps by decreasing their synthesis and/or increasing their excretion through the intestinal trace, and these effects increase with the amount of garlic used. It has also been suggested that the increased HDL associated with garlic feeding may enhance the removal of cholesterol from arterial tissue. The hypocholesterolemic effects of garlic have been ascribed to its oily substance called allitin (diallyl disulfide) by Pushpendran et al.[82] They used synthetic DADS and showed its hypocholesterolemic effects in experimental animals. DADS is a deoxy form of allicin (Equation 11).

F. EFFECTS ON TISSUE LIPIDS

Garlic extract and oil significantly lowered liver cholesterol and triglycerides in normal and streptozotocin diabetic rats.[29,32,66] Garlic and onion oils were more effective than the antilipidemic agent clofibrate against lipid deposition on the aorta in rabbits.[31] Garlic oil was more effective than onion oil in this respect.[31,34,74,75] Supplementation of the diet with garlic decreased cholesterol and phospholipids and resulted in less atherosclerosis as assessed by visual grading. Decreased atheromatous lesions have been found consistently in rabbits fed with high cholesterol diets supplemented with garlic.[72,73,77] Garlic, onion, and their oils largely prevented the hyperlipidemic effects of high fat, sucrose, and alcohol-supplemented diets in animals, and the effect of garlic was more than that of onion.[27,32,83-87]

Of various sulfur-containing amino acids isolated from garlic, SACS has been shown to exert significant antilipidemic effects.[33] This and its decomposition products like allicin, DADS, and related polysulfides have been ascribed with the property of combining with −SH groups, the functional part of many enzymes (e.g., CoA), and oxidizing reduced pyridine nucleotides, viz., NADPH, which are necessary for the biosynthesis of fatty acids, choles-terol, triglycerides, and phospholipids.[32,71,85] The lipid-lowering effect of garlic sulfur com-pounds may thus be attributed to the inactivation or oxidation of the –SH compounds and NADPH in the body (see Equations 4 to 9). In both *in vivo* and *in vitro* tests garlic feeding of animals reduced the conversion of acetate into cholesterol by liver tissue.[65,76]

G. ANTIPLATELET-AGGREGATION EFFECTS

Garlic and its products significantly reduce the aggregation of blood platelets.[43,51,88,89] The antiplatelet-aggregation effect is caused by the polysulfides present in garlic, e.g., allicin, as discussed under the similar effects of onion.[43]

The effect of garlic extract on human platelet aggregation was studied by three inde-pendent groups.[52,88,89] They demonstrated *in vitro* that garlic extract inhibited platelet ag-gregation induced by epinephrine or collagen and the effect was dose related. Oral administration of garlic also decreased platelet aggregation. It was suggested that garlic may inhibit some aspects of thrombus formation associated with platelets.[88] Ariga et al.[52] reported that methyl allyl trisulfide in the oil of garlic actually possesses the main antiplatelet ag-gregation property.

H. EFFECTS ON OTHER BLOOD COAGULATION PARAMETERS

The aforementioned study conducted by Sainani et al.[37,90] showed that three groups of population differing in their use of onion and garlic also differed in their blood coagulation parameters. The group with total abstinence from onion and garlic had significantly higher plasma fibrinogen, shorter clotting time, and poorer fibrinolytic activity (blood clot lyses) when compared to two groups using onion and garlic. The group which consumed small amounts of onion and garlic had significantly higher plasma fibrinogen and significantly poorer fibrinolytic activity than the group that consumed onion and garlic liberally, but there was no significant difference in the clotting time. Bordia et al.[91] studied the effect of garlic oil (extract from 1 g garlic per 1 kg body weight) on serum fibrinolytic activity in patients with myocardial infarction. Patients with previous myocardial infarction showed a significant increase of 83% in fibrinolytic activity over control patients after 3 months of feeding with garlic oil. Patients with acute myocardial infarction fed with garlic oil showed an increase in fibrinolytic activity of 63 and 95.5% above the postinfarction value at 10 and 20 d, respectively, and the values were significantly above those of the placebo-treated control group. Similar effects were noticed when fried or raw garlic was consumed.

Increased plasma fibrinogen and decreased coagulation time and blood fibrinolytic ac-tivity are associated with a high-fat diet. All of these increase the risk of thrombosis.[50] Garlic feeding prevented the rise of plasma fibrinogen and the fall in clotting time and fibrinolytic activity,[31,72,73] suggesting that components in garlic may prevent thrombotic disorders.[31,92,93]

I. OTHER EFFECTS

According to Saxena et al.[94] feeding of garlic oil to rats counteracted the decrease in physical endurance induced by isoproterenol. On autopsy, it was found that feeding garlic oil significantly reduced the severity of the myocardial lesions induced by isoproterenol. Other properties such as hypotensive[95] and antifungal effects[96,97] have also been reported for garlic. Tumor immunity in mice was induced by injecting them with tumor cells treated with extract of garlic. This may be caused by sulfhydryl blocking agents present in garlic.[98,99]

Synthetic DADS the major component of garlic oil significantly increased both NADH- and NADPH-dependent microsomal monooxygenases and glucose-6-phosphatase and aden-osine triphosphatase activities in suckling rats.[100] Garlic oil has also been shown to have considerable insecticidal properties.[57] Allicin has been shown to be effective against a variety of human pathogens,[58] bacteria, and fungi.[101] Antibacterial, antihelminthic, antiprotozoal, and anticarcinogenic properties have also been ascribed to garlic.[102] Garlic has been shown to be effective against organisms such as *Salmonolla typhimurium* and *E. coli*.[103] Garlic oil is larvicidal for a number of species of mosquitoes.[104] The active principle is a mixture of two rather simple organic compounds,viz., DADS and diallyl trisulfide.[105] These principles inhibit protein synthesis in mosquito larvae.[57]

Prolonged feeding of high levels of raw garlic to rats has resulted in anemia, weight loss, and failure to grow.[106] Uncontrolled use for onion may also lead to anemia[107,108] as all these vegetables contain compounds which remove cystein from a system and inhibit protein synthesis.[109] Therefore, only the customary amounts of onion and garlic, of about 50 to 80 g and 7 to 10 g/d, respectively, are advised.[37,90] The following tables show the medicinal effects of onion and garlic oils in different experimental conditions. As with onion, garlic may not be as active as hypoglycemic drugs for the control of diabetes. However, these vegetables may enhance the effect of insulin (see tables) and reduce hyperlipidemia, obesity, and resistance to insulin treatment. Also, these vegetables may prevent the incidence of atherosclerosis and related diseases of the heart.

Different biological effects of garlic and onion principles are being reported from time to time. Both beneficial and unpleasant effects of these vegetables and their extracts are due to their sulfur compounds. They contain a large number of various sulfur amino acid peptides and their derived products.[9,12] Bactericidal, allergic, antifungal, antitumor, and larvicidal effects can be explained by interactions of these sulfur compounds with thiol substances in these living systems. Despite the evidence for the considerable medical benefits of garlic and onions and preparation derived from them, my group and others have found difficulties in obtaining funding to further this work.

ACKNOWLEDGMENT

This chapter is dedicated to the services of the founder and retiring Prof. P. A. Kurup of this department with whom I worked for two decades on oral hypoglycemics.

REFERENCES

1. **Nadkarni, K. N.,** *Allium cepa* Linn and *Allium sativum* Linn, in *The Indian Materia Medica,* 3rd ed. (Part 1), Puranik, M. V. and Bhatkal, G. R., Eds., Popular Book Depot, Bombay, 1954, 63.
2. **Chopra, R. N., Nayar, S. L., and Chopra, I. C.,** *Allium* (Liliaceae), in *Glossary of Indian Medicinal Plants,* Chopra, R. N., Ed., Council of Scientific and Industrial Research, New Delhi, 1956, 11.
3. **Collip, J. B.,** Glucokinin. An apparent synthesis in the normal animal of a hypoglycemia producing principle. Animal passage of the principle, *J. Biol. Chem.,* 58, 163, 1923.
4. **Janot, M. M. and Laurin, J.,** Hypoglycemic action of the bulbs of *Allium cepa* Linn., *C. R. Acad. Sci. (Paris),* 191, 1098, 1930.
5. **Manjunath, B. L.,** *Allium cepa* Linn and *Allium sativum* Linn, in *The Wealth of India. Raw Materials,* Vol. 1, Manjunath, B. L., Ed., Council of Scientific and Industrial Research, Government of India, Delhi, 1948, 56.
6. **Augusti, K. T.,** *Allium cepa* Linn and *Allium sativum* Linn, in *Orally Effective Hypoglycemic principles from plant sources,* Ph.D. thesis, University of Rajasthan, Jaipur, India, 1963, 52.
7. **Platenius, H.,** A method for estimating the volatile sulphur content and pungency of onions, *J. Agric. Res.,* 51, 847, 1935.
8. **Augusti, K. T.,** Gas chromatographic analysis of onion principles and a study on their hypoglycemic action, *Indian J. Exp. Biol.,* 14, 110, 1976.
9. **Virtanen, A. I.,** Studies on organic sulphur compounds and other labile substances in plants, *Phytochemistry,* 4, 207, 1965.
10. **Johnson, A. E., Nurstem, H. E., and Williams, A. A.,** Vegetable volatiles — a survey of components identified. I, *Chem. India,* 1, 556, 1971.

11. **Lukes, T. M.,** Thin-layer chromatography of cysteine derivatives of onion flavor compounds and the lacrimatory factor, *J. Food Sci.*, 36, 662, 1971.
12. **Virtanen, A. I. and Matikkala, E. J.,** The isolation of S-methyl cysteine sulfoxide and s-n-propyl cysteine sulfoxide from onion (*Allium cepa*) and the antibiotic activity of crushed onion, *Acta Chem. Scand.*, 13, 1898, 1959.
13. **Kolthoff, T. M., Stricks, W., and Kapoor, R. C.,** Equilibrium constants of exchange reactions of cystine with glutathione and with thioglycolic acid both in the oxidized and reduced state, *J. Am. Chem. Soc.*, 77, 4733, 1955.
14. **Collip, J. B.,** Glucokinin, a new hormone present in plant tissue, *J. Biol. Chem.*, 56, 513, 1923.
15. **Collip, J. B.,** Glucokinin. II, *J. Biol. Chem.*, 57, 65, 1923.
16. **Laurin, J.,** Hypoglycemic action of the bulbs of *Allium cepa* Linn, *C. R. Acad. Sci. (Paris)*, 192, 1289, 1931.
17. **Kreitmair, H.,** Pharmacological trials with some domestic plants, *E. Merck's Jahresber.*, 50, 102, 1936.
18. **Brahmachari, H. D. and Augusti, K. T.,** Hypoglycemic agent from onions, *J. Pharm. Pharmacol.*, 13, 128, 1961.
19. **Brahmachari, H. D. and Augusti, K. T.,** Effects of orally effective hypoglycemic agents from plants on alloxan diabetes, *J. Pharm. Pharmacol.*, 14, 617, 1962.
20. **Mathew, P. T. and Augusti, K. T.,** Isolation of hypo- and hyperglycemic agents from *Allium cepa* Linn, *Indian J. Exp. Biol.*, 11, 573, 1973.
21. **Mathew, P. T. and Augusti, K. T.,** Hypoglycemic effects of onion, *Allium cepa* Linn on diabetes mellitus — a preliminary report, *Indian J. Physiol. Pharmacol.*, 19, 213, 1975.
22. **Augusti, K. T.,** Chromatographic identification of certain sulfoxides of cysteine present in onion (*Allium cepa* Linn) extract, *Curr. Sci.*, 45, 863, 1976.
23. **Katzen, H. M. and Tietz, F.,** Studies on the specificity and mechanism of action of hepatic glutathione insulin transhydrogenase, *J. Biol. Chem.*, 241, 3561, 1966.
24. **Black, S.,** Reduction of sulfoxide and disulfides, in *Methods in Enzymology*, Vol. 5, Colowick, S. P. and Kaplan, N. O., Eds., Academic Press, New York, 1962, 992.
25. **Augusti, K. T., Roy, V. C. M., and Semple, M.,** Effect of allyl propyl disulphide isolated from onion (*Allium cepa* L.) on glucose tolerance of alloxan diabetic rabbits, *Experimentia*, 30, 1119, 1974.
26. **Augusti, K. T. and Benaim, M. E.,** Effect of essential oil of onion (allyl propyl disulphide) on blood glucose, free fatty acid and insulin levels of normal subjects, *Clin. Chim. Acta*, 60, 121, 1975.
27. **Wilcox, B. F., Joseph, P. K. and Augusti, K. T.,** Effects of allyl propyl disulphide isolated from *Allium cepa* Linn on high-fat fed rats, *Indian J. Biochem. Biophys.* 21, 214, 1984.
28. **Karawya, M. S., Abdel Wahab, S. M., Elolemy, M. M., and Farrag, N. M.,** Diphenylamine, an antihyperglycemic agent from onion and tea, *J. Nat. Prod. (Lloydia)*, 47, 755, 1984.
29. **Augusti, K. T. and Mathew, P. T.,** Effect of long term feeding of the aqueous extracts of onion and garlic on normal rats, *Indian J. Exp. Biol.*, 11, 239, 1973.
30. **Sharma, K. K., Chowdhury, N. K., and Sharma, A. L.,** Long term effect of onion on experimentally-induced hypercholesterolemia and consequently decreased fibrinolytic activity in rabbits, *Indian J. Med. Res.*, 63, 1629, 1975.
31. **Bordia, A., Arora, S. K., Kothari, L. K., Jain, K. C., Rathore, B. S., Rathore, A. S., Dube, M. K., and Bhu, N.,** The protective action of essential oils of onion and garlic in cholesterol-fed rabbits., *Atherosclerosis*, 22, 103, 1975.
32. **Adamu, I., Joseph, P. K., and Augusti, K. T.,** Hypolipidemic action of onion and garlic unsaturated oils in sucrose fed rats over a two month period, *Experientia*, 38, 899, 1982.
33. **Itokawa, Y., Inoue, K., Sasagawa, S., and Fujiwara, M.,** Effect of S-methyl cysteine sulphoxide, S-allyl cysteine sulfoxide and related sulfur containing amino acids on lipid metabolism of experimental hypercholesterolemic rats, *J. Nutr.*, 103, 88, 1973.
34. **Jain, R. C.,** Onion and garlic in experimental cholesterol induced atherosclerosis, *Indian J. Med. Res.*, 64, 1509, 1976.
35. **Augusti, K. T.,** Effect on alloxan diabetes of allyl propyl disulphide obtained from onion, *Naturwissenschaften*, 61, 172, 1974.
36. **Olatunde, R. M., Goji, I. A., Joseph, P. K., and Augusti, K. T.,** Biochemical efffects of allyl propyl disulphide isolated from *Allium cepa* L. on streptozotocin diabetic rats, Abstr. 13th Int. Congr. Biochemistry, August 25-30, Amsterdam, The Netherlands, 1985, 397.
37. **Sainani, G. S., Desai, D. B., Gorhe, N. H., Natu, S. M., Pise, D. V., and Sainani, P. G.,** Effect of dietary garlic and onion on serum lipid profile in Jain Community, *Indian J. Med. Res.*, 69, 776, 1979.
38. **Singh, M. and Kanakaraj, P.,** Hypocholesterolemic effect of onion extract on cholesterol enriched erythrocytes, *Indian J. Exp. Biol.*, 23, 456, 1985.
39. **Mittal, M. M., Mittal, S., Sarin, J. C., and Sharma, M. L.,** Effect of feeding onion on fibrinolysis, serum cholesterol, platelet aggregation and adhesion, *Indian J. Med. Sci.*, 28, 144, 1974.
40. **Baghurst, K. I., Raj, M. J., and Truswell, A. S.,** Onion and platelet aggregation, *Lancet*, 2, 101, 1977.

41. **Makheja, A. N., Vanderhoek, J. Y., and Bailey, J. M.,** Effects of onion (*Allium cepa*) extract on platelet aggregation and thromboxane synthesis, *Prostaglandins Med.,* 2, 413, 1979.
42. **Makheja, A. N., Vanderhoek, J. Y., Bryant, R. W., and Bailey, J. M.,** Altered arachidonic acid metabolism in platelets inhibited by onion or garlic extracts, in *Advances in Prostaglandin and Thromboxane Research*, Vol. 6., Samuellson, B., Ramwell, P. W., and Paoletti, R., Eds., Raven Press, New York, 1980, 309.
43. **Makheja, A. N., Low, C. F., and Bailey, J. M.,** Biological nature of platelet inhibitors from *Allium cepa, Allium sativum* and *Auricularia polytrica,* Abstr. 8th Int. Congr. Thrombosis and Haemostasis, Toronto, July 1981.
44. **Vatsala, T. M. and Singh, M.,** Changes in shape of erythrocytes in rabbits on atherogenic diet and onion extracts, *Atherosclerosis,* 36, 39, 1980.
45. **Gupta, N. N., Mehrotra, R. M. L., and Sircar, A. R.,** Effect of onion on serum cholesterol, blood coagulation factors and fibrinolytic activity in alimentary lipaemia, *Indian J. Med. Res.,* 54, 48, 1966.
46. **Menon, I. S., Kendal, R. V., Dewar, H. A., and Newell, D. J.,** Effect of onions on blood fibrinolytic activity, *Br. Med. J.,* 3, 351, 1968.
47. **Sharma, K. K., Gupta, S., and Dwivedi, K. K.,** Effect of raw and boiled onion on the alterations of blood cholesterol, fibrinogen and fibrinolytic activity in man during alimentary lipaemia, *Indian Med. Gaz.,* 16, 479, 1977.
48. **Augusti, K. T., Benaim, M. E., Dewar,H. A., and Virden, R.,** Partial identification of the fibrinolytic activators in onion, *Atherosclerosis,* 21, 409, 1975.
49. **Water, D. Adam, O., Weber, J., and Zeigeltrum, T.,** Antiasthmatic effects of onion (*Allium cepa*) extracts. Detection of benzyl isothiocyanate and other isothiocyanates (mustard oils) as antiasthmatic compounds of plant origin, *Eur. J. Pharmacol.,* 107, 17, 1985.
50. **Lau, B. H. S., Adetumbi, M. A., and Sanches, A.,** *Allium sativum* (garlic) and atherosclerosis: a review, in *Nutrition Research,* Vol. 3., Pergamon Press, 1983, 119.
51. **Skinner, F. A.,** Antibiotics. Plants which yield antibiotically active preparations, in *Modern Methods of Plant Analysis,* Vol. 3., Paech, K. and Tracey, M. V., Eds., Springer-Verlag, Berlin, 1955, 655.
52. **Ariga, T., Oshiba, S., and Tamada, T.,** Platelet aggregation inhibitor in garlic, *Lancet,* 2, 150, 1981.
53. **Prat, M. A.,** Algunas consideraciores sobre al accion antibiotica del preparades, *Gac. Vac,* 11, 184, 1949.
54. **Sharma, V. D., Sethi, M. S., Kumar, A., and Rarortra, J. R.,** Antibacterial property of *Allium sativum* Linn in vivo and in vitro studies, *Indian J. Exp. Biol.,* 15, 466, 1977.
55. **Kumar, A. and Shrma, V. D,** Inhibitory effect of garlic (*Allium sativum* L.) on enterotoxigenic *Escherichia coli, Indian J. Med. Res.,* 76, 66, 1982.
56. **Lilianna, J., Rafinski, T., and Wrocinski, T.,** Investigations on the antibiotic activity of *Allium sativum* L., *Herba Pol.,* 12, 3, 1966.
57. **George, K. C., Amonkar, S. V., and Eapen, J.,** Effect of garlic oil on incorporation of amino acids into proteins of *Culex pipiens quinquefasciatus,* Say larvae, *Chem. Biol. Interact.,* 6, 169, 1973.
58. **Singh, K. V. and Shukla, N. P.,** Activity on multiple bacteria of garlic (*Allium sativum*), *Fitoterapia,* 55, 313, 1984.
59. **Willis, E. D.,** Enzyme inhibition by allicin, the active principle of garlic, *Biochem. J.,* 63, 514, 1956.
60. **Cavallito, C. J. and Bailey, J. H.,** Allicin, the antibacterial principle of *Allium sativum.* I. Isolation, physical properties and antibacterial action, *J. Am. Chem. Soc.,* 66, 1944, 1950.
61. **Laland, P. and Havrevold, O. W.,** The active principles of onions (*Allium sativum*) which lowers blood sugar per os, *Z. Phys. Chem.,* 221, 180, 1933.
62. **Brahmachari, H. D. and Augusti, K. T.,** Orally effective hypoglycemic agents from plants, *J. Pharm. Pharmacol.,* 14, 254, 1962.
63. **Mathew, P. T. and Augusti, K. T.,** Studies on the effect of allicin (diallyl disulphide-oxide) on alloxan diabetes. I. Hypoglycaemic action and enhancement of serum insulin effect and glycogen synthesis, *Indian J. Biochem. Biophys.,* 10, 209, 1973.
64. **Jain, R. C., Vyas, C. R., and Mahatma, O. P.,** Hypoglycemic action of onion and garlic, *Lancet,* 2, 1491, 1973.
65. **Chang, M. L. W. and Johnson, M. A.,** Effect of garlic on carbohydrate metabolism and lipid synthesis, *J. Nutr.,* 110, 931, 1980.
66. **Farva, D., Goji, I. A., Joseph, P. K., and Augusti, K. T.,** Effects of garlic oil on streptozotocin-diabetic rats maintained on normal and high fat diets, *Indian J. Biochem. Biophys.,* 23, 24, 1986.
67. **Zaman, Q. A. M., Banoo, H., Choudhury, S., Choudhury, S. A. R., and Khaleque, A.,** Effect of garlic oil on serum cholesterol and blood sugar levels in adult human volunteers in Bangladesh, *Bangladesh Med. J.* 10, 6, 1981.
68. **Mahanta, R. K., Goswami, R. K., Kumar, D., and Goswami, P.,** Effect of *Allium sativum* (garlic) on blood lipids, *Indian Med. Gaz.,* April, 157, 1980.
69. **Rasiah, S. V.,** Insulin-sparing action of garlic oil in streptozotocin diabetic rabbits, Dissertation submitted to the University of Maiduguri, Nigeria for the award of B.Sc. in Biochemistry to Miss Rasiah; supervisors Joseph, P. K. and Augusti, K. T., 1985, 19.

70. **Tempel, K. H.,** *Med. Ernaehr.*, 3, 197, 1962.
71. **Augusti, K. T. and Mathew, P. T.,** Lipid lowering effect of allicin (diallyl disulphid-oxide) on long term feeding to normal rats, *Experientia*, 30, 468, 1974.
72. **Bordia, A. K. and Verma, S. K.,** Garlic on the reversibility of experimental atherosclerosis, *Indian Heart J.*, 30, 47, 1978.
73. **Bordia, A. and Verma, S. K.,** Effect of garlic feeding on regression of experimental atherosclerosis in rabbits, *Artery*, 7, 428, 1980.
74. **Jain, R. C. and Konar, D. B.,** Effect of garlic oil in experimental cholesterol atherosclerosis, *Atherosclerosis*, 29, 125, 1978.
75. **Kritchevsky, D.,** Effect of garlic oil on experimental atherosclerosis, *Artery*, 1, 319, 1975.
76. **Kritchevsky, D., Tepper, S. A., Morrisey, R., and Klurfeld, D.,** Influence of garlic oil on cholesterol metabolism in rats, *Nutr. Rep. Int.*, 22, 641, 1980.
77. **Chi, M. S., Koh, E. T., and Stewart, T. J.,** Effects of garlic on lipid metabolism in rats fed cholesterol or lard, *J. Nutr.*, 112, 241, 1982.
78. **Augusti, K. T.,** Hypocholesterolaemic effect of garlic (*Allium sativum* Linn), *Indian J. Exp. Biol.*, 15, 489, 1977.
79. **Bordia, A. and Bansal, H. C.,** Essential oil of garlic in prevention of atheroscloerosis, *Lancet*, 2, 1491, 1973.
80. **Bordia, A.,** Effect of garlic on blood lipids in patients with coronary heart disease, *Am. J. Clin. Nutr.*, 34, 2100, 1981.
81. **Jain, R. C.,** Effect of garlic on serum lipids, coagulability and fibrinolytic activity, *Am. J. Clin. Nutr.*, 30, 1380, 1977.
82. **Pushpendran, C. K., Devasagayam, T. P. A. Banerji, A., and Eapen, J.,** Cholesterol-lowering effect of allitin in suckling rats, *Indian J. Exp. Biol.*, 18, 858, 1980.
83. **Sebastian, K. l., Zacharias, N. T., Philip, B., and Augusti, K. T.,** The hypolipidemic effect of onion (*A. cepa* Linn) in sucrose-fed rabbits, *Indian J. Physiol. Pharmacol.*, 23, 27, 1979.
84. **Zacharias, N. T., Sebastian, K. L., Philip, B., and Augusti, K. T.,** Hypoglycemic and hypolipidemic effects of garlic in sucrose fed rabbits, *Indian J. Physiol. Pharmacol.*, 24, 151, 1980.
85. **Sodimu, O., Joseph, P. K., and Augusti, K. T.,** Certain biochemical effects of garlic oil on rats maintained on high fat-high cholesterol diet, *Experientia.*, 40, 78, 1984.
86. **Shoetan, A., Augusti, K. T., and Joseph, P. K.,** Hypolipidemic effects of garlic oil in rats fed ethanol and a high lipid diet, *Experientia*, 40, 261, 1984.
87. **Bobboi, A., Augusti, K. T., and Joseph, P. K.,** Hypolipidemic effects of onion oil and garlic oil in ethanol-fed rats, *Indian J. Biochem. Biophys.*, 21, 211, 1984.
88. **Bordia, A.,** Effect of garlic on human platelet aggregation in vitro, *Atherosclerosis*, 30, 355, 1978.
89. **Boullin, D. J.,** Garlic as a platelet inhibitor, *Lancet*, 1, 776, 1981.
90. **Sainani, G. S., Desai, D. B., Gorke, N. H., Natu, S. M., Pise, D. V., and Sainani, P. G.,** Dietary garlic, onion and some coagulation parameters in Jain community, *J. Assoc. Physicians India.*, 27, 707, 1979.
91. **Bordia, A. K., Joshi, H. K., Sanadhya, Y. K., and Bhu, N.,** Effect of essential oil of garlic on serum fibrinolytic activity in patients with coronary artery disease, *Atherosclerosis*, 28, 155, 1977.
92. **Bordia, A., Verma, S. K., Khabia,B. L., Vyas, A., Rathore, A. S., Bhu, N., and Bedi, H. K.,** The effectiveness of active principle of garlic and onion on blood lipids and experimental atherosclerosis in rabbits and their comparison with clofibrate, *J. Assoc. Physicians India*, 25, 509, 1977.
93. **Bordia, A., Verma, S. K., Vyas, A. K., Khobya, B. L., Rathore, A. S., Bhu, N., and Bedi, H. K.,** Effect of essential oil of onion and garlic on experimental atherosclerosis in rabbits, *Atherosclerosis*, 26, 379, 1977.
94. **Saxena, K. K., Gupta, B., Kulshrestha, V. K., Srivastava, R. K., and Prasad, D. N.,** Effect of garlic pretreatment on isoprenaline-induced myocardial necrosis in albino rats. *Indian J. Physiol. Pharmacol.*, 24, 233, 1980.
95. **Malik, Z. A. and Siddiqui, S.,** Hypotensive effect of freeze-dried garlic (*Allium sativum*) sap in dog, *J. Pakistan Med. Assoc.*, 31, 12, 1981.
96. **Fromtling, R. A. and Bulmer, G. S.,** In vitro effect of aqueous extract of garlic (*Allium sativum,*) on the growth and viability of *Cryptococcus neoformans*, *Mycologia*, 70, 397, 1978.
97. **Amer, M., Taha, M., and Tosson, Z.,** The effect of aqueous garlic extract on the growth of dermatophytes, *Int. J. Dermatol.*, 19, 285, 1980.
98. **Weisberger, A. S. and Pensky, J.,** Tumor inhibition by a sulfhydryl blocking agent related to an active principle of garlic (*Allium sativum*), *Cancer Res.*, 18, 1301, 1958.
99. **Fugiwara, M. and Natata, T.,** Induction of tumour immunity with tumor cells treated with extract of garlic *Allium sativum*, *Nature*, 216, 83, 1967.
100. **Devasagayam, T. P. A., Puspendran, C. K., and Eapen, J.,** Diallyl disulphide induced changes in microsomal enzymes of suckling rats, *Indian J. Exp. Biol.*, 20, 430, 1982.

101. **Rao, R. R., Rao, S. S., and Venkataraman, P. R.,** Investigations on plant antibiotics. I. Studies on allicin, the antibacterial principle of *Allium sativum* (Garlic), *J. Sci. Ind. Res. India*, 5, 31, 1946.
102. **Stoll, A. and Seebeck, E.,** Chemical investigations on allicin, the specific principle of garlic, in *Advances in Enzymology*, Vol. 11, Nord, F. F., Ed., Interscience, New York, 1951, 377.
103. **Johnson, M. G. and Vaughn, R. H.,** Death of *Salmonella typhimurium* and *Escherichia coli* in the presence of freshly reconstituted dehydrated garlic and onion, *Appl. Microbiol.*, 17, 903, 1969.
104. **Amonkar, S. V. and Reeves, E. L.,** Mosquito control with active principle of garlic *Allium sativum, J. Econ. Entomol.*, 63, 1172, 1970.
105. **Amonkar, S. V. and Banerji, A.,** Isolation and characterisation of larvicidal principle of garlic, *Science*, 174, 1343, 1971.
106. **Nakagawa, S., Masamoto, K., Sumiyoshi, H., Kunihiro, K., and Fuwa, T.,** Effect of raw and extracted-aged garlic juice on growth of young rats and their organs after peroral administration, *J. Toxicol. Sci.*, 5, 91, 1980.
107. **Sebrell, W. H.,** An anemia of dogs produced by feeding onions, *U.S. Public Health Rep.*, 24, 1175, 1930.
108. **Thorp, F. and Harsfield, G. S.,** Onion poisoning in horses, *J. Am. Vet. Med. Assoc.*, 94, 52, 1939.
109. **Augusti, K. T.,** Cysteine-onion oil interaction, its biological importance and the separation of interaction products by chromatography, *Indian J. Exp. Biol.*, 15, 1223, 1977.

Chapter 6

GARLIC *Allium sativum* L.

H. Takagi

TABLE OF CONTENTS

I. INTRODUCTION

World production of garlic (*Allium sativum* L.) ranks in importance second to onions among the *Allium* species. Garlic is grown in many countries, though at present China, Egypt, India, Spain, South Korea, and Turkey are the principal producers.[1]

Fresh garlic is widely used in cooking, and dehydrated products are quite common as a condiment and in the food industry. (See chapter on ''Processing of *Allium*, Use in Food Manufacture''.) In some countries, particularly in Asia, the leaves of young, nonbulbing plants and also freshly emerged young flower stalks are used as food. It is widely held that garlic has some therapeutic values, and many folk-medicinal uses of garlic have developed (See chapter on "Medicinal and Therapeutic Effects".)

Garlic is relatively resistant to diseases and pests, and the bulbs store well. Therefore, only a few problems exist with garlic culture, mainly those concerned with virus diseases (see chapter on "Virus Diseases"), nematodes (see chapter on "Nematodes"), mites (see chapter on "Insect Pests"), and some bulb rotting diseases (see chapter on "Storage Diseases"). In some situations, the sprouting of lateral buds before clove formation, i.e., secondary growth of cloves, is also a problem. In addition, cracking of bulbs at harvest and the formation of abnormal bulbs, such as single-cloved bulbs, few-cloved bulbs, or bulbs which contain small cloves developing from lateral axes (secondary cloves), may be problems.

There are many publications on cultural practices for garlic. Fewer reports deal with the physiology of growth and development.[2] This chapter summarizes mainly knowledge on the physiology of plant growth, bulb and inflorescence formation, and dormancy. Other aspects of garlic culture (such as pests, carbohydrate biochemistry, flavor biochemistry, therapeutic effects, and other properties) are discussed elsewhere in this book. Details on agronomic practice are described by Jones and Mann.[3]

II. TAXONOMY AND CLASSIFICATION

A. Origin

A truly wild species of garlic *A. sativum* L. is not yet known, but *A. longicuspis* Rgl. is considered to be the most closely related species to the cultivated crop and is considered to be the garlic's wild ancestor.[3,4] Since *A. longicuspis* is endemic to central Asia, it is held that garlic originated there.

It used to be thought that garlic was totally sterile, but recently, many fertile plants were collected in the bazaars of five towns on the northern side of the Tien Shan mountains in Soviet central Asia.[5] Since this area is close to where *A. longicuspis* grows naturally, it is a possibility that these fertile "garlic" clones include some representatives of the wild species *A. longicuspis*. Etoh states that it is difficult to distinguish between *A. longicuspis* and *A. sativum*.[5] The fertile plants from central Asia are useful indicators of the center of origin of cultivated garlic, and they may in the future be useful for garlic breeding.

B. VARIETAL DIFFERENTIATION

Since the cultivated forms of garlic are all sterile and have been propagated only vegetatively, it might be expected that they would show little intraspecific variation. However, there are many of strikingly distinct clones in regions where garlic has long been grown, e.g., China[6] and elsewhere. Taxonomists have recognized at least four botanical varieties within *A. sativum* L., namely, *A. sativum* L. var. *sativum*; *A. sativum* L. var. *ophioscorodon* (Link) Doll.: *A. sativum* L. var. *pekinense* (Prokh) Maekawa; and *A. sativum* L. var. *nipponicum* Kitamura.

Jones and Mann[3] could not satisfactorily classify many garlic cultivars into clearly defined botanical varieties, because many show combinations of characteristics from two or more varietal groups. It appears, then, that the above botanical classification has little advantage. The distinguishing between garlic cultivars according to the following criteria is relatively simple and may be of practical use:

1. Morphological characteristics: bolting type, number and size of (the primary) cloves, number of leaf-axils forming (the primary) cloves, number of secondary cloves formed in a lateral bud (or a primary clove), bulb weight, color of the outer protective leaf of the cloves, number of protective leaves, width and length of foliage leaves, plant height, and tenderness of the green leaves.

2.   Physiological and ecological characteristics: time of bulbing and maturity, low-temperature and long-day requirements for bulb formation, winter hardiness, and bulb dormancy.

Besides the above characteristics, the taste of the cloves may also be used in practical classification. Also, adaptation to agroclimatic zones may be of help. This is a complex characteristic determined by the above-mentioned environmental requirements and is possibly the most important factor for cultural practice, since garlic cultivars adapted to one area are sometimes worthless as a crop in another region.[7]

III. MORPHOLOGY, GROWTH, AND DEVELOPMENT

The morphology of garlic was first described by Mann,[8] and later Yamada[9] provided details on inflorescence organogenesis. Shah and Kothari[10-12] described the histogenesis of cloves, inflorescences, and shoot apices. Here the morphological information necessary for discussing garlic growth and, in particular, the development of the inflorescences and bulbs is given (see chapter on "Morphology and Anatomy").

A. DESCRIPTION OF GROWTH AND DEVELOPMENT
1. *Shoot Growth*
In temperate regions of Asia, garlic is usually planted in autumn, but in cooler areas, e.g., north China, where plants cannot survive the severe winter cold, the cloves are planted in the spring. After autumn planting, garlic grows vegetatively, and neither inflorescence nor lateral buds are normally formed until early spring.

2. *Reproductive Growth*
Garlic plants from temperate regions are vernalized by low winter temperatures. Cultivars that bolt in Japan initiate inflorescences when the mean air temperature reaches about 8°C in early spring. Inflorescence initiation begins with the differentiation of the spathe (an involucral bract enclosing the umbel), and at about the same time lateral-bud primordia are initiated in the axils of the youngest two (or sometimes three) foliage leaves (Figure 1).

3. *Clove and Bulb Growth*
Each of the bud primordia forms between two and six growing points, each of which develops into a lateral bud. These lateral buds develop later into cloves. When air temperatures reach about 10°C, the growing points start to form leaf initials. As air temperatures rise further, each lateral bud initiates leaves successively which grow vigorously. However, the first and second leaves of each lateral bud differentiate in a particular way, this being initiated and promoted by long photoperiods. Both these leaves develop thickened sheaths and produce no blades. The first (oldest) sheath ceases thickening when the second leaf begins to increase in size. As the second leaf grows, the first one grows to enclose it. The growth of the first leaf ceases about 2 weeks before the foliage senesces entirely, and it finally develops into a dry protective skin for the swollen second sheath. The second leaf sheath grows and swells rapidly under warm long-day conditions prevailing late in the spring, thus forming the main storage leaf (Figure 2). In some cultivars, both the first and the second leaves of a lateral bud develop into dry protective skins, and the third leaf-sheath swells to form the storage leaf. Normally, however, the third leaf develops only slightly, and remains enclosed within the storage leaf until the clove sprouts. After the initiation of a lateral bud, the fourth leaf plus a few more leaf initials in the growing point become dormant.

Clove formation in nonbolting cultivars differs only a little from the process described above. In cv. "California", lateral-bud primordia form in the axils of the youngest six to

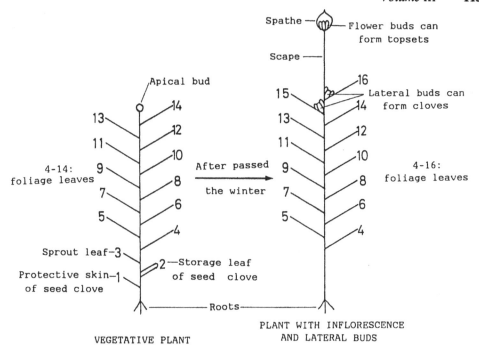

FIGURE 1. Diagrammatic representation of (left) a vegetative plant before cold induction, (right) a reproductive plant after cold induction, in the bolting garlic.

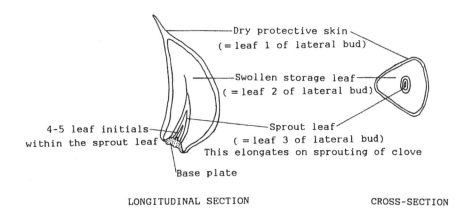

FIGURE 2. Outline drawings of median longitudinal and cross sections of the garlic clove.

eight foliage leaves, beginning in the oldest one. When this lowest axil initiates lateral buds, the growing point is still initiating foliage leaves. The apex then follows one of two alternative developmental paths.[9] It may initiate an incomplete leaf, lacking a blade, and then cease growth and activity and finally degenerate. Alternatively, the terminal growing point may behave similarly to that of a lateral bud. That is, it may form a clove and become dormant.

B. THE VARIATION OF BOLTING WITH GENOTYPE

The degree of inflorescence development varies considerably with genotype, and garlic can be classified into three groups according to the extent of flower stalk development.

Complete bolting — These produce a long, thick flower stalk, bearing many flowers and topsets (bulbils). These normally form cloves in the axils of the youngest two foliage leaves.

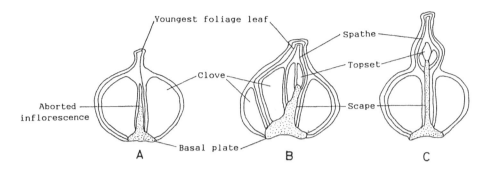

FIGURE 3. Diagrammatic representation of the three types of incomplete bolting. The longitudinal sections of the bulbs are shown. (A) An inflorescence aborts; (B) topsets and a scape remain within the bulb since the scape elongates slightly; (C) topsets and a scape remain enclosed by the pseudostem.

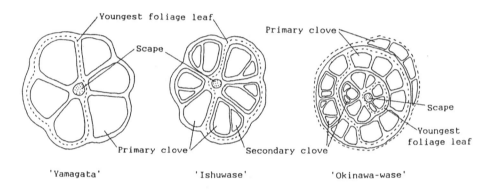

FIGURE 4. Diagrammatic representation of cross sections of the cloves of cvs. "Yamagata", "Ishuwase" and "Okinawa-wase". Number of nodes forming cloves, number of cloves per node, and number of secondary cloves vary with genotype. Number of nodes forming cloves of cvs. "Yamagata", "Ishuwase", and "Okinawa-wase" are 2, 2, and 5,[13] respectively.

Incomplete bolting — These usually produce flower stalks, but they are thinner and shorter than the complete bolting genotypes. The stalk and topsets may often remain enclosed by the pseudostem. In the extreme, the complete flower stalk plus topsets remain within the bulb (Figure 3). A few, relatively large topsets are produced per scape. There is usually no flower formed. The extent of flower stalk development varies considerably even among plants of the same clone. Some of these genotypes form cloves in the axils of the youngest three or more foliage leaves, and others normally form cloves in the axils of the youngest two (Figure 4).[13]

Nonbolting — These do not normally form flower stalks; rather, cloves are produced, but occasionally scapes may develop. These form cloves in the axils of the youngest five or more foliage leaves. When incomplete-bolting cultivars produce very short scapes within the bulb, they may appear like nonbolting types.

C. RELATION BETWEEN LATERAL BUD FORMATION AND INFLORESCENCE FORMATION

In most bolting cultivars, lateral buds are normally initiated at two or three consecutive nodes, just below the node at which the inflorescence is initiated. Inflorescence initiation is shortly followed by lateral-bud initiation, and it appears that lateral-bud formation is induced by a weakening of apical dominance caused by inflorescence initiation.

Since nonbolting cultivars occasionally form inflorescences, it is thought that these cultivars show some of the characteristics enabling inflorescence initiation, but in attenuated form. Lateral-bud initiation in nonbolting cultivars is nearly contemporaneous with inflorescence initiation in bolting cultivars. For example, lateral-bud primordia start to initiate in cv. "California" almost simultaneously with spathe initiation in bolting cultivars.[9] Therefore, although inflorescence formation is not seen, the early physiological processes involved in inflorescence initiation may occur.

D. THE FORMATION OF SECONDARY CLOVES

In the common cultivars, a lateral bud forms only a single storage leaf in most cases. However, in some cultivars, a lateral bud may often form two or more storage leaves (Figure 4). In such cases, secondary axillary buds form in the axils of some of the first leaves of the primary lateral buds and these secondary axillary buds each form a storage leaf (Figure 5). As a consequence, a cluster of small cloves termed secondary cloves develops from each primary lateral bud.

E. SPROUTING OF LATERAL BUDS BEFORE CLOVE FORMATION

Leaves of a lateral bud do not normally grow their blade during a period of clove formation. However, if external conditions become disadvantageous to storage-leaf formation — short days, heavy manuring with nitrogen, etc. — for about one or more months starting a few weeks before lateral-bud formation, some of the first leaves of each lateral bud often grow their blades and develop into green foliage leaves (Figure 6). This growth is termed "secondary growth" of a clove.[14-17] In an extreme case, a lateral bud forms an inflorescence and some small cloves.[18]

IV. FACTORS AFFECTING BULB INITIATION AND DEVELOPMENT

A. INTRODUCTION

There are many reports that bulbing in garlic is promoted by previous exposure of cloves or growing plants to low temperatures and by long days.[7,8,14,19-21] However, there are few reports concerning the interactions between the temperature effect and the daylength effect in bulbing of garlic.[22]

Bulbing of garlic consists of two processes. The first process is formation of lateral buds in the axils of some of the youngest foliage leaves. This process normally occurs with inflorescence formation. The second process is the transition of lateral buds into storage leaves. In this process, the second (or the third) leaf of each lateral bud develops into a storage leaf, and the sprout enclosed within the storage leaf gradually becomes dormant as the storage leaf grows. The external conditions favoring the first process of bulbing are not always the same as those favoring the second process. Therefore, if one intends to examine strictly the effects of external conditions on bulbing, one should examine the two processes of bulbing separately. We will also discuss the two processes of bulbing separately. First, we will consider the transition of lateral buds into storage leaves.

B. DIFFERENTIATION OF A LEAF PRIMORDIUM INTO A STORAGE LEAF

Normally, the storage leaf develops from the second or, in some cases, from the third leaf initiated by a lateral bud. This may vary with environmental conditions. Garlic plants grown under 24-h photoperiods, with sufficient cold treatment, can form a storage leaf from the first leaf initial on a lateral bud.[22] Here, no protective skin is developed, since this normally develops from the first leaf initial. In contrast, plants exposed to low temperatures followed by short days may develop several foliage leaves on the lateral bud, and, as a result, the bud sprouts and the storage leaf forms from the third or even later initiated

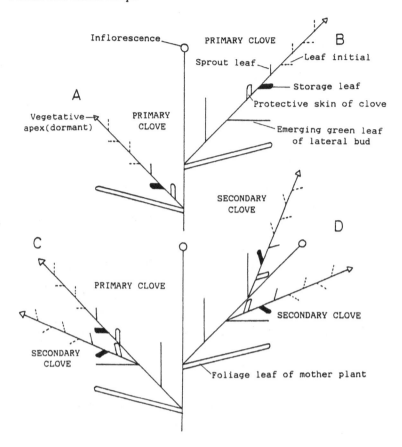

FIGURE 5. Diagrammatic representation of secondary growth of cloves and formation of secondary cloves. (A) A clove undergoing no secondary growth; no leaf of the lateral bud develops its leaf blade; leaf 2 develops into a storage leaf. (B-D) A clove undergoing secondary growth: (B) a clove forming no secondary clove; leaf 1 or leaf 1 to 2 or more emerge (develop into green foliage leaves); leaf 2 or more develops into a storage leaf; (C) a clove forming a secondary clove or more but no inflorescence; the lateral bud initiates some foliage leaves and secondary lateral-buds (developing into secondary cloves), and ends in forming a terminal clove; this cluster of small cloves is akin to an intermediate bulb; (D) a clove forming both secondary cloves and an inflorescence; the lateral bud initiates some foliage leaves and secondary lateral buds, and ends in forming an inflorescence; this cluster of small cloves with a scape is akin to a normal bulb.

leaf.[14,15,18] When a temperate-region cultivar is not exposed to low temperatures or long photoperiods, leaf primordia develop into foliage leaves, and the cultivar will form neither an inflorescence nor storage leaves.[22]

Plants grown from small bulbils or very small cloves, often produce single-cloved bulbs (Figure 7). Such bulbs are also formed when neither an inflorescence nor a lateral bud is formed. The formation of single-cloved bulbs involves only the second process of bulbing (see Section IV.A.). When a plant forms a single-cloved bulb, a leaf primordium in the apical bud develops into a storage leaf without the formation of an inflorescence and lateral buds. When such a single-cloved bulb develops, the leaf number on the main axis which forms the storage leaf is determined by the extent of cold induction and by photoperiod. Earlier formed leaf initials develop into the storage leaf with long periods of low temperature and long photoperiods. Later formed leaf initials develop into the storage leaf with a brief cold treatment and/or shorter photoperiods. The results in this section are based mainly on experiments with plants forming single-cloved bulbs.

FIGURE 6. Secondary growth of cloves in cv. ''Hoki''. (A) An intact plant; (B) the plant with foliage leaves of the main axis removed.

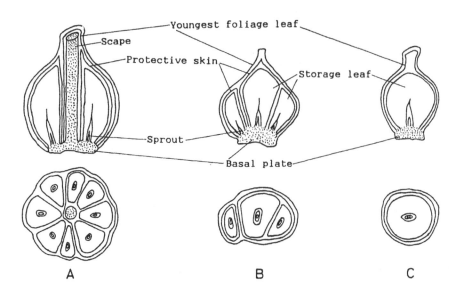

FIGURE 7. Outline drawings of longitudinal sections (top) and cross sections (bottom) of (A) a normal multicloved bulb, (B) an intermediate bulb, and (C) a single-cloved bulb in garlic.

C. FACTORS INDUCING STORAGE-LEAF DIFFERENTIATION

1. *Long Photoperiods and Low Temperatures*

Temperate-region garlic cultivars do not bulb under the long photoperiods of spring unless previously exposed to low temperatures in the winter. Equally, bulbing is very delayed if they are not exposed to long photoperiods in spring, even if they have been previously

TABLE 1

Effects of Temperature During 60-d Storage Period before Planting, Photoperiod, and Cultivar on the Number of Days from Planting to Forming Storage Leaves[a] in Garlic Grown at 20 to 25°C[22]

Cultivar	Photoperiod (h)	Storage temperature (°C)			
		5	10	15	20
		Period from planting to storage-leaf formation (d)			
Ishuwase	8	—	—	68	124
	16	<40[b]	<40	<40	47
	24	—	—	<40	—
Okinawa	8	—	—	V[c]	—
	16	<40	<40	103	131
	24	—	—	<40	—
Kanko	8	—	—	V	—
	16	<40	<40	117	V
	24	—	—	47	—
Yamagata	8	75	89	V	V
	16	<40	47	131	V
	24	—	—	61	75
Hoki	8	(V)[d]	V	V	—
	16	47	54	V	V
	24	—	—	75	96

[a]   Five plants were lifted and examined at about 7-d intervals from 40 d after planting.
[b]   Storage leaves were formed in less than 40 d from planting.
[c]   No storage leaf was formed. The plants remaining vegetative for 227 d after planting.
[d]   No storage leaf was formed for 124 d after planting.

exposed to low temperatures in the winter.[7,9,14,21] Therefore, it seems that both low temperatures and long photoperiods are required for the formation of storage leaves by such cultivars. However, this is not strictly true, since in some genotypes very long photoperiods, e.g., 24 h, can substitute for normally inadequate low-temperature treatments, e.g., 20°C (Table 1). Furthermore, given sufficient cold treatment, some cultivas readily form storage leaves under 8-h photoperiods (Table 1).

The above relationship between the low temperature effect and the photoperiod effect may be explained as follows. A photoperiod longer than a critical value is considered to be the main factor inducing storage-leaf formation. Different cultivars may have different critical photoperiods. The longer the cold treatment, the shorter the critical day-length required for storage-leaf induction.

2. Effect of Temperature and Its Duration
a. Optimum Cold Treatment

Between 2 and 4°C are the most effective storage temperatures for temperate-garlic cultivars in order to stimulate the formation of storage leaves after planting out at 20°C (Table 2). However, the optimum temperature in the strict sense varies with cultivar, being 4°C in cv. "Yamagata" and 2°C in cv. "Hoki".

b. Duration of Low-Temperature Treatment

The effects of cold treatment vary with temperature, the duration of the treatment, the subsequent photoperiod, and cultivar. The Japanese cool-temperate cv. "Yamagata" will not form a storage leaf when grown at temperatures above 20°C with 16-h photoperiods

TABLE 2
Effects of Temperatures during a 30-d Storage Period and Cultivar, on Storage-Leaf Induction in Garlic Grown at 20°C or More Under Natural Photoperiods (January to July, in Japan).[24]

Cultivar	Storage temperature °C	Plants forming storage leaves (%)[a]	Leaves initiated by the clove sprout at planting time (number)[b]	Foliage leaves formed before the storage leaf (number)[c]
Yamagata	−4	0	7.2	—
	−2	0	7.3	—
	0	93	7.3	11.9
	2	100	7.3	10.7
	4	100	7.5	9.8
	6	100	7.3	11.3
	15	0	7.6	—
Hoki	−4	0	6.0	—
	−2	0	6.0	—
	0	9	6.0	—
	2	69	6.0	12.2
	4	67	6.0	13.3
	6	23	6.0	14.7
	15	0	6.0	—

[a] Every plant either formed a single-cloved bulb or still remained vegetative approximately 6 months after planting.

[b] For cvs. "Yamagata" and "Hoki" the S.E. values of the means were 0.2 and 0.0, respectively.

[c] In all cases except cv. "Hoki" at 6°C, the S.E. values were <5%, of the mean. For the latter, the S.E. was 9% of the mean. Replication approximately 13.

unless cloves are given a cold treatment before planting. However, 1 week of cold treatment at 5°C before planting was sufficient for storage-leaf formation as single-cloved bulbs (Figure 8). With longer duration at 5°C up to 3 weeks, the storage leaf was formed by successively earlier (lower leaf number) leaf initials (Figure 8). With a duration of 5°C for more than 3 weeks, this effect continued but less dramatically. Nevertheless, the development of a single-cloved bulb is accelerated the longer the cold treatment at 5°C, up to about 3 months druation.[25] When large cloves of cv. "Yamagata" were stored at 4 to 6°C for 4 months or more, the formation of new storage leaves (as single-cloved bulbs) commenced within the sprouting cloves, even during storage.[26] In such cases, the growth of the developing cloves is very slow. The stimulating effect of low temperature on formation of storage leaves of cv. "Yamagata" seems to reach a maximum after 3 to 4 months, when cloves are stored at optimum temperatures (4 to 6°C) (see also Section IV.C.2.a). However, at 0 to 2°C a storage period of 6 months or more was needed to induce the maximum effect.[26] During storage at 0 to 2°C, storage leaves were not formed within the clove, even after 7 months.[26]

c. Critical Temperature for Cold Treatment

The limits of temperature which are effective for storage-leaf induction vary with cultivar, storage period, and photoperiod after planting. In cv. "Yamagata", the upper limit of inductive temperature is between 17 and 20°C, when plants are grown under 16-h photoperiods, and is between 10 and 15°C when plants are exposed to 12-h photoperiods, irrespective of the duration of storage.[22] The lower limit of inductive storage temperature is −2°C in the case of storage for 4 weeks followed by growth under 16-h photoperiods. Storage of cloves at −4°C causes little or no induction, no matter how long the storage period.[26]

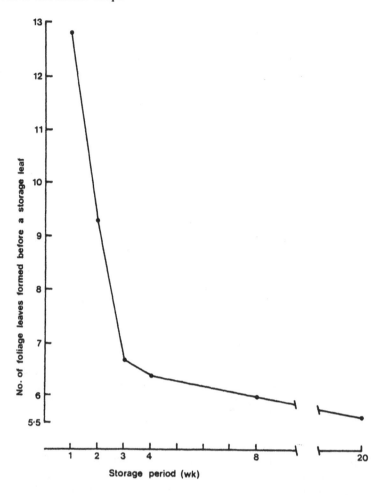

FIGURE 8.   Effect of duration of storage at 5°C on storage-leaf induction in cv. "Yamagata" grown at 20 to 25°C under 16-h photoperiod. All plants formed single-cloved bulbs. Replication approximately 12.[22]

3. *Genotypic Differences in Requirement for Cold Treatment and Photoperiod*

As shown in Table 1, cv. "Ishuwase" formed storage leaves (as single-clove bulbs) under 8-h photoperiods, even when the plants were not exposed to temperatures lower than 20°C between harvest and replanting. The cold treatment and photoperiod requirement of this cultivar may be quantitative rather than absolute. In contrast cv. "Hoki", hardly formed a storage leaf under 8-h photoperiods, even if exposed to low temperatures of 5°C for 2 months before planting (Table 1), or even if overwintered outdoors (Table 3). Hence, the long photoperiod requirement of cv. "Hoki" may be absolute (qualitative) under natural photoperiods since it cannot form a storage leaf under 16-h photoperiods (close to the longest natural daylength in most countries in the temperate zone), unless exposed to a temperature of 10°C or less; in addition, this cultivar may have an absolute cold requirement. The cultivar "Yamagata" may also have an absolute (qualitative) cold requirement, but it is less than that of cv. "Hoki" (Table 1). Although cv. "Ishuwase" has less cold and photoperiod requirement than the other mentioned cultivars, its storage-leaf formation was nevertheless greatly delayed unless it was exposed to either low temperatures or long photoperiods (Table 1). When this cultivar was planted at the beginning of autumn in a temperate region, it was not until the following spring that the plant formed storage leaves. However, Shah and Kothari[10] report that bulbing started after 45 d in a white-skinned cultivar planted in India

TABLE 3
Effect of Short-Day Treatment[a] of Field-Grown Over-Wintered Garlic Plants on the Formation of Storage Leaves by Lateral Buds[15]

Cultivar	Photoperiod (h)	Clove weight per plant (g)	Buds forming storage leaves (%)[b]	Leaves before the storage leaf in each lateral bud (number)[c]
Yamagata	8	6.1	100	4.4
	ND[d]	33.8	100	1.0
Hoki	8	—[e]	11	—
	ND	29.1	100	1.4

[a] Treated from March 1 to July 2. Number of replicate plants of ND and 8-h photoperiods, 34 and 5, respectively.
[b] Percentage of the total number of lateral buds that formed storage leaves.
[c] In normal growth this number is 1.0, this being the outer protective skin.
[d] ND = natural daylength.
[e] The foliage remained green in the middle of July when that of ND plants had senesced.

at the beginning of October, and that there, garlic is a crop with a growth duration of only about 4 months.[10] Tropical cultivars seem to form storage leaves without exposure to either low temperature or long photoperiod. However, the bulbing of some garlic cultivars from Thailand and Hong Kong is promoted by long photoperiods.[27] Therefore, even tropical cultivars may have a small quantitative cold and long photoperiod response.

4. Effects of Temperature and Daylength on Storage-Leaf Growth

The optimum temperature range for growth of storage leaves is 17 to 26°C.[22] At lower temperatures, storage-leaf growth slows, and it is very slow at 9°C or less. At 28°C or more, growth is inhibited.[28] The longer the photoperiod, the more rapidly storage leaves grow and the earlier they mature. However, the final bulb size does not necessarily increase with increase of photoperiod.

5. The Reversal of Cold-Treatment Induction by High Temperatures

If cloves which have been exposed to cold treatment are grown with some water stress at temperatures of 30°C or more, the inductive effect of low temperatures is reduced or disappears (Table 4). This reversion effect of high temperature was greater the higher the treatment temperature and the longer the duration of this treatment. However, the longer the period of cold treatment, the more difficult it becomes to reverse the cold-treatment effect by subsequent high temperatures. When previously low temperature-treated plants, which have then been treated with high temperature, are given a further cold period, storage-leaf induction again occurs. Thus, the effect of high and low temperatures on storage-leaf induction are, to some extent, reversible. This high/low temperature reversibility is akin to vernalization, devernalization, and revernalization in flower induction. If the plants are well watered during the high-temperature treatment and are growing rapidly, the high-temperature treatments are less effective in reversing the low-temperature effect (Table 5). However, if water supply is limited during the high-temperature exposure, the high-temperature effect is stronger, and it becomes apparent at lower temperatures than is the case for well-watered plants. When stored cloves are subjected to temperatures even of 20°C, the effect of previous cold treatment on clove formation is slightly reversed.[22]

6. Other Factors Affecting Storage-Leaf Formation
a. Dormancy

The effects of low temperature on inducing later storage-leaf formation are less when

TABLE 4

**Effects of High Temperatures After Cold Induction on Storage-Leaf Formation in
cv. "Yamagata" Grown at 20 to 25°C Under 16-h Photoperiod[22]**

Duration of cold storage at 5°C (d)	Temperature imposed after planting		Plants forming storage leaves[b] (% of total)
	Air temperature[a](°C)	Duration (d)	
30	—	—	100
30	25	30	93
30	30	20	100
30	30	30	17
30	30	40	0
30	35	10	100
30	35	20	0
60	—	—	100
60	30	30	70
60	30	40	0
90	—	—	100
90	30	30	66
90	30	40	52
30	30	30[c]	100

[a] Soil temperatures were always 1°C below air temperatures.

[b] Every plant either formed a single-clove bulb or remained vegetative. Control plants stored at 20°C in place of 5°C for 30 d did not form a storage leaf. Replication approximately 10.

[c] Plants were cold induced at 5°C for 30 d following the high-temperature treatment.

applied to dormant cloves. The deeper the dormancy of cloves, the less the effectiveness of cold treatment.[23]

b. Spectral Quality of Supplementary Light Used to Extend Photoperiods

When long photoperiods are provided by extending the natural daylength by artificial illumination, the effect on storage-leaf induction by incandescent lamps is much more effective than that from fluorescent ones.[30] When the natural daylength is extended by light of different spectral composition, far-red (approximately 730 nm) radiation has a promotive effect on storage-leaf formation, and blue (approximately 440 nm) has less effect, but red (approximately 660 nm) has little or no effect.[30] This resembles the response of onion.[31] It seems that the effect of the light from incandescent lamps is caused by the far-red component of the light spectrum and that the lesser effect of the light from some fluorescent lamps is caused by their lesser emission of far-red wavelengths.

c. Water Supply

When cold-induced cloves were planted and grown under low water supply, storage-leaf formation was delayed relative to that with high water supply. However, the restricted water treatment caused storage leaves to form at lower leaf positions (number) (older leaves) on the main axis.[32] Following storage at 2°C for 3 to 5 months, storage leaves were formed within the cloves subsequently kept at 6°C. However, these storage leaves formed much later than those of plants stored at 2°C for 3 to 5 months and then planted at approximately 20°C (Table 6). Storage leaves formed at lower leaf positions (numbers) in plants placed in 6°C storage than those planted after storage at 2°C (Table 6). This may have been because the differentiation of leaf primordia into foliage leaves is inhibited by shortage of water. Hence, in the stored cloves, even early formed (low leaf number) leaf primordia are promoted to differentiate into storage leaves when water is in short supply. It should be noted, however, that the growth rate of storage leaves is very low during storage.

TABLE 5
Effect of Soil Moisture During High-Temperature Treatment[a] on Leaf-Blade Growth and Storage-Leaf Formation by cv. "Yamagata" Grown at 20 to 25°C under 16-H Photoperiods[29]

Watering regime	Soil temperature (°C)	Expanded leaves at the end of the hot treatment (number)	Plants forming storage leaf[b] (%)
Daily irrigation	29.5	4.1	100
Once, at planting	31.5	0.0	50

[a] After cold induction at 5°C for approximately 5 weeks, cloves were planted and treated at 30°C for 4 weeks under 16-h photoperiods.
[b] Plants either formed a single-cloved bulb or remained vegetative. Replication 10.

TABLE 6
Formation of a New Storage Leaf Within a Clove During Low-Temperature Storage of cv. "Yamagata" in Comparison with Storage-Leaf Formation in Growing Plants[26]

Pretreatment[a] duration (d)	Leaves in the sprouted cloves at the end of pretreatment (number)	Storage at 6°C	Storage at 20°C	Growth at 20 to 25°C; Photoperiod 12 h	Growth at 20 to 25°C; Photoperiod 24 h
90	7.0	5.5	V	8.5	6.9
150	—	5.2	V	8.3	6.5

Number[b] of foliage leaves before the storage leaf[c]

[a] Pretreatment temperature was 2°C.
[b] In all cases the SE values were <5%, of the mean. Replication approximately 13.
[c] Plants either formed a single-cloved bulb or remained vegetative (V).

d. Nitrogen Nutrition

A high N supply to growing plants stimulates the growth of foliage leaves, but partially inhibits the differentiation of leaf primordia into storage leaves. Consequently, at high N levels, storage leaves form at higher leaf positions (number) on the main or lateral axes.[33] In contrast, N shortage promotes storage-leaf differentiation. The effect of nitrogen on storage-leaf differentiation is much smaller than that of photoperiod.

e. Plant Growth Regulators

Benzyladenine (BA) at 50 ppm applied as a spray during the period of inflorescence formation in the spring inhibits storage-leaf formation and slightly promotes the initiation of lateral buds in field-grown plants.[34] In BA-treated plants, the lower leaf initials on lateral buds are promoted to form foliage leaves. Hence, in such plants, storage leaves are formed at a high leaf number on the lateral axes. In *in vitro* cultures of the excised buds of cloves, 10 ppm BA inhibits bulb formation.[35] Because BA has a striking effect in breaking bulb dormancy in garlic (see Section VI.D), the basic effect of BA seems to be inhibit the onset of dormancy. As a result, BA promotes vegetative growth and inhibits storage-leaf formation.

Ethephon (2-chloroethyl-phosphonic acid), an ethylene-releasing agent, at 4 ppm inhibits storage-leaf formation in *in vitro* excised-bud cultures.[36] Ethephon also inhibits storage-leaf formation by intact plants when they are sprayed, or when cloves are immersed in a 240- to 480-ppm solution before planting.[34] Because ethephon has no effect on the breaking of bulb dormancy (see Section VI.D), its inhibiting effect on storage-leaf formation seems to be unrelated to the effect of BA. Contrary to the above results, there is a report from India that the application of 250 to 750 ppm ethephon to plants at the 4-leaf stage increased plant growth and clove yield.[37]

Gibberellic acid (GA) strongly induces the initiations of lateral buds (clove primordia) and bulbil primordia when overwintered plants are sprayed with a 200- to 400-ppm solution (Figure 9). Furthermore, in the plants before inflorescence induction, GA induces the initiation of lateral buds.[34] GA strongly inhibits the growth of storage leaves.

Naphthaleneacetic acid (NAA) at 0.5 to 4 ppm promotes the growth of storage leaves in *in vitro* excised-bud cultures.[36] However, 8 ppm or more NAA inhibits the growth of storage leaves.[35] NAA at 50 ppm or more inhibits storage-leaf formation in intact plants.[34]

V. FACTORS AFFECTING INFLORESCENCE INITIATION AND DEVELOPMENT

A. RELATIONSHIP BETWEEN INFLORESCENCE FORMATION AND BULB FORMATION

Completely and incompletely bolting cultivars (see Section III.B) commonly form inflorescences, but may fail to form normal inflorescences when subjected to abnormal environmental conditions. There are two cases in which no inflorescence is formed. In both cases, a clove is formed (terminal clove) and replaces the inflorescence. In one case lateral cloves are also formed, while in the other case only a terminal clove is formed, resulting in formation of a single-cloved bulb (see Section IV.B). In the fomer case, the terminal clove may in reality be a storage inflorescence because a rudiment of an inflorescence is occassionally seen on the surface of such terminal cloves.[22] When storage leaves are strongly induced in the early stages of inflorescence initiation, the direction of development of the young inflorescence primordium may change. However, the terminal bud may never initiate an inflorescence but develop directly into a clove. A bulb which forms no inflorescence but rather a terminal clove and lateral cloves usually has fewer cloves than a normal bulb which has formed both an inflorescence and lateral cloves.

Bulbs with terminal cloves and few lateral cloves can be thought as intermediate between normal multicloved and single-cloved bulbs (Figure 7). Hence, in this paper, such bulbs are termed "intermediate bulbs", and it is postulated that incomplete inflorescence formation occurs when such intermediate bulbs form (see Section V.C.1).

B. FACTORS INDUCING INFLORESCENCE INITIATION
1. *Low-Temperature Requirement*

Because temperate cultivars form no inflorescences without exposure to low temperatures, it is thought that they have a qualitative (absolute) requirement for vernalization. The vernalization requirements of tropical cultivars are not yet known. They may be vernalized at higher temperatures than temperate garlic, or they may have only a quantitative vernalization requirement.

2. *Optimum Temperatures and Critical Temperatures*

The most effective clove storage temperatures for inducing inflorescences are rather lower than the optima for inducing storage leaves. They vary with cultivar, being − 2 to + 6°C in cv. "Hoki", and − 2 to + 2°C in cv. "Yamagata" (Table 7). Both the lower and the upper limits of storage temperature effective in inducing inflorescences are also lower than those for storage-leaf induction, and again they vary with cultivar. The lower limit for cv. "Hoki" is − 4°C or slightly less, and that for cv. "Yamagata" is between − 2 and − 4°C. The upper limit storage temperature effective for inflorescence induction for cv. "Yamagata" is about 10°C.[22] When growing plants are induced, the upper limit temperatures

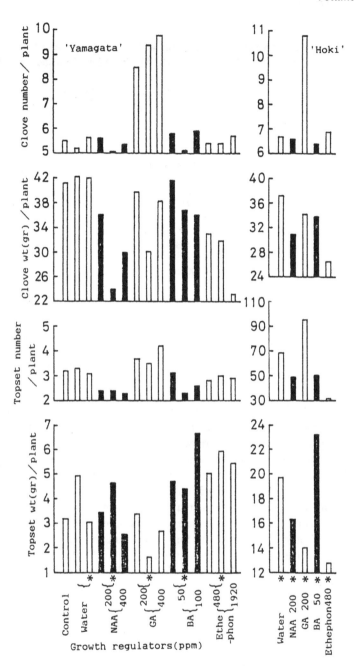

FIGURE 9. Effect of foliar spray of naphthaleneacetic acid (200 and 400 ppm), gibberellins (200 and 400 ppm), benzyladenine (50 and 100 ppm), and ethephon (480 and 1920 ppm) on field-grown over-wintered cvs. "Yamagata" and "Hoki" on numbers and weights of cloves and topsets. Sprayed from March 16 to April 27 (six sprays) or to May 25 (ten sprays) once in every week. Replication 10 to 25.[34] *Sprayed to May 25.

are 10 to 13°C in cv. "Yamagata", 13°C in cv. "Shanghai", about 16°C in cvs. "Ishuwase" and "Saga", and less than 10°C in cv. "Hoki".[39]

3. Effect of Cold-Treatment Duration

When cloves of cv. "Hoki" were stored at 2°C before planting, the longer the storage period, the fewer the leaves formed before the inflorescence. This applied to storage periods

TABLE 7
Effects of Storage Temperature and Duration of Storage on Inflorescence Formation in Garlic[a] Grown at 13°C Under 12-h Photoperiods[38]

Storage duration (d)	Storage temperature (°C)					
	−4	−2	0	2	4	6
	Plants forming inflorescences (%)[b]					
	cv. "Yamagata"					
60	0/21[c]	100	79/86	64/86	23/54	22/56
90	0/17	86/93	71/86	36/50	14/14	33/33
120	—	67/74	—	62/62	0/8	0/8
	cv. "Hoki"					
60	93/93	100	100	100	100	100
120	—	—	100	100	70/100	100
180	—	—	82/91	100	100	14/14

[a] Clove weight was 6 to 9g.
[b] Plants formed either a complete inflorescence or an incomplete inflorescence or a single-cloved bulb. Replication approximately 12.
[c] Percentage of complete and the percentage of complete plus incomplete inflorescences are shown to the left and right of the / sign, respectively.

of up to 6 months (Table 8). In cv. "Yamagata", a similar effect occurs with a storage period of up to 4 months duration. When a storage period of 5 months or more was given, the percentage of plants forming an intermediate bulb or a single-cloved bulb increased, and consequently the percentage of plants forming a normal inflorescence decreased. As a result, it was difficult to evaluate the inflorescence-inducing effect. It is generally recognized that, regardless of cultivar, the percentage of bulbs forming inflorescences decreases, and the percentage forming single-clove bulbs increases, the longer the storage beyond some particular duration. This is true when cloves are stored at temperatures higher than about 2°C, in the case of cv. "Yamagata". However, this temperature limit will vary with cultivar and storage period.[22] This can probably be explained as follows. First, cloves stored at 4 to 6°C for long periods very slowly develop single-cloved bulbs with little sprout growth during storage (see Section IV.C.6.c), but cloves stored at temperatures of about 2°C or less do not form storage leaves during the storage period. In other words, if the conditions necessary for storage-leaf formation have been fulfilled, storage leaves can form even during storage, but this is difficult when the storage temperature is about 2°C less. Second, no inflorescences are formed during storage, even if cloves are stored for very long periods, e.g., 5 to 8 months.[41] This is true whatever the storage temperature. Therefore, it may often be the case in cv. "Yamagata" that cloves stored at temperatures of 4 to 6°C for longer than some critical duration are about to form, or have already formed, single-cloved bulbs by planting time. Consequently, the percentage of inflorescence formation decreases sharply as the period of storage at such temperatures increases beyond this critical duration. The same effect was seen in cv. "Hoki", but here the storage temperature that reduced the percentage of inflorescence formation under long storage was somewhat higher than for cv. "Yamagata". Also the period of storage needed before the percentage of inflorescence formation was lowered was considerably longer than for cv. "Yamagata". The duration of the latter period tends to increase as storage temperature decreases. In other words, the longer the storage period the lower the temperature which will cause a decrease in the percentage of plants forming inflorescences (Tables 7 and 8).[42] If growing plants are cold treated for a very long period,

TABLE 8
Effect of Duration of Storage at 2°C on the Number of Leaves Formed
Before the Inflorescence in Garlic[a] Grown at 13°C Under 12-h
Photoperiods[40]

Storage period (d)				
60	90	120	150	180
Number of leaves formed before the inflorescence				
cv. "Yamagata"				
9.2 ± 0.2(9)[c]	8.0 ± 0.3(5)	7.6 ± 0.2(8)	S[b]	S[b]
cv. "Hoki"				
10.4 ± 0.2(11)	—	8.6 ± 0.2(10)	—	7.4 ± 0.1(14)

[a] Clove weight was 6 to 9 g.
[b] Plants formed either an intermediate bulb or a single-cloved bulb.
[c] Mean ± SE; replicates in brackets.

inflorescences will form during the low-temperature period, and in this case, the percentage of plants forming inflorescences is not reduced, because low temperatures during the growth phase (as opposed to during storage) are favorable to inflorescence formation, but unfavorable to storage-leaf formation (see Section V.C.2).

C. FACTORS AFFECTING INFLORESCENCE FORMATION AFTER INDUCTION BY LOW TEMPERATURES

1. *Introduction*

A plant in which inflorescences have been induced does not always go on to form an inflorescence. If such plants are grown under unsuitable conditions for inflorescence formation, they may form single-cloved bulbs or intermediate bulbs (see Sections IV.B and V.A). Low-temperature treatment induces both inflorescence and storage-leaf formation. Therefore, a plant induced for inflorescences is capable of either course of development at the terminal bud. If the induced plant forms a storage leaf in the terminal bud, it will never subsequently form either an inflorescence or a lateral bud, but rather it develops into a single-cloved bulb, with the terminal bud enclosed by the storage leaf, and finally the terminal bud will become dormant. However, when inflorescence formation occurs before storage-leaf formation, lateral buds form, and subsequently these form storage leaves. Therefore, since the percentage of inflorescences formed from induced plants varies considerably with environmental conditions after the cold treatment, these conditions can have a considerable effect on the type of bulb formed, i.e., the number of cloves; namely, whether it is a single-cloved, an intermediate, or a normal multicloved bulb. The main factors affecting the percentage of inflorescences formed after cold induction are the temperature, photoperiod, light intensity, water supply, and nitrogen nutrition encountered subsequent to cold treatment, and also the plant size and the cultivar.

2. Temperature

Temperatures after inflorescence induction strongly influence inflorescence formation. Cold-induced cloves of cv. "Yamagata" had a much lower percentage of inflorescence formation if they were grown at temperatures above about 13°C (Table 9). In 12-h photoperiods the upper limits for inflorescence formation can be taken as approximately 13°C.

TABLE 9
Effect of Temperature at which Plants Were Grown
and the Photoperiod on Inflorescence Formation by cv.
"Yamagata"[a] in Plants in which Inflorescences had
been Induced by Low Temperatures Before Planting[22]

Growing temperature (°C)	Photoperiod (h)		
	12	10	8
	Plants forming inflorescence[b] (%)		
8 weeks cold treatment[c] before planting			
13	81/81	89/89	88/88
15	44/50	61/72	94/94
17	18/24	25/31	39/44
12 weeks cold treatment[c] before planting			
13	13/13	50/50	83/83
15	6/6	12/12	56/56
17	0/0	6/6	7/7

[a] Clove weight was 5.1 to 7.1 g.
[b] Plants formed either an inflorescence or a single-cloved bulb. Percentage of complete/incomplete plus complete inflorescences. Replication 13 to 18.
[c] Cold induction at 5°C.

However, when cv. "Yamagata" was grown under 8-h photoperiods, the upper limit temperature for inflorescence formation was approximately 15°C. Thus, the upper limit temperature for inflorescence formation tends to increase as photoperiod decreases and vice versa. This upper limit temperature for inflorescence formation varies with cultivar. It is about 20°C in cv. "Hoki" under 12-h photoperiods.[38] When plants are grown at low temperatures after induction, a high percentage of plants form inflorescences even if other environmental factors are not favorable for inflorescence formation; for example, after long cold storage of cloves before planting or under long photoperiods (Table 10). The lower the growing temperature after the cold induction, the higher the percentage of plants to form an inflorescence. However, at these lower temperatures, inflorescence development progresses more slowly than at higher temperatures. For example, at 13°C, inflorescences are formed within a month of cold-treatment induction, but in plants grown at 5°C, this takes 2 months or more under 24-h photoperiods.[22]

3. Photoperiod

When plants are grown under 8-h photoperiods following cold induction, they form two more leaves before the inflorescence develops than plants grown under 12-h photoperiods.[22] Therefore, short photoperiods after induction inhibit subsequent inflorescence formation. However, because short photoperiods inhibit storage-leaf formation more strongly, they ultimately favor inflorescence formation, and they result in a higher percentage of bolting plants. Although the number of leaves formed before the inflorescence is almost equal in plants grown under photoperiods ranging from 12 to 24 h, inflorescences form earlier as the photoperiod increases.[22] The longer the photoperiod after cold induction, the more inflorescence formation is accelerated; however, storage-leaf formation is even more strongly accelerated. Therefore, storage-leaf formation tends to occur earlier than inflorescence for-

TABLE 10
Effect of the Duration of a Cold Induction of Cloves and the Subsequent Growing Temperatures and Photoperiods on Inflorescence Formation by cv. "Yamagata"[a22]

Photoperiod (h)	During cold[b] of storage (d)					
	0	30	60	90	120	150
	Percentage of plants forming inflorescences[c]					
Growth at 13°C						
16	—	0	35	0	13	0
8	50	94	100	64	19	0
Growth at 9°C						
16	—	100	100	58	0	12
8	100	100	94	85	36	24
Growth at 5°C						
16	100	100	100	100	41	11

[a]   Clove weight was 6.5 to 10 g.
[b]   Cloves were cold induced at 5°C.
[c]   Percentage of complete inflorescences. Plants formed either an inflorescence of a single-cloved bulb. Few plants formed incomplete inflorescences. Replication 12 to 19.

mation as photoperiods lengthen. This prevents inflorescence formation in some plants, and the percentage bolting decreases as the photoperiod increases (Table 11). This effect is magnified when other conditions besides photoperiod are unsuitable for inflorescence formation.

4. Light Intensity
When garlic plants are grown at very low light intensities after inflorescence induction, the percentage of plants forming inflorescences is reduced.[43] However, this is counteracted by also growing the plants at low temperatures. At 5°C, 100% of plants can flower even in darkness.

5. Water Supply and Nitrogen Nutrition
When the induced cloves are grown with low soil moisture for 2 weeks after planting, inflorescence formation is inhibited, and single-cloved bulbs result.[32] As previously mentioned, inflorescence formation does not occur during the storage of cloves. This may be because inflorescence formation needs an external water supply after induction, which is not available to stored cloves.

When small plants are starved of nitrogen for the month preceding inflorescence initiation, the precentage of single-cloved bulbs increases in most cultivars, and the percentage of those forming inflorescences decreases (Table 12). This indicates that nitrogen deficiency promotes storage-leaf formation.

6. Plant Size
When cloves are induced by low temperatures, only those above a particular size can produce flower-stalks (Table 11). The minimum size needed to form an inflorescence depends on the photoperiod during growth, the duration of vegetative growth before cold induction,

TABLE 11
Effect of the Weight of Cloves at Planting, the Stage at Which Cold Induction is given, and Photoperiod on Inflorescence Formation in cv. "Yamagata"[22]

Photoperiod (h)	Percentage of plants forming inflorescence[a]				
	Chilled before planting then grown in a greenhouse[b]				
	Weight of clove at planting (g)				
	0.5—1.1	1.5—2.1	2.5—3.1	3.5—4.1	4.5—5.1
16	0/0	0/0	0/0	71/82	92/92
12	0/0	11/11	78/78	90/90	86/86

Photoperiod (h)	Chilled after planting and grown in the field[c]		
	Weight of clove at planting (g)		
	0.5—1.0	1.1—2.0	2.1—3.0
18	5/10	67/78	100
ND[d]	44/83	90/100	100

[a] Percentage of complete/incomplete plus complete inflorescences.
[b] Cloves were stored at 5°C for 62 d, and then planted in pots in an unheated greenhouse on November 27. The nautral daylength extended was extended using incandescent lamps. Replication 10 to 20.
[c] Cloves were planted in the field on September 17, The natural day length was extended from March 1. Replication 12 to 23.
[d] ND = natural daylength.

TABLE 12
Effect of Nitrogen Starvation and Cultivar on Inflorescence Formation by Small Garlic Plants[a] After Cold Induction[44]

Nitrogen application[b]	Percent of plants forming:		
	Complete inflorescence	Incomplete inflorescence	Single-cloved bulb
	cv. "Hoki"		
+	100	0	0
−	100	0	0
	cv. "Ishuwase"		
+	28	25	47
−	17	3	80
	cv. "Yamagata"		
+	81	3	16
−	53	0	47

[a] Clove weight was 1.5 to 2.0 g. Cloves were planted in pots in an unheated greenhouse on October 28. Replication approximately 39.
[b] Low N plants were deprived of N from February 24. Inflorescences formed between the end of March and early April.

TABLE 13
**Effect of the Weight of Cloves at Planting and the Duration of
Vegetative Growth[a] Before Cold Induction[b] on Inflorescence Formation
by cv. "Yamagata" Grown at 13°C Under 12-h Photoperiods[22]**

Clove weight (g)	Duration of vegetative growth before cold induction (d)				
	0	30	60	90	120
	Percentage of plants forming inflorescences[c]				
1. 1—1.5	—	0/0[d]	12/12	18/18	19/19
2. 1—2.5	5/15	0/0	61/61	89/89	89/89
3. 1—3.5	0/19	36/43	100	—	—
4. 1—4.5	14/64	79/79	91/91	86/86	—

[a] Grown at 20 to 25°C under natural short-days.
[b] Cold induced at 5°C for 2 months under 10-h photoperiods.
[c] Plants formed either a complete inflorescence, an incomplete inflorescence, or a single-cloved bulb.
[d] Percentage of complete/complete plus incomplete inflorescences.

and the other growing conditions. The minimum size for forming an inflorescence decreases as the photoperiod shortens (Table 11) and as the period of vegetative growth before cold induction increases (Table 13).

Interestingly, the percentage of plants forming inflorescence when grown under controlled conditions was usually considerably lower than in field-grown plants (Table 11). This was probably due to differences in cold-induction treatment, light spectral quality, soil conditions, plant population density, and possibly some other environmental factors.

7. *Genotypic Differences*

The three cultivars mentioned in Table 12, usually flower 100% when grown outside in Japan. However, under conditions less favorable to inflorescence formation, they had very different percentage bolting (Table 12). Cv. "Hoki" almost always produced a higher percentage of flowering plants than the others (Tables 7 and 12). Cv. "Hoki" can form inflorescences at higher temperatures than the others after cold induction (see Section V.C.2). Cv. "Ishuwase" is the least easily induced to form an inflorescence of these three cvs., yet it develops much sturdier scapes than cv. "Yamagata". Therefore, the ease with which inflorescences can be induced to form may not always be related to the ultimate size and vigor of the flower stalk.

D. FACTORS AFFECTING THE GROWTH AND DEVELOPMENT OF INFLORESCENCES

The growth and development of the inflorescence and the lateral cloves are concurrent during normal bulb development. Accordingly, the inflorescence and cloves may compete with each other for photosynthates. If the growth of one is promoted, the other tends to be inhibited. Long photoperiods and warm temperatures promote clove growth more strongly than inflorescence growth. If long days and warm days occur immediately after the initiation of an inflorescence, it may abort and degenerate, turning into a thin membranous rudiment. If plants are grown under short photoperiods and/or at low temperatures immediately after inflorescence initiation, both clove growth and flower-stalk development are retarded. However, in this case, because clove growth is more strongly inhibited, inflorescence growth is better, and its final size is larger than at higher temperatures and longer photoperiods. After a certain stage of inflorescence development, long photoperiods and warm temperatures no longer inhibit its growth, and large well-formed inflorescences result.

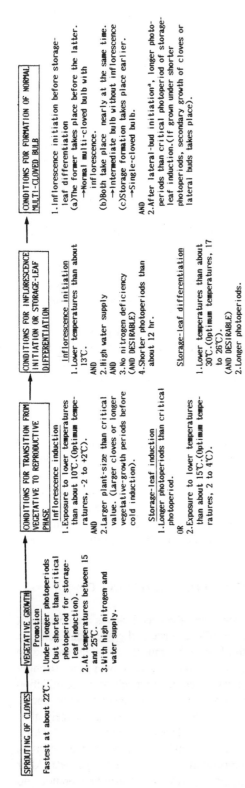

FIGURE 10. Conditions for bulbing and flowering in temperate-zone bolting garlic. The critical temperatures vary with cultivar and the other conditions. *Inflorescence initiation shortly follows lateral-bud initiation.

E. SUMMARY ON PHYSIOLOGY OF BULB AND INFLORESCENCE FORMATION

The effects of environmental conditions on bulb and inflorescence formation in garlic are rather complex. Figure 10 illustrates the effects of the external and internal factors on bulbing and flowering in temperate bolting cultivars.

VI. DORMANCY AND STORAGE OF BULBS

A. THE COURSE OF DORMANCY UNDER NATURAL CONDITIONS

Immediately after harvest, garlic cloves are deeply dormant.[45] The number of days to sprout emergence of cloves planted under optimal conditions for growth, varies with the degree of maturity of the cloves and with storage temperatures (Figure 11). Cloves lifted when the foliage is still green require more time to sprout than mature cloves lifted when the foliage has senesced. On this evidence it appears that the more immature the cloves are, the deeper their dormancy. However, if dormancy is estimated in terms of the growth rate of cloves while still attached to the mother plant, cloves from plants whose foliage has almost totally senesced can be considered the most dormant. Those from more immature plants where the foliage is still green are less dormant, since the storage leaf and sprout leaves are still growing.

In plants with near totally senescent foliage, the growth of storage leaves in the clove has almost ceased. Furthermore, the increase in number and length of the sprout leaves enclosed by the storage leaf, ceases at this time (Figures 12 to 14). The growth of sprout leaves or of the new roots does not occur for about 2 weeks, when cloves are planted under optimal conditions for growth, i.e., at 20°C and with plentiful water (Figures 13 and 14 and Table 14). In field-grown bulbs, the growth of sprout leaves within the cloves ceases when rapid senescence of the foliage occurs. This happens in late June to mid-July for cv. "Yamagata" in Japan (Figure 12). The growth inhibition of cloves is not caused by external conditions directly (e.g., rising air- and ground temperatures) but by an internal (physiological) control which slows down the growth of the cloves. At this time, conditions in Japan are ideal for vegetative growth, the mean air temperature being about 22°C and the soil being well watered. Thus, as garlic plants mature and their foliage senesces, the growth rate of the cloves declines to almost zero. When the foilage is almost totally senesced, the cloves are truly dormant.

When cloves of cv. "Yamagata" were planted in growth chambers at 20°C just after leaf senescence, they rooted and resumed growth of vegetative leaves after 2 weeks, and resumed leaf initiation 1 to 2 weeks later still (Figures 13 and 14 and Table 14). Subsequently, the growth rate of sprouts and roots accelerates, and the cloves visibly sprout 7 to 8 weeks after transplanting. Thus, cloves of cv. "Yamagata" planted under ideal conditions for growth, have a 2-week period of total inactivity as regards measurable growth. They may be considered as truly dormant during these 2 weeks.

When cloves from plants that had just senesced were transplanted to the field, growth is not resumed for about 5 weeks. Since, as we have seen, at 20°C growth could resume in only 2 weeks, we must assume that the conditions in the field were not ideal for the resumption of growth, probably because the temperatures (mean field temperatures were 26.5°C) were too high.

Cloves stored in well-ventilated sheds after total leaf senescence resumed sprout growth even later than those immediately transplanted to the field. The rooting of these stored cloves was much delayed (Table 14). When such stored cloves were planted at approximately 2-week intervals in 20°C, cloves planted immediately after lifting and those stored for 2 weeks did not differ in the number of days for sprout emergence. Cloves stored for 4 weeks sprouted considerably faster (Figure 15). This indicates that the cloves were in total true dormancy

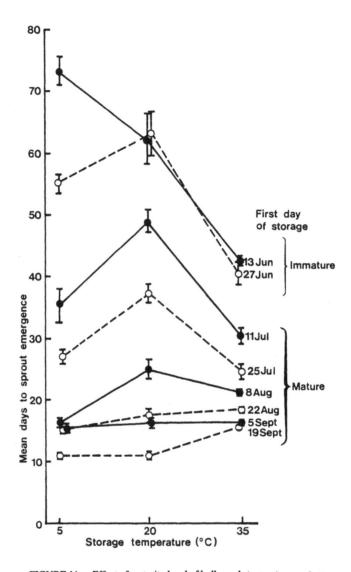

FIGURE 11.   Effect of maturity level of bulbs and storage temperature on the sprouting of cloves of cv. "Yamagata" after planting at 20°C. Immature bulbs were lifted on the first day of the 3-week controlled temperature storage period. Mature bulbs were lifted on July 11 when foliage was almost fully senesced. They were then stored at natural temperatures until given 3-week controlled temperature storage before planting at 20°C.[22] Replication approximately 20. Vertical bars represent SE.

for about 2 weeks after lifting, but that dormancy was in part imposed by the storage conditions in those stored for 4 weeks. Dormancy, as measured by time to visible sprouting, declined steadily and reached a minimum about 8 weeks after lifting (Figure 15). The rate at which dormancy declines depends strongly on storage temperature (see Section VI.C.2) and on cultivar. The dormancy of cv. "Yamagata" stored at natural temperatures disappeared by mid-September, whereas for cv. "Hoki" this took until mid-October.[46]

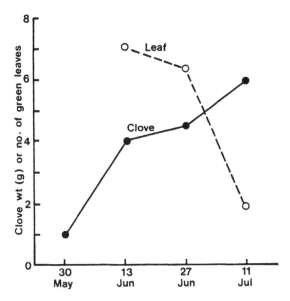

FIGURE 12. The increase in clove weight and the decrease in number of green leaves on the mother plant during bulbing when cv. "Yamagata" was grown under field conditions.[22] Replication approximately 20.

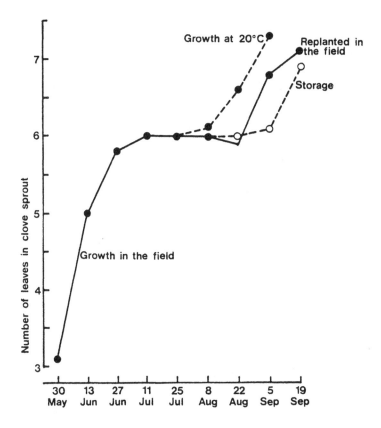

FIGURE 13. The number of leaves and leaf initials (excluding the skin and storage leaf) within a clove sprout of cv. "Yamagata" from an early stage of bulbing until 10 weeks after a July 11 harvest. After harvest, bulbs were replanted in the field, grown at 20°C, or stored at natural temperatures.[22] Replication 6.

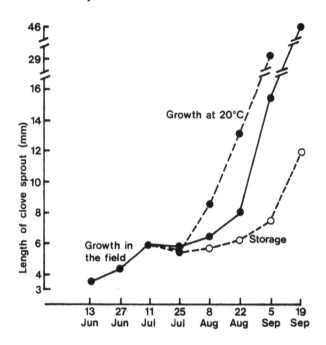

FIGURE 14. The increase in length of the clove sprout with time in the experiment detailed in Figure 13.[22]

TABLE 14
Sprouting and Root Growth of Cloves[a] of cv. "Yamagata" Which, After Harvest, Remained in the Field, was Grown at 20°C, or Stored at Ambient Temperature.[22] The Date of Near Complete Senescence of Foliage and Clove Harvest was July 11. No Sprouting Occurred in Stored Cloves Before September 19. Those at Controlled Temperature, 35% Sprouted by September 5, and in the Field, 47% Sprouted by September 19.[22]

	Root length (mm)		
Observation date	Field	Controlled temperature 20°C	Storage
July 25	<0.5	<0.5	<0.5
August 8	<1.0	12.6 ± 3.4[b]	<0.5
August 22	19.4 ± 6.6	54.2 ± 8.7	≤0.5
September 5	62.8 ± 6.3	136.8 ± 8.5	≤0.5
September 19	—	—	≤1.0

[a]   Clove weight was 5.0 to 7.0 g.
[b]   ± SE. Replication approximately 20.

B. OPTIMUM TEMPERATURES FOR SPROUTING AND ROOTING OF CLOVES

1. *Optimum Temperature for Rooting*

In cloves which are no longer truly dormant, roots emerge within a few days under a wide range of temperatures, when they are adequately supplied with water, although extremes of high or low temperatures can delay such rooting.[22] With dormant cloves, however, the time for root appearance is strongly temperature dependent. The optimum temperature for root emergence from such dormant cloves is 15°C.[47] Rooting is greatly delayed at higher temperatures than 20°C or lower temperatures than 10°C. The elongation rate of roots after

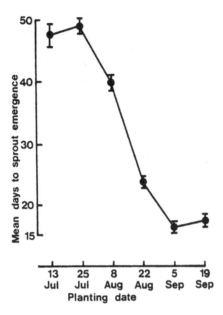

FIGURE 15. Effect of planting date on the time from planting to sprouting at 20°C in cloves of cv. "Yamagata" stored at natural temperatures after harvest on July 11.[22] Replication approximately 20. Vertical bars represent SE.

their emergence increases with temperatures, within the range of 0 to 25°C. However, the number of new roots to emerge is maximal at 15 to 20°C, and decreases at higher or lower temperatures. As a result, the development of the root system as a whole is fastest at 20°C[47].

2. Optimum Temperature for Sprouting

The optimum temperature for the sprouting of cloves that are no longer dormant is about 22°C[22]. Over the range of 15 to 30°C, differences in the rate of sprout emergence are small, but at 10°C, this rate is much lower. The relationship between temperature and sprouting time for dormant cloves is similar. However, with dormant cloves, the time to sprouting is much longer and even a temperature of 5°C about or below the optimum considerably delays sprouting.[22,47]

C. TEMPERATURE AND THE BREAKING OF DORMANCY

As discussed above, garlic cloves are in a deep state of dormancy immediately after harvest, and the dormancy of the cloves gradually diminishes during storage. The rate at which dormancy is lost depends upon storage temperature. Moreover, the effective temperatures for diminishing dormancy depends on the phase of dormancy of the cloves.

1. Effect of Temperature on Breaking Dormancy in Cloves in Different Phases of Dormancy

When cloves of cv. "Yamagata" were harvested between June 13 and 27, when they were still growing and were therefore immature (i.e., in a predormant phase), a storage temperature of 35°C was much more effective than 5 or 20°C in promoting rapid sprouting after replanting (Figure 11). When cloves from fully mature plants were harvested (i.e., truly dormant cloves), a 3-week period of storage at 35°C prior to replanting again led to the fastest sprouting. However, in this case, storage at 5°C was almost as effective (Figure 11). This shows that the low-temperature (5°C) effect on dormancy breaking increases in importance as the cloves approach the true dormant state.

During postdormancy (i.e., when dormancy is declining in depth) storage at 35°C no longer accelerates sprouting, as can be seen from cloves lifted between August 8 and 22 in Figure 11. However, storage at 5°C still caused some acceleration of sprouting (Figure 11). At the end of this postdormant phase (September 5 lifting of cloves), storage at 5, 20, or 35°C has similar effects on the sprouting rate (Figure 11). By the time innate dormancy has totally disappeared (September 19, lifting of cloves), 35°C storage actually delayed sprouting after planting (Figure 11). Mann and Lewis[45] also report that sprouting is delayed the higher the storage temperature, within a range, e.g., 15 to 30°C, in cloves which are stored for 4 months or more before planting.

Cloves which have emerged from dormancy have roots which are almost protruding, or in some cases already doing so. High-temperature storage (approximately 35°C) of such cloves often causes necrosis of such young roots. Dormant cloves are not damaged by such high temperatures. It seems that as dormancy weakens, the resistance of cloves to damage by such high temperatures diminishes, and those cloves that are on the point of rooting, are easily damaged. This may be why 35°C storage delays sprouting when applied at this stage.

2. *Effect of Temperature and Storage Duration on Dormancy*

When truly dormant cloves were stored at temperatures between 0.5 and 40°C for between 5 and 70 d, only 5 to 7 days of storage at temperatures of 35 to 40°C considerably accelerated sprouting upon replanting at 20°C (Figure 16).[48] With storage at 5 to 10°C, more than 2 weeks of storage were needed before such an effect on sprouting was noticeable (Figure 16).[48] Alternating the storage temperature between 35°C (day) and 5°C (night) was similar in its effect to a constant 35°C.[48] With storage for longer than 4 weeks, 5 to 10°C storage was more effective in promoting sprouting than 35 to 40°C (Figure 16). There may be two reasons for this. First, storage at 35 to 40°C may cause some injury, and second, 35 to 40°C storage is effective in overcoming deep dormancy, but has little effect on cloves where dormancy has partially declined (i.e., during postdormancy, see Section VI.C.1) In addition, Mann and Lewis[45] reported that cloves from bulbs stored for between 4 and 24 weeks at 5 and 10°C sprouted earlier than those stored at 15 and 20°C.

The influence of temperature on overcoming clove dormancy may be summarized as follows (Figure 17). In order to overcome deep dormancy in cloves immediately after harvest, storage at 35 to 40°C for 2 weeks or less is best, whereas with 3 or more weeks of storage, a temperature of 5 to 10°C is optimal. However, storage for 2 to 3 weeks at the higher temperature followed by a similar period at the lower temperature may be even better (see Section VI.C.3). It can be seen that a storage temperature of 5 to 10°C is suitable for promoting the sprouting of postdormant cloves irrespective of the duration of the storage period (Figure 16).

3. *Effect of Storage at Two Different Temperatures for Successive 3-Week Periods on Dormancy*

Freshly harvested, fully mature cloves were stored for 3 weeks at temperatures of 35, 20, or 5°C, and then transferred for a further 3 weeks of storage at another of these temperatures. Storage at 35°C followed by 20°C or 5°C led to the most rapid sprouting after planting at 20°C (Figure 18). Thus, high-temperature storage followed by low temperature (5°C) was more effective in promoting sprouting than any constant temperature. This is because high temperatures (35°C) are most effective for breaking true dormancy, and 5°C is most effective in promoting sprouting later (in postdormancy), (see Section VI.C.1.) Nevertheless, 35°C followed by 20°C promoted sprouting slightly better than 35°C followed by 5°C (Figure 18).

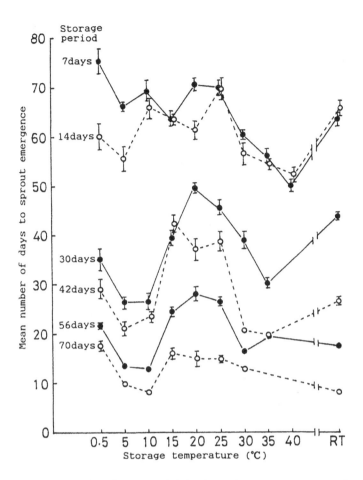

FIGURE 16. Effect of storage period duration and storage temperature on the time to sprouting after planting cloves of cv. ''Yamagata'' at 20°C. Bulbs were harvested on July 3 and stored at controlled temperatures from July 16 onwards.[48] RT represents room temperature. Replication approximately 20. Vertical bars represent SE.

4. *Effect of High Temperatures During the Last Month of Bulb Growth on Clove Dormancy*

When mother plants are grown at 22 to 30°C during the last month of bulb development, the resulting cloves sprout sooner after replanting at 20°C than those from plants grown at 11 to 20°C (Figure 18). This indicates that the high temperatues late in growth inhibit, or partly reverse, the development of dormancy in cloves.

5. *Conditions for Prolonged Storage of Bulbs*

Methods for prolonged storage of garlic bulbs have been widely studied. The main techniques for prolonged storage are low temperatures,[22,48-51] spraying with maleic hydrazide (MH) on mother plants before harvest,[52-55] gamma irradiation of harvested bulbs,[54,56,57] and controlled atmosphere storage,[58] Here, low-temperature storage only is discussed. When partially dormant cloves were stored at −2, 0.5, 5, and 20°C for 3.5, 6, or 8 months, −2°C storage was much more effective than the other regimes in preventing rooting and weight losses, and in prolonging dormancy (Table 15). Cloves stored at −2°C for 8 months showed no rooting or rotting. At the end of storage, the appearance was similar to that of fresh bulbs.

A temperature of −2°C is much more effective than 0.5°C for prolonging storage life

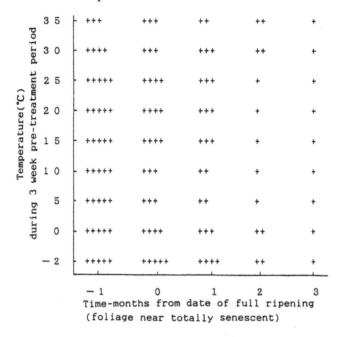

FIGURE 17.   Effect of temperatures applied for 3 weeks at various stages of bulb maturation and clove storage on subsequent rate of sprouting of cv. "Yamagata" planted at 20°C.[22] Key: time to visible sprouting of plants grown at 20°C: +, 0 to 1 week; + +, 2 to 3 weeks; + + +, 4 to 5 weeks; + + + +, 6 to 7 weeks; + + + + +, >8 weeks.

(Table 16). However, undried cloves placed immediately after lifting into −2 to −3°C may suffer freezing injury.[50] Susceptibility to such freezing injury decreases with time as bulbs are air-dried after harvest. Bulbs dried naturally in the open air in summer for 1.5 months or longer do not suffer freezing damage at −3 to −2°C. Thoroughly dried bulbs are not injured even by storage at −8°C[50]. The more the start of −2°C storage is delayed after harvest, the lower the proportion of cloves injured by freezing. If cloves have emerged completely from dormancy when they are stored at −2°C, rooting can occur. This was observed after 7 months of storage.[50] It is best to start at −2°C storage, while the cloves are still partially dormant (i.e., during postdormancy). These findings are corroborated by the results of Andorosova,[51] which show that the optimum storage temperature for garlic is −3 to −1°C.

D. CHEMICAL CONTROL OF DORMANCY

The sprouting of dormant cloves is strongly promoted by immersing them in BA solution before planting (Figures 19 and 20). However, rooting is somewhat inhibited by such a treatment. The breaking of dormancy has been shown to correlate with a rise in endogenous cytokinin activity[60,61] and in endogenous gibberellin-like activity.[62-64] There are conflicting reports on the influence of exogenous gibberellins. Gibberellin application has been reported to promote the sprouting of cloves,[65] but others found no such promotion.[66] The present author also observed no promotion of sprouting or rooting by exogenous application of GA_3 (Figure 19). Exogenous auxin has been reported to promote sprouting,[65] but again, the present author did not find such an effect (Figure 20). However, indole-butyric acid did promote rooting somewhat (Figure 20). Exogenous ethylene (ethephon) does not promote the sprouting or rooting of dormant cloves (Figure 19). Methyl disulfide, one of the constituents of the sap extracted from garlic, promotes the rooting of dormant cloves exposed

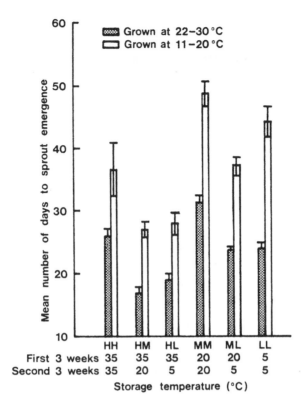

FIGURE 18. Effect of exposure of mother plants to high temperatures during the last month of bulb growth, and of storage for 3 weeks at one temperature (the upper figure) followed by 3 weeks at another temperature (the lower figure) on the time to sprouting after planting cloves of cv. "Yamagata" at 20°C. Treatments commenced 4 d after harvest.[22] Replication approximately 20. Vertical bars represent SE.

to its vapor before planting (Figure 19). This compound can also break dormancy in some flowering shrubs, bulbous plants, and other perennial plants.[67] Cytokinins, gibberellins, auxins, and growth inhibitors all may be involved in regulating the dormancy of garlic bulbs. In addition, one or more of the many sulfur compounds (see chapter on "Flavor Biochemistry") in garlic plants may also be involved in the regulation of dormancy.

TABLE 15
Effect of Temperatures During Prolonged Storage on Rooting and Weight Loss During Storage, and on Sprouting After Planting at 20°C by Cloves of cv. "Hoki" Harvested on July 18 and Placed into Controlled Temperature Storage on August 13[48]

Storage period (d)	Storage temperature (°C)				
	RT[a]	20	5	0.5	−2
Plants rooting during storage (%)					
105	90	0	100	0	0
175	100	50	—	100	0
245	100	—	—	100	0
Weight loss during storage, % of initial fresh weight					
105	7	4	6	0	1
175	13	8	—	1	1
245	23	—	—	5	3
Mean number days for sprout emergence after planting					
105	6.6 ± 0.7[b]	23.6 ± 2.0	7.4 ± 0.2	17.5 ± 1.9	31.9 ± 1.4
175	3.4 ± 0.1	9.1 ± 1.3	—	6.0 ± 0.2	54.9 ± 3.2
245	4.5 ± 0.2	—	—	4.5 ± 0.2	—

[a]   RT = room temperature.
[b]   ± SE. Replication approximately 20.

TABLE 16
Effect of Low Temperature Storage for 7 Months on Sprout and Root Growth by Cloves of c.v. "Yamagata" During Storage Between August 15 and March 19. On August 15, 6 Leaves were Counted in the Young Sprout.[48]

Storage temperature (°C)	Length of sprout (mm)	Leaves per sprout (number)	Length of root (mm)
0.5	21.6 ± 0.6[a]	6.6 ± 0.2	3.8 ± 0.4
−2	9.2 ± 1.7	6.3 ± 0.1	No roots

[a]   ± SE. The four cloves of almost the same size were examined in each storage.

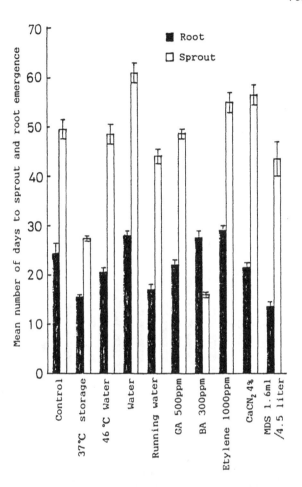

FIGURE 19. Effect of high temperatures and chemical application on sprouting and rooting after planting at 17°C, of cloves of cv. "Yamagata" harvested on July 4 and treated on July 11. Treatment durations were as follows: storage at 37°C for 2 weeks; hot water for 1 h; running water for 1 d; ethylene for 4 d at 5h/d; methyl disulfide (MDS) for 8 h; the remaining treatments with the chemicals for 5 h.[59] Replication approximately 14. Vertical bars represent SE.

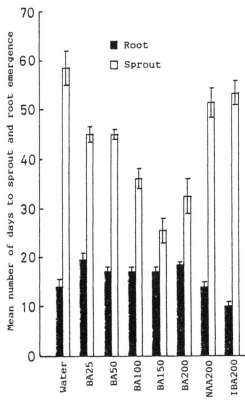

FIGURE 20.   Effect of benzyladenine, naphthaleneacetic acid, and indolebutyric acid on sprouting and
rooting after planting cloves of cv. "Hoki" at 17°C.
Cloves were harvested in mid-July and treated with the
chemicals for 5 h on August 4.[59] Replication approximately 12. Vertical bars represent SE.

REFERENCES

1. *Production Yearbook,* Vol. 39, Food and Agriculture Organization, Rome, 1985.
2. **Muller, J. J. V., Sperry, S., and Vieira, L. M. F.,** *Bibliografia Internacional de Alho* (International Bibliography of Garlic), Doc. No. 17, Empresa Catarinense de Persquisa Agropecuria S. A., Florianopolis, Brazil, 1982.
3. **Jones, H. A. and Mann, L. K.,** *Onions and Their Allies,* Leonard Hill Books, London. 1963, 37.
4. **Vvedensky, A. I.,** The genus *Allium* in USSR, (transl.) *Herbertia,* 2, 65, 1946.
5. **Etoh, T.,** Fertility of the garlic clones collected in Soviet Central Asia, *J. Jpn. Soc. Hortic. Sci.,* 55, 312, 1986.
6. **Shirahama, K.,** *A Comprehensive Survey of China Vegetable Cultivars. I. Phanerogams and Cryptogams* (Japanese), Private Press, Tokyo, 1985, 702.
7. **Mann, L. K. and Minges, P. A.,** Growth and bulbing of garlic (*Allium sativum* L.) in response to storage temperature of planting stocks, daylength, and planting date, *Hilgardia,* 27, 385, 1958.
8. **Mann, L. K.,** Anatomy of the garlic bulb and factors affecting bulb development, *Hilgardia,* 21, 195, 1952.
9. **Yamada, Y.,** Studies on the culture of (*Allium sativum* L. var. *japonicum* Kitamura), *Agric. Bull. Saga Univ.,* 17, 1, 1963.

10. **Shah, J. J. and Kothari, I. L.**, Histogenesis of garlic clove, *Phytomorphology*, 23, 162, 1973.
11. **Kothari, I. L. and Shah, J. J.**, Structure and organization of shoot apex of *Allium sativum* L., *Isr. J. Bot.*, 23, 216, 1974.
12. **Kothari, I. L. and Shah, J. J.**, Histogenesis of seed stalk and inflorescence in garlic, *Phytomorphology*, 24, 42, 1974.
13. **Ishibasi, Y., Ogawa, T., and Matubara, N.**, Ecological and morphological classification of garlic cultivars (Japanese), *Bull. Nagasaki Agric. For. Exp. Stn.*, (Sect. Agric.). 15, 95, 1987.
14. **Aoba, T.**, Studies on bulb formation of garlic plants. I. The effect of size of mother-bulb, day-length and strains on developmental process of clove and inflorescence, *J. Jpn. Soc. Hortic. Sci.*, 35, 284, 1966.
15. **Takagi, H. and Aoba, T.**, Studies on bulb formation of garlic. V. Secondary growth of garlic clove (Japanese), Abstr. Jpn. Soc. Hortic. Sci. Autumn Meet., 1972, 132.
16. **Abe, T., Yoshitake, T., and Takahashi, K.**, Studies on secondary growth of garlic plants (Japanese), *Bull. Iwate Hortic. Exp. Stn.*, 6, 21, 1985.
17. **Moon, W. and Lee, B. Y.**, Studies on factors affecting secondary growth in garlic (*Allium sativum* L.) I. Investigations on environmental factors and degree of secondary growth, *J. Korean Soc. Hortic. Sci.*, 26, 103, 1985.
18. **Takagi, H.**, Morphology of secondary growth of garlic (Japanese), *Abstr. Jpn. Soc. Hortic. Sci. Tohoku Reg.*, 1983, 41.
19. **Yamada, Y.**, On the relation between the cultivating temperature and effects of cold storage of bulb in Japanese garlic (*Allium sativum* var. *japanicum* Kitamura). I (Japanese), *Agric. Bull. Saga Univ.*, 8, 23, 1959.
20. **Yamada, Y.**, On the relation between the cultivation temperature and effects of cold storage of bulb in Japanese garlic (*Allium sativum* var. *japanicum* Kitamura). II (Japanese), *Agric. Bull. Saga Univ.*, 9, 79, 1959.
21. **Aoba, T.**, Studies on the bulb formation in garlic plants. II. On the effect of low temperature, *J. Yamagata Agric. For. Soc.*, 33, 35, 1971.
22. **Takagi, H.**, Studies on bulb formation and dormancy of garlic plants, *Bull. Yamagata Univ. Agric. Sci.*, 8, 507, 1979.
23. **Takagi, H. and Aoba, T.**, Studies on bulb formation of garlic plants. VII. Effects of temperature and daylength on formative induction, formation and growth of storage leaf, *Bull. Yamagata Univ. Agric. Sci.*, 7, 423, 1977.
24. **Takagi, H.**, unpublished data, 1978a.
25. **Aoba, T. and Takagi, H.**, Studies on the bulb formation in garlic plants. III. On the effect of cooling treatments of seed-bulbs and day-length during the growing period on bulbing. *J. Jpn. Soc. Hortic. Sci.*, 40, 240, 1971.
26. **Takagi, H.**, unpublished data, 1979a.
27. **Ogawa, T., Mori, N., and Matsubara, N.**, The studies on the ecological distribution and bulbing habit of garlic plants (Japanese), *Nagasaki Agric. For. Exp. Stn.*, 3, 3, 1975.
28. **Takagi, H.**, unpublished data, 1987.
29. **Takagi, H.**, unpublished data. 1980.
30. **Takagi, H. and Aoba, T.**, Studies on the bulb formation in garlic plants. V. Effects of kind of supplemental light on growth, and bulb and inflorescence formations, *Bull. Yamagata Univ. Agric. Sci.*, 7, 401, 1976.
31. **Terabun, M.**, Studies on the bulb formation in onion plants. I. Effects of light quality on the bulb formation and the growth, *J. Jpn. Soc. Hortic. Sci.*, 34, 196, 1965.
32. **Takagi, H.**, unpublished data, 1978b.
33. **Takagi, H.**, unpublished data, 1981.
34. **Takagi, H. and Aoba, T.**, Studies on bulb formation in garlic plants. VI. Effects of growth regulators on shoot growth and bulb formation, *J. Yamagata Agric. For. Sci.*, 33, 39, 1976.
35. **Terabun, M., Furukawa, Y., Inagaki, N., and Maekawa, S.**, Bulb formation of garlic in organ culture (Japanese), Abstr. Jpn. Soc. Hortic. Sci. Autumn Meet., 1986, 214.
36. **Takagi, H. and Aoba, T.**, Studies on bulb formation of garlic. IX. *In vitro* culture of the excised buds of cloves (Japanese), Abstr. Jpn. Soc. Hortic. Sci. Autumn meet., 1975, 154.
37. **Tewari, J. P., Awasthi, D. N., Kanaujia, J. P., and Joshi, K. R.**, Effect of growth retardants on the growth and yield of single clove garlic, *Prog. Hortic.*, 16, 199, 1984.
38. **Takagi, H.**, Varietal difference of optimum temperature for inflorescence induction and of upper limit temperature of inflorescence formation in garlic (Japanese), Abstr. Jpn. Soc. Hortic. Sci. Autumn Meet., 1980, 138.
39. **Takagi, H.**, unpublished data, 1978b.
40. **Takagi, H.**, unpublished data, 1979b.
41. **Takagi, H.**, unpublished data, 1977.
42. **Takagi, H. and Aoba, T.**, Studies of spring planting culture of garlic, *J. Jpn. Soc. Hortic. Sci.*, 51, 318, 1982.

43. **Takagi, H.**, unpublished data, 1982.

44. **Takagi, H.**, unpublished data, 1983.

45. **Mann, L. K. and Lewis, D. A.**, Rest and dormancy in garlic, *Hilgardia*, 26, 161, 1956.

46. **Takagi, H. and Aoba, T.**, Studies on bulb formation of garlic. VIII. Storage temperature and diminishing of dormancy of cloves (Japanese), Abstr. Jpn. Soc. Hortic. Sci. Spring Meet., 1975, 140.

47. **Takagi, H.**, Effects of storage and planting temperatures on rooting and root growth in garlic bulbs after planting, *J. Yamagata Agric. For Soc.*, 44, 51, 1987.

48. **Takagi, H. and Aoba, T.**, Studies on the bulb formation in garlic plants. IV. Effects of storage temperatures and humidities on dormancy, sprouting and rooting of bulbs, *J. Yamagata Agric. For. Sci.*, 32, 71, 1975.

49. **ASHRAE Guide and Data Book Comm.**, *ASHRAE Guide and Data Book Applications, 1971*. American Society Heating, Refrigerating, Air-conditioning Engineers, New York, 1971, 389.

50. **Takagi, H.**, Prevention from freezing injury in long cold storage of garlic bulbs, in *Studies from the Institute of Horticulture Kyoto University*, Vol. 9, Tomana, T. and Ashahira, T., Eds., Yokendo, Tokyo, 1979, 151.

51. **Androsova, O. G.**, Studies on storage methods for garlic, *Sb. Nauchn. Tr. Khar k. Skh. Inst.*, 291, 75, 1983; *Hortic. Abstr.*, 54, 8069, 1984.

52. **Katsumata, K.**, Cultivars and cultures of garlic in warm regions (Japanese), *Agric. Hortic.*, 41, 1628, 1966.

53. **Lee, W. S.**, On the retardation of garlic sprouting in storage by MH-30 application, *Agric. J. Kyungpook Natl. Univ.*, 1, 4, 1968.

54. **El-Oksh, I. I., Abdel-Kader, A. S., Wally, Y. A., and El-Kholly, A. F.**, Comparative effects of gamma irradiation and maleic hydrazide on storage of garlic, *J. Am. Soc. Hortic. Sci.*, 96, 637, 1971.

55. **Om, H. and Awasthi, D. N.**, Effect of maleic hydrazide sprays on the storage of garlic, *Prog. Hortic.*, 9, 63, 1978.

56. **Watanabe, T. and Tozaki, H.**, The Co-60 irradiation of garlic to prevent sprouting and its influence to alliin-lyase activities, *Food Irradiat.*, 2, 106, 1967.

57. **Park, N. P., Choi, E. H., and Kim, Y. M.**, Effect of gamma-radiation on sprout inhibition of garlic bulbs and changes in their chemical composition, *J. Korean Agric. Chem. Soc.*, 12, 83, 1969.

58. **Cessari, A. and Tonini, J.**, Entreposage de l'ail en atmosphere artificielle, Inst. Intern. du Froid/Intern. Inst. Refrig. Comm. 4-5, Bologne, Paris, 1966; *Hortic. Rev.*, 1, 375, 1979.

59. **Takagi, H.**, Breaking of summer dormancy in asatsuki and garlic bulbs (Japanese), Abstr. Jpn. Soc. Hortic. Sci. Spring Meet., 1987, 398.

60. **Solomina, V. F.**, Changes in the activity of endogenous cytokinins during the relative dormancy of garlic bulbs, *Skh. Biol.*, 11, 571, 1976; *Hortic. Abstr.*, 47, 4476, 1977.

61. **Rakhimbaev, I. R. and Solomina, V. F.**, The activity of endogenous cytokinins during garlic storage at low temperature, *Vestn. Skh. Nauki Kaz.*, 2, 46, 1980; *Hortic. Abstr.*, 50, 8925, 1980.

62. **Rakhimbaev, I. R. and Ol'shanskaya, R. V.**, Changes in endogenous gibberellins during the transition of garlic bulbs from dormancy to active growth, *Fiziol. Rast.*, 23, 76, 1976; *Hortic. Abstr.*, 46, 9253, 1976.

63. **Argüello, J. A., Bottini, R., Luna, R., de Bottini, G. A., and Racca, R. W.**, Dormancy in garlic (*Allium sativum* L.) cv. Rosado Paraguyayo. I. Levels of growth substances in "seed cloves" under storage, *Plant Cell Physiol.*, 24, 1559, 1983.

64. **Argüello, J. A., de Bottini, G. A., Luna, R., and Bottini, R.**, Dormancy in garlic (*Allium sativum* L.) cv. Rosado Paraguyayo. II. The onset of the process during plant ontogeny, *Plant Cell Physiol.*, 27, 553, 1986.

65. **Srivastava, R. P. and Adhikari, B. S.**, Influence of growth substances on the germination of onion and garlic, *Allahabad Farmer*, 42, 103, 1968; *Hortic. Abstr.*, 39, 4893, 1969.

66. **Pyo, H. K. and Lee, B. Y.**, A physiological and ecological study on post-harvest garlic, *J. Korean Soc. Hortic. Sci.*, 14, 25, 1973.

67. **Hosoki, T., Hiura, H., and Hamada, M.**, Breaking of dormancy in horticultural plants by sulfur compounds (Japanese), Abstr. Jpn. Soc. Hortic. Sci. Spring Meet., 1984, 376.

Chapter 7

GARLIC AGRONOMY

J. L. Brewster and H. D. Rabinowitch

TABLE OF CONTENTS

I. INTRODUCTION

The general culture of garlic was described by Jones and Mann.[1] Since their review there has been considerable development in garlic agronomy, e.g., weed control using both pre- and postemergence herbicides is now well developed and widely used.[2] Much of the new information is pertinent only to certain localities, but some generally applicable principles have been established or reinforced by recent studies. The methods of culture for garlic used in California, where the majority of the U.S. crop is grown, are well documented,[3] and Shinde and Sontakke[2] have also recently reviewed the culture of garlic.

II. PREPLANTING STORAGE TEMPERATURE

As is clear from the physiology reviewed in the previous chapter, the temperature and duration of bulb storage have a profound effect on the pattern of growth and bulb development after planting the cloves. The optimum storage regime for a particular cultivar depends on the climate encountered subsequent to planting out, particularly the temperature and photoperiods. Cold storage for a long period may result in early bulbing and hence low yields, if regrowth resumes in long photoperiods.[4-9] Storage for prolonged periods at low temperatures (4°C) can also cause sprouting of side shoots, secondary clove formation, and this results in undesirable, low quality, "rough" bulbs.[3] On the other hand, lack of cold storage may result in excessively prolonged leaf-blade growth or failure to bulb, if regrowth resumes in a climate with short photoperiods. Thus, in some conditions, and with some strains, storage at approximately 0°C gives good yields and maturity, whereas 18°C storage results in delayed maturity.[10]

In practice, farmers have selected cultivars that are well adapted to their region and which bulb satisfactorily after storage at the local ambient temperatures.[1] In California bulbs for planting are normally stored in unheated warehouses. In these conditions, the average optimum temperature for storage is about 8°C, and temperatures should not fluctuate outside the range 4 to 15°C.[3] In Israel, planting material with attached tops is stored through the summer, at ambient temperatures, under well-aerated shades. A standard precooling treatment at 14°C for 2 weeks results in a 2- to 4-week earlier harvest.[11]

III. PLANTING

A. PLANTING DATES

Garlic is extremely hardy and survives long periods at temperatures below zero centigrade. Consequently, in temperate regions it may be planted in autumn or in spring. Studies on successional sowing dates in temperate regions consistently show that autumn planting gives higher yields than midwinter or spring planting.[12-17] In Kalyani, India, mid-November planting gave higher yields than late December planting.[18] In Bangladesh, late October planting gave higher yields than November or mid-December planting.[19] At Pukekohe, New Zealand, although yields were highest from a late autumn planting, the occurrence of secondary bulbing from this planting was much greater than from later sowings. Consequently, the quality of bulbs from the early plantings was lower. For the best yields of marketable bulbs a planting date approximately 1 month after midwinter was optimal.[12]

In California the time of planting varies with locality. In the warm desert valleys of southern California planting is between September and November and harvesting occurs from May to June. In the cooler Salinas and Gilroy regions planting occurs from January to March for harvest between September and November.[3]

B. PLANTING METHODS

Individual cloves do not store well and so separation into individual cloves is left until just before planting. However, separation of large cloved bulbs can be done up to 1 month before planting.[20] Cloves are then sorted, and small (less than 0.7 g), infected, and damaged cloves are discarded.[21] Machinery for mechanically separating and grading cloves from the bulbs prior to planting has been developed.[22] In Israel, cloves are planted at a depth of 3 cm in September, and in late autumn a 2-cm planting depth is common.[21] Machine planting using a pneumatic planter gave more even planting depths and interplant spacing than the widely used tulip planting machines.[23] The pneumatic planter gave higher emergence percentages, yields, and quality. If cloves are damaged, as can happen with machine grading and planting, their subsequent multiplication is reduced.[24] When different clove orientations were tested, those planted upright emerged earlier than those planted sideways which were, in turn, earlier emerging than those planted upside down. The percentage emergence and bulb size were lower with upside-down planting, but differences in bulb quality were slight.[25] When cloves were oriented with their backs along the row, 82% of leaves were oriented along the row, and this was more convenient for interrow cultivation. Bulb size and multiplication were not affected by such clove orientation.[26]

C. EFFECT OF CLOVE SIZE AND TYPE OF PLANTING MATERIAL

Numerous studies show that both yield and the size of harvested bulbs increases with planted clove weight.[27-33] The weights of the cloves used in such experiments have ranged from less than 1 g[30] to 8 or 9 g.[32,33] Recommendations for optimal size depend on cultivar and the intended use of the crop. A clove weight around 4 to 5 g has been recommended for planting in Chile and in Poland.[29,33] Clove weights of 3.4 to 4 g; 1.5 to 2 g, and 1.3 g are recommended in Israel for cvs "Shani", "Gat", and "Alpha", respectively.[21] Cloves taken from larger bulbs gave better yields than those from smaller bulbs.[19,34] For cloves of equal weight, the position of the clove on the base plate of the mother bulb did not effect growth and productivity.[35] Cloves of different weight produce plants of different growth rate and maturity date, and this may cause difficulties at harvest time.[20,35] Hence, crop uniformity can be increased and culture simplified by planting size-graded cloves.[24]

Aerial bulbils can be planted as an alternative to cloves.[36] As with cloves, the smaller the bulbil the smaller the resulting bulb, and with very small bulbils only single-cloved bulbs may result.[37] Hence, acceptable yields are obtained only at much denser plant populations than normal.[20] Such bulbils can be used to raise planting material for the following year. The physiological responses of aerial bulbils to storage temperature are like those of ordinary cloves. Thus, cold storage (0 to 5°C) results in earlier bulbing and maturity and usually smaller bulbs than warm storage (18 to 20°C).[38] Single cloves raised from aerial bulbils were more winter hardy and productive than ordinary cloves of comparable weight, and make good planting material.[39]

D. PLANT POPULATION DENSITY

The yield of bulbs has usually been found to increase with increases in plant population density in the range of approximately 17 to 100 plants per m².[15,19,28,29,32,40] However, as population density increases, the mean bulb size decreases, and frequently there is a market price premium for bulbs above a certain size. Therefore, the economic optimum planting density depends on the bulb size required. Typically, planting densities of 30 to 40 plants per m² are used to give good-sized bulbs.[40]

In California, garlic is grown on raised beds (1 m wide, 15 to 20 cm high) at densities ranging from 40 to 60 plants per m². The lower densities are preferred for producing large bulbs for fresh market sale, and the higher densities are used to produce maximum yields for processing, where bulb size is less important.[3] Similarly, four to five rows on 1.4-m-wide beds at 35 to 40 plants per m² is recommended in Israel.[21]

Certain authors report maximum yields at other than the highest planting density.[41,42] In a study with plant densities ranging from 47 to 148 per m², maximum yields occurred at 89 plants per m².[42] Komissarov and Karlovich[43] reported that maximum yields occurred at 27 to 37 plants per m² with 9-g cloves, but 37 to 55 plants per m² with 3- or 5-g cloves. Thus, the planting density needed to achieve maximum yield probably depends on clove size and genotype.

Interrow spacing used in the above experiments ranged from 10 to 60 cm. The rectangularity of spacing was frequently confounded with the effect of plant population density in these studies.

IV. GROWTH IN THE FIELD

A. OPEN-FIELD CONDITIONS

After planting both the number of green leaves and plant height increase. Bulb initiation is characterized by an increase in the bulbing ratio (bulb diameter/neck diameter). At this time the number of cloves per plant increases sharply, the rate of production of green leaves slows down and finally ceases, and the plant height plateaus.[3] The fresh weight of garlic plants in trials at Davis, California reached a maximum 1 month before harvest, but dry weight increased until harvest. The dry matter content of growing plants was about 25% but increased to 30% near harvest.[3] Bulbs are harvested when leaves turn yellow, necks soften, and/or tops bend over, but the latter does not occur when a flower stalk develops.[44] At this stage the bulbing ratio is about four[3] or five.[44]

Studies in New Zealand[17] and France[45] have shown that thermal time, with a base temperature of 0°C, is appropriate for describing the field growth of garlic. In the French study, the production of each leaf required 100 degree-days, and in the New Zealand study, 131 degree-days were required per leaf under optimal levels of nitrogen. By using thermal time rather than chronological time, data from crops sown on different dates can be critically compared.[17] The patterns of growth of a crop planted in midwinter, and another planted 2 months later, were very similar when plotted against thermal time. Both crops increased their leaf-blade weight until about 2400 degree-days from emergence, after which the leaf-blade weight began to decline. For both crops, bulb dry matter declined during the first 700 to 1000 degree-days and then increased until harvest. Using this time base, the leaf-blade growth of both crops was described by a single quadratic equation relating log leaf weight to thermal time. A cubic equation described the relation between log bulb weight and time, but here the two sowing dates had significantly different relationships.[17]

Garlic can be grown on many soils, but a well-drained soil suitable for irrigated row crops is best. Heavy clay soils can be difficult to plant in winter. It may also be difficult to loosen the bulbs at harvest from such soils, and the bulbs may be misshapen.[3] In light soils, where water-holding capacity is low, fluctuations in water tension between irrigations result in irregular growth, thus causing the development of misshapen bulbs. In Israel, this is especially true with late-maturing cultivars, which bulb in the late spring or early summer months.[20]

B. POLYETHYLENE FILM MULCHING

Increases in growth and yield have been reported from Korea by mulching the crop during the winter and/or spring with polyethylene films. Mulching on February 20, followed by mulch removal on June 20, gave yields of 10.9 t/ha compared with 5.4 t/ha from unmulched control plots. However, late mulch removal (June 20 as opposed to April 20) increased the proportion of both plants with secondary bulb growth and those bolting.[46] The rapid growth made under plastic film mulches or tunnels tends to accelerate secondary growth. This causes undesirable "rough" bulbs to develop.[47]

C. NUTRITION, FERTILIZERS, AND IRRIGATION
1. Pattern of Nutrient Uptake

Successional sampling and plant analysis show that after emergence, the nitrogen concentration increases to a maximum around 4 to 5% N in whole-plant dry matter, and thereafter declines slowly throughout later growth and bulbing.[48-50] In a New Zealand study, the peak nitrogen concentration was reached 500 degree-days (base temperature 0°C) after emergence.[50] Potassium, sodium, and phosphorus concentration per unit whole-plant dry matter, all show a decline from early maximal values, particularly during bulb growth.[48,49] Potassium concentrations start to decrease sooner than phosphorus.[49] Calcium concentration in whole-plant dry matter increases throughout growth,[48,49] and magnesium concentration is fairly constant throughout growth.[48]

2. Principles of Fertilizer Application

The fertilizer requirement of a crop depends on the nutritional demand required to sustain the growth needed to attain the maximum possible yield and on the rate at which the unfertilized soil can supply nutrients. Because both these factors vary between sites and seasons, the results of straightforward fertilizer trials tend to be site and season specific.[50] As a move toward more generality, it has been suggested that fertilizer recommendations should be based on plant tissue nutrient concentrations. With garlic, however, the tissue nutrient concentrations needed to sustain maximal growth change markedly with stage of development.[50] Therefore, development stage, as well as tissue concentration, must be specified if this approach is to be used.

In a recent study in New Zealand, responses to nitrogen fertilizer were studied in relation to the growth and development of the crop. The intention was to establish some general principles upon which to base fertilizer recommendations which were of more than local applicability. The growth and development pattern of crops planted on different dates, and therefore growing in seasons of different temperature, was similar when a time scale of accumulated degree-days (thermal time, base temperature 0°C) was used. Fertilizer nitrogen increased the rates of leaf initiation and extension early in growth. The thermal time, after which bulbs initiated and leaf-blade growth plateaued, was the same for all levels of applied nitrogen. Therefore, increased early relative leaf-growth rate resulted in a higher leaf-blade weight at the start of bulbing, and this resulted in higher bulb yields.[17] The response of relative leaf-growth rate to nitrogen was greater in an early-sown crop, than a late crop. The early crop was exposed to lower temperatures during early growth than the later one. This difference in N response may have been because the mineralized nitrogen content of unfertilized soil was greater in the warmer conditions of the second sowing.

There was a linear relationship between relative leaf-growth rate and the concentration of nitrogen per unit of leaf-tissue water. The same relationship applied at all levels of fertilizer.[17] The "critical" nitrogen concentration in the dry matter (nitrogen concentration needed to achieve 90% of maximum yield) varied from 5.8% at 500 degree-days from emergence to 1.5% close to harvest. The nitrate concentration in the youngest mature leaf, 500 degree-days after emergence, was highly correlated with final yield, as was the mineral nitrogen content of the top 15 cm of soil at this time. Either of these measurements, at this stage of growth, could be a useful guide to the level of nitrogen topdressing required to achieve maximum yield. Bulb yields increased linearly with the logarithm of the fertilizer application rate. Despite the similarities in growth pattern of the crops from the two planting dates and the qualitative similarities in their responses, the actual yield responses to applied nitrogen varied in a way which was not fully explicable in terms of the measurements recorded in this study. Nevertheless, by using growth analysis in fertilizer studies, a clearer understanding of responses to nitrogen resulted than was possible from analysis of the final yields alone. Furthermore, there is the useful technique of basing recommendations for topdressing on precisely defined plant or soil diagnostic criteria at 500 degree-days after emergence.

3. Optimum Fertilizer Application Rates

In view of the site- and even season-specific nature of fertilizer responses, it is not surprising that in different regions, different rates of application for maximal yields have been reported. Optimal levels of N, P, and K reported by various workers are given in Table 1.

In California, about half the N is applied at planting, and the rest as a topdressing or in irrigation water in early spring.[3] The best yields from an application of N, P, K, and borax in Brazil were obtained by banding the fertilizer below, and five cm to the side, of the planted cloves.[61] Banding fertilizer above or very close to the cloves reduced yields, probably because of a toxic effect of the dissolved fertilizer salts when present in high concentrations. Banding phosphorus below, and to the side of, the garlic rows just before planting is also the recommended procedure in California.[3]

High levels of nitrogen fertilizer have been reported to increase the percentage of plants showing secondary growth.[47,50] Such plants give rise to "rough" surfaced bulbs, which is undesirable for quality.

4. Irrigation

Like the common onion (*Allium cepa*) and other edible alliums, garlic has a rather sparse and shallow root system, with roots limited to the top 60 cm of soil.[62] Consequently, as for these other crops, the upper layers of the soil must be kept moist to attain maximum growth rates and yields.[3] Therefore, garlic crops are normally irrigated. In California and in Israel, the best yields occur in crops that are kept at near-field capacity, using sprinklers, drip, or furrow irrigation.[3,21,44] In South Korea, yields were increased by 84% when 30 mm of water was applied at 5-day intervals.[63] Later in crop growth, a slightly drier soil was optimal than was the case earlier on. In southern Argentina, five irrigations of 20 mm or 30 mm were needed to achieve high yields and good quality bulbs.[64] In Egypt, the yields were compared of crops receiving between 50 and 150 kg/ha N, which were watered when the available soil water had declined by 20, 40, 60, 80, or 100%. The maximum yields were attained at the highest rate of nitrogen along with irrigation at the least water deficit.[65] In contrast, in a study in southern Brazil, where irrigation was applied when 0, 25, 50, 75, and 95% of the available soil water had been removed, irrigation at 75% along with a nitrogen topdressing gave the highest yields.[66] Irrigation treatments given at the 50% level or above, combined with nitrogen topdressing, accelerated sprouting of stored bulbs.[67] A similar more rapid wastage in storage of bulbs from highly irrigated crops, compared with those given less irrigation, was reported from Egypt.[68] The recommendation in Israel and in California is to cease the irrigation of crops for fresh market sale 3 weeks before harvest, when the shoots start to bend over. Late irrigation can result in rotting, discoloration of the bulb skins, and exposure of the outer cloves, thus decreasing their market value. In crops for processing, where appearance is less important, light irrigation can continue until most of the shoots have collapsed.[3] Salinity of irrigation water can reduce yields. Glasshouse studies in Argentina indicated that irrigation with solutions of electrical conductivity greater than 3.0 dS m^{-1} may significantly reduce yields.[69]

V. PESTS, DISEASES, AND WEEDS

A. PESTS

Onion thrips (*Thrips tabaci*) is a major pest which causes severe foliage damage in warm weather (see chapter on "Insect Pests"). Young garlic plants are heavily damaged by the common onion fly maggot (*Hylemia antiqua*). In Israel, this occurs mainly in the fall, but

TABLE 1
Recent Reports of N, P, or K Application Rates Needed for Maximum Yields of Garlic

Country	Fertilizer application rate needed for maximum yield (kg/ha)		
	Nitrogen(as N)	Phosphorus(as P_2O_5)	Potassium(as K_2O)
Argentina[51]	200	—	—
Chile[52]	150	—	—
Chile[53]	150	0	—
Costa Rica[54]	—	(no effect of P on yield) 300	—
India[55]	60	60	120
Israel[44]	80—120 (base) 2kg/d topdressing[a]	260—350	320—380
Italy[56]	180	—	—
Italy[57]	160	0	160
Mexico[58]	240	—	—
New Zealand[59]	200	—	—
New Zealand[50]	120	—	—
U.S. (California)[3]	84—168	84—112(up to 224 in desert areas)	—
USSR (Ukraine)[60]	60	60 (plus 40 t/ha farmyard manure)	120

[a] Applied in irrigation water or before every second irrigation.

a second-wave attack may occur in early spring. Garlic is also badly damaged by stem and bulb nematodes (*Ditylenchus dipsaci*). The pest can be eradicated from planting stock by presoaking bulbs for 2 h in a solution containing 1% formalin and 0.1% detergent and then immersing them in this solution at 49°C for 20 min[3] (see chapter on "Nematode Pests of *Allium* Species"). Minor pests, which may spread from nearby or preceding onion crops to garlic, include the Japanese onion aphid (*Micromyzus fomosanus*), the eriophyd mite (*Aceria tulipae*), the brown wheat mite (*Petrobia latens*), and certain red spider mites of the genus *Tetranychus*[3] (see chapter on "Insect Pests").

B. DISEASES

Both the soil-borne fungal diseases white-rot (*Sclerotium cepivorum*) and pink-root (*Pyronochaeta terrestris*) attack garlic[3] (see chapter on "Root Diseases"). Rust (*Puccinia porri* (syn. *P. allii*)) may seriously damage the leaves, thus causing heavy yield losses (see chapter on "Onion Leaf Diseases"). Bulb rots, caused by *Botrytis allii*, occasionally occur on garlic.[3] The fungus *Penicillium corymbiferum* may cause decay of cloves in storage or in the field, and *Sclerotium bataticola* may cause a gray discoloration of the outer bulb skins[3] (see chapter on "Storage Disease of Onions"). A disorder known as "waxy breakdown", in which the clove shrinks and becomes amber, translucent, and waxy or sticky to touch, occurs in stored garlic and occasionally in the field. It may be associated with sunscald in the field, but is usually associated with poor ventilation and lack of oxygen during storage and transport.[3,20]

C. WEEDS

Because of their growth habit, garlic plants need protection against weed competition (see chapter on "Weed Control in Onions"). Selective herbicides commonly used in onion fields are also used on garlic: DCPA at 13 kg/ha or oxadiazon at 4 kg/ha are given immediately after planting to prevent weed emergence. The nonselective herbicide paraquat (methyl viologen) is applied before crop emergence to eradicate any weeds that have already emerged. The combination of proper crop rotation (fields which in the last year grew garlic, onion, gladioli, or oats should be avoided), adequate cultivation, and knowledgeable application of herbicides should ensure a field free of weeds.

VI. HARVESTING

Garlic is harvested when the foliage collapses and starts to senesce. In garlic grown in beds the bulbs are loosened by undercutting with a knife or rotating bar.[1] In warm dry climates the bulbs may be left to dry and cure in windrows in the field for about 3 weeks, using the foliage to protect the bulbs from direct sun. Where rain or dew may occur at harvest time, the tops and roots are removed at harvest, and the bulbs are cured indoors. Potato and root harvesting machines can be modified and used for garlic.[70-72] Before lifting, tops can be removed about 13 cm above the bulb using a rotary knife.[72] Topping can also be carried out at the end of the field curing. Trimmed bulbs can be stored in 2-m-deep bins or stacks provided with forced ventilation, as well as in shallow containers or sacks. Cold air (down to 0°C) of low relative humidity (RH) is best for ventilating the stacks.[71] Garlic can store well at temperatures ranging from 0 to 30°C, but the humidity should be kept at 60% RH or less. At humidities above 70% RH, surface molds and rotting occur. At 0°C it can be difficult to maintain sufficiently low RHs. Sprouting occurs most rapidly at temperatures around 5°C. At temperatures above 20°C the rate of shrinkage increases.[3] Sprouting of garlic bulbs in the field, before harvest, has been reported as a problem in parts of Brazil. The disorder was exacerbated by excessive water and nitrogen supply, and was reduced by dense planting.[73]

REFERENCES

1. **Jones, H. A. and Mann, L. K.,** *Onions and their Allies,* Interscience, New York, 1963, chap. 18.
2. **Shinde, N. N. and Sontakke, M. B.,** Bulb crops, in *Vegetable Crops in India,* Bose, T. K. and Somm, M. G., Eds., Nayon Prokash, Calcutta, 1986.
3. **Anon.,** Growing Garlic in California, Division of Agricultural Sciences, University of California, Leaflet 2948, 1976.
4. **El Motaz, M. B., Omar, F. A., El Shiaty, M. A., Arafa, A. I. G., Gheta, M. A., Shahin, A. H., and Zein, A.,** The effect of some treatments on yield and quality of Egyptian garlic. III. Breaking rest period for early crop production, *Agric. Res. Rev.,* 49, 157, 1971.
5. **Ignat'ev. M. A.,** The effect of storage temperature on garlic growth and productivity, *Tr. Chuv. Skh. Inst.,* 9, 95, 1972.
6. **Starikova, D. A.,** The effect of spring garlic storage on its growth and development, *Sel. Semenovod. Agrotekh. Polevykh Kult. v Sib. Novosib. U.S.S.R.,* 1, 140, 1976.
7. **Ledesma, A., Reale, M. I., Racca, R., and Burba, J. L.,** Effect of low temperatures and preplanting storage time on garlic clonal type Rosado Paraguayo growth, *Phyton,* 39, 37, 1980.
8. **Racca, R., Ledesma, A., Reale, M. I., and Collino, D.,** Effect of low pre-planting storage temperatures and thermo-photoperiodic conditions on bulb formation in garlic, *cv* Rosado Paraguayo, *Phyton,* 41, 77, 1981.
9. **Da Silva, J. L. O. and Alverenga, M. A. R.,** Effects of cold treatment on some agronomic characteristics of the garlic cultivar Chonan. I. Morphological characteristics, *Pesqui. Agropecu. Bras.,* 19, 1353, 1984.
10. **Polishchuk, S. F., Gorkutsenko, A. V., and Zhila, E. D.,** Effect of temperature of garlic propagating material on plant growth and development, *Sel. Semenovod.,* 50, 66, 1982.
11. **Admati, E.,** Recommendations for Garlic Cultivation for Exports, Israeli Ministry of Agriculture, 3, August 1979.
12. **Rudge, T. G.,** Garlic: time of planting, *N. Z. Commer. Grower,* 38, 28, 1983.
13. **Ledesma, A., Racca, R. W., and Reale, M. I.,** Effect of storage conditions, and planting dates on garlic (*Allium sativum* L.) *cv.* Rosado Paraguayo growth, *Phyton,* 43, 207, 1983.
14. **Rudge, T.,** July planting of garlic desirable, *N. Z. Commer. Grower,* 39, 17, 1984.
15. **Orlowski, M. and Kolota, E.,** Effect of some agrotechnical treatments on garlic yield. II. Effect of the date and density of planting on the yield, *Biul. Warzywniczy,* 27, 165, 1984.
16. **Scheffer, J.,** Red garlic — best yields from early plantings at Pukehohe, *N. Z. Commer. Grower,* 3, 25, 1985.
17. **Buwalda, J. G.,** Nitrogen nutrition of garlic (*Allium sativum* L.) under irrigation. Crop growth and development, *Sci. Hortic.,* 29, 55, 1986.
18. **Das, A. K., Sadhu, M. K., Som, M. G., and Bose, T. K.,** Effect of time of planting on growth and yield of multiple clove garlic (*Allium sativum* L.), *Indian Agric.,* 29, 177, 1985.
19. **Rahim, M. A., Siddique, M. A., and Hossain, M. M.,** Effect of time of planting, mother bulb size and plant density on the yield of garlic, *Bangladesh J. Agric. Res.,* 9, 112, 1984.
20. **Mitchnik, Z.,** Garlic, in *Israeli Encyclopaedia of Agriculture,* Vol. 2, Halperin, H., Ed., 1972, 383.
21. **Admati, E.,** Recommendations for Garlic Cultivation, Israeli Ministry of Agriculture, 4, 1977.
22. **Bartos, J. and Holik, K.,** Mechanized preparation of garlic (*Allium sativum* L.) planting material, *Sb. UVTIZ Zahradnictvi,* 12, 131, 1985.
23. **Bartos, J. and Holik, K.,** Intensification of garlic (*Allium sativum* L.) production by precision machine-planting, *Sb. UVTIZ Zahradnictvi,* 12, 195, 1985.
24. **Jarosova, J. and Bartos, J.,** Effect of grading garlic (*Allium sativum* L.) planting material by size on cultural and economic production indices, *Sb. UVTIZ Zahradnictvi,* 11, 60, 1984.
25. **Sinclair, P.,** Planting garlic — sideways, up or down?, *Farmers Newsl. Yanco Agric. Inst. Aust.,* 158, 14, 1983.
26. **Alekseeva, M. V. and Sokolova, M. G.,** Growth and productivity of winter garlic in relation to clove orientation at planting, *Hortic. Abstr.,* 55, 1063, 1984.
27. **Salvarredi, J. A. U.,** The effect of bulb size on garlic yield, *IDIA (Inf. Invest. Agric.),* 258, 46, 1969.
28. **Minard, H. R. G.,** Effect of clove size, spacing, fertilisers, and lime on yield and nutrient content of garlic, *(Allium sativum), N. Z. J. Exp. Agric.,* 6, 139, 1978.
29. **Duimovic, M. A. and Bravo, M. A.,** Effects of the weight of cloves and spacing on the yield and quality of white garlic, *Cienc. Invest. Agrar.,* 6, 99, 1979.
30. **Aliudin,** The effect of seed size and plant spacing on garlic production, *Bull. Penelitian Hortik.,* 8, 3, 1980.
31. **Lucero, J. C., Andreoli, C., Reyzabal, M., and Larregui, V.,** Influence of clove weight and planting density on garlic *cv.* Colorado, yields and quality, *An. Edafol. Agrobiol.,* 40, 1807, 1982.
32. **Duranti, A. and Cuocolo, L.,** The effect of clove weight and distance between the rows on the yield of *Allium sativum* L. *cv.* Messidrome, *Riv. Ortoflorofruttic. Ital.,* 68, 25, 1984.

33. **Orlowski, M. and Kolota, E.,** Effect of some agrotechnical treatments on garlic yield. I. Effect of clove size on garlic yield and quality, *Biul. Warzywniczy,* 27, 147, 1984.

34. **Volosky, Y. E.,** The size of the garlic clove and the type of bulb harvested, *Agric. Tec.,* 32, 32, 1972.

35. **Ershov, I. I. and Gerasimova, L. I.,** Position of cloves in the bulb and the quality of spring garlic planting material, *Sel. Semenovod.,* 9, 37, 1984.

36. **Popova, L. D.,** Utilization of garlic aerial bulbils as planting material, *Nauchn. Tr. NII Ovoshch Kh. -va,* 3, 98, 1975.

37. **Ban, C. D., Hwang, J. M., and Choi, J. K.,** Studies on growing garlic from aerial bulbils, *Res. Rep. Off. Rural Def. Hortic. South Korea,* 24, 72, 1982.

38. **Komissarov, V. A., Karlovic, S. V., and Andreeva, A. V.,** The effect of storage temperature and date of sowing aerial bulbils of garlic on the growth, development and yield of sets, *Izv. Timiryazevsk. Skh. Akad.,* 2, 121, 1969.

39. **Komissarov, V. A. and Karlovic, S. V.,** Biological and cultural characteristics of growing garlic from cloves and aerial bulbils, *Izv. Timiryazevsk. Skh. Akad.,* 1, 148, 1971.

40. **Ferraresi, A.,** Studies on sowing density in garlic in Emilia-Romagna, *Riv. Fruttic. Ortofloric.,* 47, 67, 1985.

41. **Pal, R. K. and Phogat, K. P. S.,** Effect of different spacings on the growth and yield of garlic (*Allium sativum* L.), *Prog. Hortic.,* 16, 337, 1984.

42. **Caraballo Llosas, N. and Aguila Fernandez, M.,** Influence of planting density on growth, development and yields of garlic (*Allium sativum* L.), *Cent. Agric.,* 11, 17, 1984.

43. **Komissarov, V. A. and Karlovich, S. V.,** Soil area, growth, development and yield of garlic, *Vestn. Skh. Nauki.,* 16, 73, 1971.

44. **Admati, E.,** Recommendations for Garlic Cultivation for Exports, Israeli Ministry of Agriculture, 3, February 1979.

45. **Espagnacq, L., Morard, P. and Bertoni, G.,** Determination of the growth threshold temperature of garlic (*Allium sativum*), Rev. Hortic., 275, 29, 1987.

46. **Cho, J. T., Choung, T. W., Song, Y. J., and Youn, K. B.,** The effects of a polyethylene film mulch and the time of its removal on the growth and yield of garlic, *Res. Rep. Off. Rural Dev. Hortic. South Korea,* 24, 123, 1982.

47. **Moon, W. and Lee, B. Y.,** Studies on factors affecting secondary growth in garlic (*Allium sativum* L.). I. Investigations on environmental factors and degree of secondary growth, *J. Korean Soc. Hortic. Sci.,* 26, 103, 1985.

48. **Zink, F. W.,** Rate of growth and nutrient absorption of the garlic, *Proc. Am. Soc. Hortic. Sci.,* 83, 579, 1963.

49. **Da Silva, N. et al.,** Mineral nutrition of vegetables. XI. Uptake of nutrients by a garlic crop, *Solo,* 62, 7, 1970.

50. **Buwalda, J. G.,** Nitrogen nutrition of garlic (*Allium sativum* L.) under irrigation. Components of yield and indices of crop nitrogen status, *Sci. Hortic.,* 29, 69, 1986.

51. **Lazzari, M. A., Rosell, R. A., and Landriscini, M. R.,** Absorption of ^{15}N from fertilizers in an *Allium sativum-Setaria italica* rotation in lysimeters, *Turrialba,* 34, 163, 1984.

52. **Aljaro Uribe, A. and Escaff Gacitua, M.,** Nitrogen fertilization and garlic plant population, *Agric. Tec.,* 36, 63, 1976.

53. **Ruiz, S. R. ,** Rhythm of nitrogen and phosphorus absorption and response to NP nutrition in garlic, *Agric. Tec.,* 45, 153, 1985.

54. **Ramirez, H. V. E., Lopez, G. C. A., and Loria, M. W.,** The response of garlic to phosphorus nutrition, *Proc. Trop. Reg. Am. Soc. Hortic. Sci.,* 17, 247, 1973.

55. **Das, A. K., Som, M. G., Sadhu, M. K., and Bose, T. K.,** Response to varying level of N, P, and K, on growth and yield of multiple clove garlic (*Allium sativum* L.), *Indian Agric.,* 29, 183, 1985.

56. **Cervato, A.,** Fertilizer experiments with garlic in Pizcenza province, *Ann. Fac. Agrar. Univ. Cattol. Sacro Cuore Milan,* 10, 422, 1969.

57. **Pimpini, F.,** Studies on the mineral fertilizing of garlic, *Riv. Agron.,* 3, 182, 1970.

58. **Cardenas Valdovinos, J. M.,** Nitrogen fertilization and planting layout in garlic, *Proc. Trop. Reg. Am. Soc. Hortic. Sci.,* 23, 182, 1986.

59. **Nelson, M.,** Garlic fertilizer trial, *N. Z. Commer. Grower,* 38, 28, 1983.

60. **Koltunov, V. A.,** Effect of different fertilizer rates on garlic productivity and storability, *Visn. Silskogospod. Nauki,* 11, 52, 1984.

61. **Do Amaral, F. A. L. et al.,** Fertilizer placement in a garlic crop, *Experientiae,* 11, 209, 1971.

62. **Weaver, J. E. and Brunner, W. E.,** *Root Development of Vegetable Crops,* McGraw-Hill, New York, 1927, 56.

63. **Choi, J. K., Ban, C. D., and Kwon, Y. S.,** Effects of the amount and times of irrigation on bulbing and growth in garlic, *Res. Rep. Off. Rural Dev. Hortic. Seric. (Suwon),* 22, 20, 1980.

64. **Donnari, M. A., Rosell, R. A., and Torre, L.,** Garlic productivity. II. Real evapotranspiration and water needs, *Turrialba,* 28, 331, 1978.

65. **El-Beheidi, M. A., Kamel, N. H., and Abou El-Magd, M. M.,** Effect of water regime and nitrogen fertilizer on some mineral contents, yield and amino acid contents of garlic plant. *Ann. Agric. Sci. Moshtohor,* 19, 149, 1983.
66. **Scalopi, E. J., Klar, A. E., and Vasconcellos, E. F. C.,** Irrigation and nitrogen fertilization in garlic growing, *Solo,* 63, 63, 1971.
67. **Vasconcellos, E. F. C., Scalopi, E. J., and Klar, A. E.,** The influence of irrigation and nitrogen fertilization on precocity and premature sprouting in garlic, *Solo,* 63, 15, 1971.
68. **Higazy, M. K., Shanan, S. A., Billah, M., and El-Ramadan, H. M.,** Effect of soil moisture levels on postharvest changes in garlic, *Egypt. J. Hortic.,* 1, 13, 1974.
69. **Silenzi, J. C., Moreno, A. M., and Lucero, J. C.,** Effects of irrigation with saline water on sprouting of cloves of garlic (*Allium sativum* L.) *cv.* Colorado, *IDIA (Inf. Invest. Agric.),* 433-436, 17, 1985.
70. **Lyon, M.,** The mechanization of garlic harvesting and drying, *Pepinieristes Hortic. Maraichers,* 154, 35, 1975.
71. **Peters, P., Banholzer, G., and Eysold, H.,** Preliminary results on the medium-term storage of garlic, *Gartenbau,* 31, 263, 1984.
72. **Banholzer, G., Krumbein, G., and Maut, R. J.,** Mechanization of garlic production in the agricultural plant production collective ''Progress'' Krostitz, *Gartenbau,* 32, 266, 1985.
73. **Garcia, A.,** Sprouting before harvest in garlic, *Commun. Tec. Emp. Bras. Pesqui. Agropecu. Pelotas,* 9, 3, 1980.

Chapter 8

JAPANESE BUNCHING ONION (*Allium fistulosum* L.)

Haruhisa Inden and Tadashi Asahira

TABLE OF CONTENTS

I. EVOLUTION, HISTORY, AND ECONOMIC IMPORTANCE

Allium fistulosum L. has many common names. The name "Welsh onion" is not related to Wales, but derives from the German *welsche*, meaning foreign. This was probably applied when it was first introduced into Germany near the end of the Middle Ages.[1] Other local names include the following: in China, cong; in English-speaking countries, Japanese bunching onion, Spanish onion, two-bladed onion, spring onion, green bunching onion, scallion, green trail, and Chinese small onion; in France, ciboule, oignon de Strasburg; in Germany, Röhtenlauch, Winter zwiebel; in Japan, negi; in Korea, pa; in The Netherlands, pijplook, bieslook, Indische prei; in Spain, cebolla, ceboletta; in Taiwan, chung. Tindall's[2] reference to Kikiyu onion as a common name of *A. fistulosum* in east Africa seems to be a mistake, since there is not a kikiyu onion, and the similar name Kikuyu onion, derived from the ethic group name, is probably a local variety of *A. cepa* var. *ascalonicum* (shallot).[3,4]

A. fistulosum probably originated in northwestern China, from an unknown progenitor. The closely related wild species, *A. altaicum* Pall., is still common in Mongolia and Siberia. The written history of Japanese bunching onion dates back to the 3rd century B.C. The Chinese character for *A. fistulosum* was mentioned in a Chinese script at that time.[5] However, this character also stands for most of the cultivated *Alliaceae* including the common onion, shallot, chive, and occasionally Chinese chives. The definite characteristics and agronomic practices of *A. fistulosum* were described in a Chinese book of 100 B.C.[6] *A. fistulosum* was first mentioned in Japanese literature as early as 720 A.D.,[7] probably after its introduction as plants from China. The species was brought into western Europe during or at the end of the Middle Ages, and from there into Russia.[8,9]

A. fistulosum is widely cultivated throughout the world, ranging from Siberia to tropical Asia, including China, Taiwan, Korea, Japan, Malaysia, the Philippines, and Indonesia. There are no worldwide statistics for Japanese bunching onions. Japan, Korea, China, and Taiwan grow most of the world production. In these countries Japanese bunching onion is a very important vegetable, ranking among the top ten, and is supplied to the market all year round. The annual yields and production areas, respectively, are 553,000 t and 24,000 ha in Japan (1985), 468,000 t and 19,000 ha in the Republic of Korea (1982), and 76,000 t and 5,600 ha in Taiwan (1985).

II. CLASSIFICATION OF CULTIVARS

There are many local and commercial cultivars with distinctive morphological and ecological characteristics which are adapted to a variety of climatic conditions (see Section

IV.A.1). There are different systems of cultivar classification. Among them, the system of Kumazawa[6], which divides the Japanese bunching onion into four groups: "Kaga", "Senju", "Kujyo", and "Yagura negi" — is widely accepted in Japan. This is based on utilization and ecological characteristics, especially the degree of winter growth. The general properties of each group are as follows:

"Kaga" group — The dark-green leaf blades are thick and so are the sheaths. Very little tillering occurs. Most cultivars of this group are grown for their thick pseudostems. In the winter in Japan the plants become dormant, and their above-ground parts mostly die back. Therefore, these genotypes are most suitable for overwintering. They are cultivated mainly in the north and the cold inland regions of Japan. The leading cultivars of this group are "Kaga", "Shimonita" (Figure 1A), "Matsumoto", and "Iwatsuki". "Iwatsuki" is a peculiar cultivar, which is also grown for its green tops.

"Senju" group — The plants are large, both the blades and the sheaths are thick, but the blades are tough and not so suitable as food. Tillers are few, and very long blanched leaf sheaths (pseudostems) develop and are used. The growth is slow, but continues through the winter. In many respects, this group lies between the "Kaga" and "Kujyo" groups. They are widely distributed, but cultivated mainly in central Japan, not far from Tokyo. The leading cultivars are "Senju", "Kuronobori" (Figure 1B), and "Hakusyu".

"Kujyo" group — The blades are tender and green, and their eating quality is excellent. The plants tiller easily and are mostly grown for their green tops. However, some cultivars, e.g., "Kujyofuto" and "Koshizu", are also grown to produce blanched pseudostems in the warmer regions of Japan. Plants of this group grow well at low temperatures and remain green throughout the winter, but cold resistance is low. The main production area is southwestern Japan. The leading cultivars are "Kujyofuto", "Kujyohoso" (Figure 1C), and "Koshizu".

"Yagura negi", *A. fistulosum* var. *viviparum* **Makino** — This plant develops no flowers, but in late spring it produces one to three clusters of bulblets on the scape (Figure 2A). From its shape, it is called "Yagura negi"; Yagura and negi stand for turret and Japanese bunching onion, respectively. It produces numerous tillers in spring and summer, but growth ceases in the winter. It is propagated by division of the basal cluster and by topsets, and is cultivated mainly in northern Japan for summer harvest of its green leaves. However, there is no commercial cultivation of this crop.

Here we present a modification of Kumazawa's classification system, which is supplemented by our recent work so as to apply to the Japanese bunching onion cultivated not only in Japan, but also in China, Taiwan, and Korea.[10-17] This scheme is outlined and summarized in Table 1 (see also Figure 4 later). In this system ecological groups are classified into three groups; i.e., cold region, intermediate region, and warm region. The cold region group is divided into four subgroups; i.e., Korean cold region, Chinese cold region, inland cool region, and cool region. The warm region group is divided into two subgroups; i.e., warm region, and subtropical region. In this classification system, the cultivars of the "Senju" group belong to the ecological group from the intermediate region, those of "Kujyo" group to the warm region subgroup, but those of "Kaga" group are divided between the inland cool region and cool region subgroups. The classification criteria are mainly based on winter dormancy and growth rate under low or high temperatures, plus more than 15 other characteristics, e.g., viviparousness, bolting, tillering, and the ultrastructure of the epicuticular wax layers on leaves. From multivariate analysis using these factors, we estimate that the cultivars of the "Kaga" group were introduced to Japan through the northern route from China and those of the "Senju" and "Kujyo" groups through the southern route.[15]

Genotypes with blue-green leaves and white skins are sometimes named *A. bouddhae* O. Deb.;[18,19] however, the difference does not seem sufficient to warrant a separate species name from *A. fistulosum*.[7]

FIGURE 1.   Representative plants of various *A. fistulosum* genotype in mid-December. (A) Plants of cv. "Shimonita", one of the "Kaga" group which are normally grown for their thick blanched pseudostems; (B) plants of cv. "Kuronobori", one of the "Senju" group which are grown for long blanched pseudostems; (C) plants of cv. "Kujyohoso" of the "Kujyo" group; these plants tiller profusely and are grown mostly for their tender green leaves; (D) plants of cv. "Pei Chung" of Taiwan; this plant is closely related to the "Kujyo" group, a profusely tillering cultivar which is grown for its green leaves. Notice the early bolting.

FIGURE 2.   Representative plants propagated by bulbs. (A) *A. fistulosum* var. *viviparum* Makino showing clusters of topsets forming in the scape in late spring; (B) *A. wakegi* Araki, an interspecific hybrid between *A. fistulosum* and *A. cepa* var. *ascalonicum* (shallot). This plant forms bulbs under long photoperiods and high temperatures, unlike typical *A. fistulosum*, and tillers profusely.

The properties of other related species are as follows:

A. wakegi **Araki** — This hybrid between *A. fistulosum* and *A. cepa* var. *ascalonicum* (shallot) develops a small bulb and does not form fertile seeds (Figure 2B),[20] thus differing from typical Japanese bunching onions. In Southeast Asia, China, and Japan it is propagated by planting bulbs. *A. wakegi*, which is a perennial plant, forms a bulb and becomes dormant in summer. The morphology of this plant is similar to that of *A. fistulosum*. However, the leaves are slender and 60 to 70 cm in length. One bulb produces 20 to 30 tillers. Cold resistance is low so that this plant is distributed only in the warmer regions in Japan. *A. wakegi* Araki (2n = 16) used to be considered a variety of *A. fistulosum* and was named *A. fistulosum* var. *caespitosum* Makino. (See chapter on "Taxonomy, Evolution, and History")

A. altaicum **Pall.** — *A. altaicum* is quite similar to *A. fistulosum*, except that the former has elongated bulbs which are coated with a fibrous net-like skin and the flowers have shorter pedicels than those of *A. fistulosum*. The interspecific hybrid between these two *Allium* species has high pollen and seed fertility.[21,22]

III. PLANT MORPHOLOGY

A. fistulosum is a perennial herbaceous plant, usually grown as an annual. The species name refers to the fitulous, hollow, tube-like leaves. The morphology of *A. fistulosum* is similar to that of *A. cepa*, except for the following points. The leaves are circular in cross section, as compared with the slightly flattened onion leaf. All the leaves have blades, the

TABLE 1
Classification of Cultivars of Japanese Bunching Onion in Regard to Comparative Leaf Growth at Low and High Temperatures[10-17]

Ecological group	Winter dormancy	Growth rate under low[a] temperatures	Growth rate under high[b] temperatures	Classified leading cultivars
Cold region				
Korean cold region	+ + + + +	+	+ +	"Hanlo"*, "Yeongcheon"*
Chinese cold region	+ + + +	+	+ + + + +	"Gao Jiao"**, "Bei Jing da cong"**
Inland cool region	+ + +	+ +	+ + + +	"Nanbu", "Hida", "Matsumoto"
Cool region	+ +	+ + +	+ + +	"Sapporo", "Shimonita", "Iwatsuki", "Kaga", "Zhang Qiu"**, "Zao Zhuang"**
Intermediate region	0	+ + + +	+ + + +	**"Senju" group**[c]
Warm region				
Warm region	0	+ + + + +	+	**"Kujyo" group**[d], "Zhe Jiang"**, "Shanghai"**
Subtropical region	0	+ + + + + +	+ +	"Yun Nan"**, "Pei chung"***
Viviparous type	+ + + +	+ +	+ + +	"Yagura negi"

Note: KEY: 0 = none; + = very weak; + + + = intermediate; + + + + + = very strong. * = Korean cvs. ** = Chinese cvs. *** = Taiwanese cvs. Others = Japanese cvs.

[a] Mean air temperature is lower than 5°C.
[b] Mean air temperature is higher than 25°C.
[c] Leading cultivars are "Senju", "Ishikura", "Kincho", "Hakusyu", "Kuronobori", "Akanobori".
[d] Leading cultivars are "Kujyofuto", "Kujyohoso", "Asagikujyo", "Yakko", "Kannon", "Koshizu".

basal parts — the sheaths — form a storage structure. Usually, the lateral buds in the leaf axils elongate and develop as tillers to form a vigorous clump. This tillering characteristic is more pronounced in the cultivars grown for green leaves than in those grown for long blanched pseudostems. Leaf length varies considerably with genotype, ranging from 30 to 150 cm. Unlike the common onion, it forms only a poorly developed bulb with a diameter somewhat larger than that of the neck (pseudostem). The adventitious root system is rather shallow, as most of the roots penetrate less than 30 cm in depth. The green and fistulose scape is circular in cross section, grows to 35 to 130 cm at anthesis, and is evenly inflated throughout (Figure 3), having no localized swelling so typical of the common onion. The pale-yellow perianth segments are 6 to 8 mm long whereas those of the common onion are green-white and only 4 to 6 mm. The flowers open in a consecutive order, starting at the top and continuing to the base of the umbel. In contrast, the pattern of flower opening in the common onion is quite irregular. The Japanese bunching onion flowers earlier in the spring than the common onion.

IV. PHYSIOLOGY

A. VEGETATIVE DEVELOPMENT
1. Growth
The optimum temperature for seed germination is between 15 and 25°C.[23] Germination

FIGURE 3. *A. fistulosum* cv. "Shimonita" in bloom, showing the straight nonswollen scape typical of the species.

rate is low at 30°C and is slow at temperatures below 10°C. The optimal temperature for growth ranges from 15 to 20°C. This corresponds to the temperature of late spring and midautumn in Japan. The comparative growth rates of the various ecological groups under both high and low temperatures are summarized in Table 1 (see also Figure 4). Types adapted to regions with cold winter show deep dormancy in winter, whereas those from the warmer regions have no winter dormancy (Table 1).

High temperatures affect the ultrastructure of the epicuticular wax layers on leaves.[12,14,16,17] In summer, the plants of "Kaga" and "Senju" groups, which grow well at high temperature, form dendrite-type wax, whereas those of "Kujyo" group, which do not grow well at high temperature, form cocoon-type wax. In autumn, most of the wax formed on leaves is the dendrite type, and the difference between groups becomes unclear.

2. Blanching

The most suitable temperature for production of blanched leaf sheaths (see Section VII.B.1) by earthing up (see Section VII.A.4) is about 15°C.[24] At higher temperatures, genotypes from cold regions produce thin, poor-quality pseudostems.

3. Tillering

The following is a summary of Yakuwa's[25] work on tillering. Each tiller develops from

FIGURE 4.   The effect of low temperatures on the growth of the various ecological groups in early February at Kyoto. The plants in the background are the "Kujyo" group cvs. "Asagikujyo", "Kannon", "Yakko" (from left to right). These belong to the warm region group. Those on the left-middle are the "Senju" group cv. "Akanobori" of the intermediate region group, and those on the right-middle are cv. "Bei Jing da cong" of the Chinese cold region subgroup. Those in the foreground are the "Kaga" group. Left is cv. "Sapporo" of the cool region subgroup, and right is cv. "Matsumoto" of the inland cool region subgroup.

a lateral bud formed in the leaf axil. The leaves on the tiller are initiated and grow in a plane at right angles to the original plane of phyllotaxis of the mother plant. After the appearance of two to three leaves on the tiller, however, the plane of phyllotaxis of the mother plant makes a 90° turn to become parallel to that of the daughter shoot. This pattern of tillering also occurs for secondary and tertiary tillers, and in *A. fistulosum* var. *viviparum*, *A. wakegi*, and the common onion (*A. cepa*). The number of tillers produced depends on the genotype and environment, as mentioned above. The first tiller develops mostly at the seventh node, regardless of total number of tillers produced.

Tillering is increased when the plants are transplanted after a drying treatment (see Section VII.A.2)[26-29] and after soaking with above 10-mg/l 6-benzylamino purine (BA) solution.[29,30] On the other hand, tillering is delayed and decreased under nutrient deficit conditions, especially nitrogen.[25,29]

4. Bulb Formation in *A. wakegi*

Bulb formation in *A. wakegi* is induced by photoperiods longer than 12 h and high temperatures (20 to 25°C),[31] as in the common onion (*A. cepa*). A period of low temperature, preceding the long photoperiod and high temperature phase, accelerates bulb formation.[31] *A. wakegi* plants can be classified into two groups, namely, "Japanese" and "Southern", which differ in their low temperature requirements for bulbing.[31,32] The latter bulbs earlier and has a shorter dormancy than the "Japanese" type. When bulb formation is induced by long photoperiods, endogenous levels of abscisic acid (ABA) and auxin increase.[31]

B. PHYSIOLOGY OF FLOWERING

Flowering is induced by temperatures below 13°C, when seedlings have more than 11 to 12 leaves,[33] or are more than 5 to 7 mm in pseudostem diameter.[34] The temperature and

TABLE 2
Effect of Temperature and Photoperiod on Bolting in Japanese Bunching Onion (cv. "Kaga")[33]

Treatment		No. of days from end of treatment to
Temperature and photoperiod	Duration	beginning of bolting
20°C 16 h	240	Vegetative
	15	122.3
13°C natural daylength, 10.5 to 13.0 h.	30	67.5
	60	42.8
	15	58.9
5°C 8 h	30	49.6
	45	41.3

period needed for vernalization can vary with cultivar. Certain cultivars are easy to vernalize. For example, 5 d at 5°C or 20 d at 10°C are sufficient to vernalize most of the Taiwanese cultivar "Pei Chung" population.[35] The plants of this cultivar may bolt after they have reached 4.5 mm in pseudostem diameter at Kyoto in Japan (Figure 1D) except in mid-summer.[36] Plants of cv. "Gao Jiao" cultivated in the Bei-Jing area of China can be vernalized by a treatment of 5°C for 30 d when the pseudostem diameter is only 3 mm.[16]

Yakuwa and Okimizu[33] investigated the effects of low temperature and photoperiod on bolting. Plants grown continuously under warm conditions and long days remained vegetative for more than 240 d, whereas those exposed to low temperature and short days started to flower 40 to 120 d after the termination of the treatment. A treatment of 5°C and 8-h photoperiod was more effective for flower initiation than 13°C and 10- to 13-h photoperiod (Table 2).

Plants are not vernalized when they are grown at more than 20°C, either in long-day or short-day conditions. Plants grown under approximately 13 or 18°C, however, bolt only in the case of short-days.[33] So, not only low temperature but also short photoperiods promote floral initiation in the Japanese bunching onion.

In the field at Kyoto, bolting normally occurs from March to April. Bolting is early in Korean local genotypes, rather late in the "Kaga" group, and later still in the "Senju" group.[12,13,16,17] There are some cultivars which scarcely bolt, and some are very late bolting.

Concerning flowering in the tropics, Tindall[2] states that some cultivars flower in short daylengths, but in general these conditions encourage vegetative rather than reproductive growth.

V. CYTOGENETICS AND PLANT BREEDING

A. CYTOGENETICS

The basic chromosome number of *A. fistulosum* is eight.[37] *A. fistulosum* and *A. fistulosum* var. *viviparum* Makino are known only as diploids (2n = 16).[38,39] Their karyotype formula is the same as $K_{(2n)} = 14V + 2J^T$.[20,40,41] The plants are self-compatible, but there is a high degree of cross-pollination. However, a certain percentage of self-pollination occurs. Consequently, some plants show inbreeding depression and grow poorly.

Male sterility was found by Nishimura and Shibano in 1970.[42] This is controlled by an interaction between a cytoplasmic factor S and two nuclear genes, Ms_1 and Ms_2. The genotype of the male sterile plants and their maintainer are $Sms_1ms_1ms_2ms_2$ and $Nms_1ms_1ms_2ms_2$, respectively.[43] This has been utilized to produce F_1 hybrid cultivars (see Section V.B). Experiments with such hybrids showed that a significant heterosis occurs for early growth, but this is less apparent at later stages of development.[44]

Some hybrids between *A. cepa* and *A. fistulosum* exist (see Section II and chapter on "Taxonomy, Evolution, and History"). One of them, *A. proliferum* (2n = 16), syn. *A.*

Aobanum Araki,[45] or *A. cepa* var. *viviparum* (Metzger) Alefeld, is commonly named the "top onion", "tree onion", or "Egyptian onion". The plant is viviparous and propagated by topsets (bulbils). This was previously considered by some authorities to be *A. cepa*[9] and by others to be *A. fistulosum*.[46] Now it is accepted that *A. proliferum* is a natural hybrid between these two species,[47,48] and its karyotype formula is $K_{(2n)} = 14V + J_1 + J_2^T$.[20,41,49] The tetraploid interspecific hybrid can be fertile, as demonstrated by the following examples. Jones (1950) in California, crossed *A. fistulosum* with *A. cepa* to form the amphidiploid bunching onion cv. "Beltsville Bunching".[50] This plant is fertile and propagated by seed. Another interspecific cross, which is commercially utilized, is the shallot cv. "Delta Giant", a backcross between shallot (*A. cepa* var. *ascalonicum*) with an amphidiploid of shallot ×　*A. fistulosum*.[50,51]

Tashiro[20] studied the cytogenetics of *A. wakegi* and concluded that the karyotype is $K_{(2n)} = 14V + J_1^t + J_2^T$. The basic karyotype of *A. cepa* var. *ascalonicum*(shallot) is $7V + J^t$, and that of *A. fistulosum* is $7V + J^T$. Additionally, he found that the genome of *A. wakegi* is composed of two components, one homologous with shallot and one with *A. fistulosum*. The cytoplasm of *A. wakegi* has derived from *A. fistulosum*, since hypoplasia of the anthers occurs only when *A. fistulosum* is the female parent. It was concluded that *A. wakegi* originated as an interspecific cross between *A. fistulosum* and *A. cepa* var. *ascalonicum*, the former being the female parent.

B. PLANT BREEDING

In Japan most farmers used to raise their own seed of Japanese bunching onion. Therefore, there was considerable genetic variation within established populations. Recently, breeding work has been carried out to improve cultivar performance and homogeneity. Concomitantly, the proportion of seeds purchased from seed companies has increased considerably. Following an intensive improvement program in Japan, about ten newly bred cultivars are registered annually. At present, there are over 120 registered Japanese bunching onion cultivars, including some F_1 hybrids developed utilizing male sterility. These hybrids have improved quality, heat tolerance, and resistance to bolting.

C. GENETIC RESOURCES

The Japanese bunching onion is a potential source for improvement of the common onion (*A. cepa*), especially for resistance to a number of diseases, adaptability to a wide range of climatic conditions, winter hardiness, and early flowering.[52,53] However, the sterility of the interspecific F_1 hybrid between these two *Allium* species has made progress very slow.

Collections of *A. fistulosum* exist in the Netherlands, Japan, U.K., U.S., East Germany, West Germany, and the U.S.S.R. (See chapter on "Conservation of Genetic Resources".)[52] The importance of *A. fistulosum* as a genetic resource, the identification of the important characteristics and its collection, was recognized by the International Board for Plant Genetic Resources, which ranked it second in importance in the genus *Allium*.

D. TISSUE AND MERISTEM CULTURE

Quite an effort has been invested in developing efficient techniques for maintaining and propagating male sterile lines of *A. fistulosum*. Micropropagation of plant material using callus culture is not suitable, because mutations and polyploidization are quite frequent. Fujieda and Fujita[54] using shoot meristem culture developed the following method. Small disks of stem base containing the shoot tips plus a few leaf primordia are explanted and cultured at 20°C under continuous fluorescent light at an intensity of 3 klx. The culture medium is based on the Murashige and Skoog (MS) and contains NAA at 0.5 mg/l and 1 to 2 mg/l kinetin. Regeneration of the adventitious buds is evident within 1 month of culture. These are subdivided and transplanted into a hormone-free medium. After 25d, the plantlets are large enough for transfer and hardening. Thus, hundreds of young plants can be obtained from a single plant wihtin less than 3 months.

The possible use in breeding programs of somaclonal variation induced by callus culture has been discussed.[51,55,56]

VI. PLANT PROTECTION

There are few major pest and disease problems with *A. fistulosum*. However, chemical protection of the commercial crop is still needed for high-quality crops.

A. INSECTS

Onion thrips (*Thrips tabaci* Lind.) is one of the most important pests. The cutworm (*Agrotis fucosa* Butler) feeding on the soft tissues of young seedlings may cause serious damage in nursery beds and direct-seeded fields. The mite (*Rhizoqlyphus echinopus* Fumouze et Robin) can damage the plants when cropping is repeated year after year in the same field. The stone leaf miner fly (*Dizygomyza cepae* Hering) and the stone leek miner (*Acrolepia alliella* Semenov et Kuznetsov) attack leaf blades.

B. FUNGAL DISEASES

Rust (*Puccinia allii* [de Candolle] Rudolphi) causes serious damage to the leaves. The yellowish-orange uredial lesions formed on the leaves and the chlorotic leaf spots markedly reduce market value. The outbreak of the disease varies from year to year, depending on climatic conditions. The rather low temperatures and heavy rains of both autumn and spring tend to promote rust occurrence during these seasons. In addition, deficiency of nutrients, especially that of nitrogen, stimulates the development of this disease (see chapter on "Leaf Diseases").

Purple blotch, or alternaria leaf spot (*Alternaria porri* [Ellis] Ciferri), which shows characteristic concentric spots on the leaves, causes heavy losses under moist conditions. Downy mildew (*Peronospora destructor* [Berkeley] Caspary) has not caused serious damage in recent years. Phytophthora blight (*Phytophthora nicotianae* var. *parasitica* [Dastur] Water-house), leaf spot (*Pleospora herbarum* [Persoon] Rabenhorst), black spotted leaf blight (*Septoria alliacea* Cooke), siroiro-eki-byo (*Phytophthora porri* Foister), botrytis leaf spot (*Botrytis squamosa* Walker, *Botrytis cinerea* Persoon) and fusarium wilt (*Fusarium oxysporum*) are other fungal diseases which may affect the crop (see chapter on "Leaf Diseases"). White rot (*Sclerotium cepivorum* Berkeley) may damage the plants in the fields under continuous cropping and causes a serious problem because of the long persistence of this pathogen in the soil (see chapter on "Root Diseases").

C. VIRUS DISEASES

Onion yellow dwarf virus (OYDV) is the most important virus. More than 50 aphid species are potential vectors of this virus.[57] The important carriers are the green peach aphid (*Myzus persicae* Sulzer) and the corn leaf aphid (*Rhopaloiphum maidis* Fitch). OYDV causes mosaic-type symptoms, including chlorotic mottle, chlorotic streaking, stunting, and distorted flattening of leaves. Other common symptoms are yellowing-type symptoms, including yellowing and stunting of the whole plant, an increase in tillering, and thinner leaves. The "Kujyo" group tends to be rather resistant to this virus, compared with the other groups. (See chapter on "Virus Diseases".)

Other viruses pathogenic to Japanese bunching onion are cycas necrotic stunt virus (CNSV) and tomato spotted wilt virus (TSWV).

D. WEED COMPETITION AND HERBICIDES

Weeds cause losses of both yield and quality in Japanese bunching onion fields. Much effort for weeding by hand is required during the summer, since herbicides are rarely used. The control of weeds by the permitted herbicides, trifluralin at 3 l/ha or bensulide + prometryn at 8 l/ha may be incomplete. Pendimethalin at 5 l/ha gives good weed control without damage to the crop.[58] However, this herbicide is not yet registered for Japanese bunching onion.

E. AIR POLLUTION

Physiological disorders caused by air pollution have a detrimental effect on leaf quality. Leaf-tip necrosis is caused by such air pollutants as sulfur dioxide, nitrogen monoxide, nitrogen dioxide, particulate matter, and some other oxidants. The various oxidants have the greatest effect on leaf-tip necrosis.[59] Cultivars differ in their tolerance to air pollutants. Cv. "Hakusyu" of the "Senju" group is more sensitive than others.[59]

VII. AGRONOMIC PRACTICES AND CULTURAL SYSTEMS

A. AGRONOMIC PRACTICES

Japanese bunching onion is cultivated mainly in open fields. Since the most suitable conditions for growth prevail in the autumn and spring (temperatures in the range of 10 to 25°C), these are the common sowing seasons in Japan. The agrotechniques applied by the Japanese farmers depend on the desired end product (see Section VII.B). This may be blanched pseudostems, green tops, or young shoots. Farmers select cultivars which are best adapted to produce the desired product, and best adapted to the regional environment.

1. Raising Seedlings

Seeds are sown in early spring for winter and spring crops, and in the autumn for next summer to winter harvests. In the former case, sowing too early may cause damage by drought and virus diseases in summer. In the latter case, sowing too early results in a high percentage of bolting, whereas sowing too late results in much winter damage to the young seedlings. Two to 4 kg of seeds per hectare are used in nurseries and 8 to 16 kg/ha for direct sowing in the field. The weight and numbers of seeds per 1 l are about 0.4 kg and 170,000, respectively. The seeds are broadcast or drilled in nursery beds, in shallow trenches, or in rows.

2. Transplanting

Transplants are placed in furrows, and roots and bases are then lightly covered by soil (Figure 5). The depth of furrow is about 15 cm for pseudostem production and about 5 cm for green-top production. The distance between rows and the within the row spacing are 55 to 85 and 3 to 15 cm, respectively, depending on the tillering tendency of the cultivar used. In summer, some cultivars of the "Kujyo" group are transplanted immediately after a drying treatment of 1 to 2 weeks. This increases the tillering and resistance to heat and drought, and results in high production.[26,28,29]

3. Soils and Fertilizers

Soils must be well aerated, and excessive soil moisture should be avoided. Well-drained loams or sandy loams containing high organic matter are optimal, especially for the production of blanched sheaths. A soil pH between 5.7 and 7.4[60] and soil rich in phosphate are suitable. Japanese bunching onion prefers ammonium nitrogen to nitrate nitrogen.

Two hundred to 300 kg N, 100 to 200 kg P, and 150 to 200 kg K per hectare are commonly used. This fertilizer is split over three to four application times.

4. Earthing Up

Long, blanched pseudostems are produced by mounding soil around the lower leaf bases

FIGURE 5.   The transplanting of Japanese bunching onions. Seedlings are placed in furrows ready to be covered with soil at their base.

to a height of more than 30 cm. This is done gradually in three or so stages, the first time 50 d after transplanting, the last time 20 d before harvesting for summer harvest and 40 d for winter harvest. Otherwise growth is retarded due to poor aeration.

5.   Harvesting
At harvest, plants are lifted by hand, trimmed, and packed into bundles which may then be boxed (Figures 6 and 7). Harvesting is very laborious, especially in blanched pseudostem production (see Section VII.B.1), so mechanized lifters have been developed and are being used. It takes 200 to 1000 h/ha for digging and lifting mechanically depending on the power, whereas it takes 3000 to 5000 h by hand.[61] Trimmers which peel the outer leaves using high-pressure water or air are also used.

6.   Seed Production
A suitable seed production area must have dry weather between blooming and seed ripening. Seeds from selected plants are sown from mid-March to late May. The plants of weakly tillering genotypes are transplanted in late July to August at a distance of 7 to 8 cm within the row. Those of strongly tillering cultivars are planted 20 to 25 cm apart in September. The distance between rows is 70 cm. For the seed production of F_1 hybrids the planting ratio of female plants to male plants is 3 to 1.[62] The total amount of fertilizer applied is 200 kg N, 260 kg P, and 200 kg K per hectare. Half of the fertilizer is applied before planting, and the rest is applied in two doses, one in November (early stage of flower initiation) and the other in February to March (stage of floral development).[63]

Seeds are harvested the next spring. About 40 d after full bloom, the umbels are cut with about 30 cm of scape attached, and the seeds are allowed to ripen and dry for about 20 d. The harvesting techniques are similar to those used for the common onion. Seed yields in Japan are 500 to 800 kg/ha. In 1985, the seed production area and seed yields of Japanese bunching onion in Japan were 108 ha and 68 t, respectively, as compared with 164 ha and 79 t for the common onion.

FIGURE 6.   Harvested bundles of Japanese bunching onion of the long green leaf type, or "Kujyo" group. The foreground three are cv. "Kujyofuto", the white sheaths of which have been produced by slight earthing up. Those in the background are cv. "Kujyohoso".

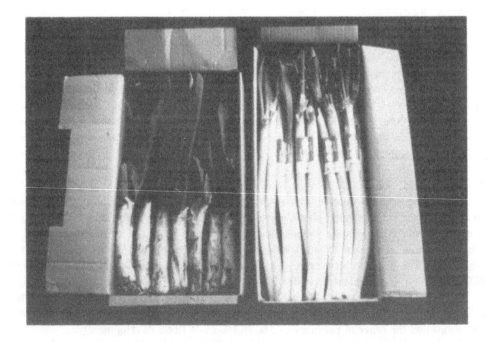

FIGURE 7.   Harvested bundles of white-sheaths type. Thick or long blanched sheaths are produced by progressive earthing up. Those on left are cv. "Shimonita" of the "Kaga" group. Their sheaths are not so long, but thick, very sweet, and delicious. Those on right are cv. "Hakusyu" of the "Senju" group. The green tops of both cultivars are tough and are discarded.

FIGURE 8. The production of blanched pseudostems using cv. "Akanobori" of the "Senju" group, at Saitama prefecture near Tokyo, in February. Soil is progressively mounded around the pseudostems as the plants grow. Note that the middle ridge is already harvested.

B. CULTURAL SYSTEMS
1. Blanched Pseudostem Production
The plants may be sown in the spring or early autumn. Spring-sown plants are transplanted in the summer and are harvested in the winter. Autumn-sown plants are transplanted in the next spring to early summer and are harvested in the summer to winter. The transplants are earthed up progressively as they grow. This type of production occurs in regions with deep cultivated soil and in cooler regions which are best blanching. (See Section IV.A.2 and Figure 8.) Cultivars of the "Kaga" and "Senju" groups are mostly used for this market, and mainly the latter group since these produce very long (more than 35 cm), nice-looking sheaths which retain their qualities even after long shipment. In warmer regions of Japan, some cultivars of the "Kujyo" group are cultivated by this system, and the green tops as well as the blanched pseudostem are harvested. In Taiwan, a few Japanese cultivars of the "Senju" group are used for short term crops. Thses are sown early in the autumn and harvested about 120 d later.

2. Green Tops (Leaf Blade) Production
Green tops are harvested all year-round. The growing period varies from 1 month to 1 year, depending on the size of plant to be harvested (between 40 and 90 cm), season, price movements of the market, the utilization of greenhouse, cultivar, and region. Since freshness is important, this crop is grown near centers of population. In recent years, precooling at 0°C and air transportation have been used to bring a fresh, high-quality product from remote regions. Cultivation occurs in southwestern Japan and in the area around Osaka. The cultivars of the "Kujyo" group are mostly used because of the high quality of their tender juicy leaves and their tendency for fast tillering (Figure 9). Cv. "Iwatsuki", which belongs to the "Kaga" group, is also used for green tops in the central area of Japan.

FIGURE 9.   Plants of cv. ''Kujyohoso'' of the ''Kujyo'' group being grown for green tops, at Kyoto in June.

3. Production Using Vegetative Stock

Nonbolting or bolting-resistant cultivars are grown for harvest in late spring, when other cultivars bolt. In the year prior to harvest, divided stocks (vegetative shoots) are temporarily planted in late May and then redivided and replanted in September. *A. fistulosum* var. *viviparum* is also grown using this schedule.

4. Production of Seedlings

There is a market for young seedlings with 15- to 30-cm-long leaves which are harvested 1 to 2 months after sowing and sold in bunches. The crop is grown all year-round, chiefly near main population centers. The main cultivars used for this crop belong to the ''Kujyo'' group or cv. ''Iwatsuki'' (''Kaga'' group). Hydroponic cultural systems are sometimes used for seedling and for green tops production (Figure 10). In addition, there is a specialized market from restaurants for 5- to 7-cm-long seedlings for use in soups and garnishing.

5. Perennial Production

In the tropics, perennial cultivation is quite common. Tops are cut, but plant bases are left for tillering and further growth, and are repeatedly harvested in this way.

VIII. COMPOSITION, NUTRITIONAL, AND THERAPEUTIC PROPERTIES

A. FLAVOR

The typical odor of the Japanese bunching onion derives mainly from volatile allyl sulfides (see chapter on ''Flavor Biochemistry''), but it is not very strong. Though the odor of this plant is somewhat different from other alliaceous plants, it is related to that of the common onion because of a similarity of chemical properties.[7] The leaf blades contain high levels of vitamins A, B_2, and C as compared with other leafy vegetables, and they are low in calories.[64] Details of its nutritional values are given in Table 3. Like garlic and other

FIGURE 10.   The hydroponic production of young *A. fistulosum* seedlings intended for sale when about 30 cm long.

Alliaceae, it contains alliin, a precursor of allicin which plays an important role in the intake of thiamin (vitamin B_1) and also has a strong antimicrobial effect. In addition, it is effective in removing the smell from meat and fish. Hence, Japanese bunching onion is favored by the Japanese as an invigorating and healthy food which is readily available all year-round.

B. CARBOHYDRATE BIOCHEMISTRY

Mizuno and Kinpyo[65] found the following free sugars and storage carbohydrates in Japanese bunching onion: glucose, fructose, sucrose, maltose, and fructose-oligosaccharides (see chapter on "Carbohydrate Biochemistry"). Leaf blades are rich in rhamnose, galactose, glucose, arabinose, and xylose. These are usually found as components of glycosides. Additionally, galacturonic acid and mannose were also detected in ethanol and water-extractable polysaccharides. The eating qualities of *A. fistulosum* improve under low temperatures because both sugar and protein content increase. The leaves contain also small quantities of alpha-cellulose and lignin (this depends on the age of the tissue), but starch is not found. Mucilages increase in bolting plants. These are well-hydrated gels of cellulose, hemicellulose, protopectin, and water-soluble pectin.[66]

C. PROCESSING

Japanese bunching onion has not been used in a processed form until quite recently, when a dehydration industry started. The product is mainly used as an additive to preprocessed food such as instant noodles.

D. THERAPEUTIC PROPERTIES

The therapeutic properties of this crop are detailed in some ancient Chinese medical books, i.e., it improves eyesight and the functioning of the internal organs, enhances metabolism, and prolongs life. In addition, it is effective in aiding digestion, perspiration, recovery from common colds, headaches, wounds, and festering sores.

TABLE 3
Chemical Composition and Nutritional Properties of Japanese Bunching Onion per 100 g of Edible Fresh Weight[64]

	Plant Tissue	
	Pseudostem	Green tops
Macronutrients (g)		
Proteins	1.1	1.7
Fat	0.1	0.2
Digestible carbohydrates	6.7	5.4
Vitamins		
A (IU)	85	480
B_1 (μg)	40	60
B_2 (μg)	60	100
Niacin (μg)	300	400
C (mg)	14	33
Minerals (mg)		
Ca	47	80
Fe	0.6	1.0
Na	1.0	1.0
K	180	200
P	20	38
Others		
Energy (kcal)	27	25
Water (g)	91.6	92.0
Waste (%)	5	5

REFERENCES

1. **Purseglove, J. W.,** *Tropical Crops: Monocotyledons,* Longman, London, 1972, 52.
2. **Tindall, H. D.,** *Vegetables in the Tropics,* Macmillan, London, 1983, 23.
3. **Kahangi, E. M.,** personal communication, 1987.
4. **Yazawa, S.,** Vegetable consumption and production by farmers, in *Agriculture and Soils in Kenya,* Hirose, S., Ed., Nihon University, Tokyo, 1987, 40.
5. **Kitamura, S.,** The origin of cultivated plants in China (in Japanese), *Touhougakuhou,* 19, 76, 1950.
6. **Kumazawa, S.,** *Vegetable Crops (in Japanese),* Yokendo, Tokyo, 1956, 325.
7. **Aoba, T.,** *Vegetables in Japan — Fruit vegetables and Allium* (in Japanese), Yasakashobo, Tokyo, 1982, chap. 2.
8. **Helm, J.,** Die zu Würz- und Speisezwecken Kultivierten Arten der Gattung *Allium* L., *Kulturpflanze,* 4, 130, 1956.
9. **Jones, H. A. and Mann, L. K.,** *Onions and their allies,* Leonard Hill, London, 1963, chap. 19.
10. **Inden, H., Yamasaki, A., and Asahira, T,** Comparative leaf growth at high temperatures in Japanese bunching onion (in Japanese), Abstr. Jpn. Soc. Hortic. Sci. Spring Meet., 1985, 180.
11. **Inden, H., Yamasaki, A., and Asahira, T.,** Comparative leaf growth at low temperatures in Japanese bunching onion (in Japanese), Abstr. Jpn. Soc. Hortic. Sci. Autumn Meet. 1985, 152.
12. **Yamasaki, A.,** Comparative ecological studies on cultivars of Japanese bunching onion (in Japanese), Master's thesis, Kyoto University, 1986.

13. **Inden, H., Yamasaki, A., and Asahira, T.,** Bolting characteristics of cultivars of Japanese bunching onion (in Japanese), Abstr. Jpn. Soc. Hortic. Sci. Spring Meet., 1986, 194.
14. **Inden, H., Yamasaki, A., and Asahira, T.,** Seasonal changes of the ultrastructure of the epicuticular wax layers on leaves of some Japanese bunching onion cultivars (in Japanese), Abstr. Jpn. Soc. Hortic. Sci. Autumn Meet., 1986, 184.
15. **Inden, H., Yamasaki, A., and Asahira, T.,** Multivariate analysis on cultivars of Japanese bunching onion (in Japanese), Abstr. Jpn. Soc. Hortic. Sci. Spring Meet., 1987, 274.
16. **Inden, H., Goto, T., and Asahira, T.,** unpublished data, 1987.
17. **Inden, H., Nishimura, K., and Asahira, T.,** unpublished data, 1987.
18. **Araki, E.,** Specilegia florae Nipponiae, *J. Jpn. Bot.,* 25, 205, 1950.
19. **Aoba, T.,** Introductory route of Japanese vegetables (in Japanese), *Agric. Hortic.,* 45, 320, 1970.
20. **Tashiro, Y.,** Cytogenetic studies on the origin of *Allium wakegi* Araki (in Japanese with English summary), *Bull. Fac. Agric. Saga Univ.,* 56, 1, 1984.
21. **Inada, I. and Iwasa, S.,** Interspecific hybrids between *A. fistulosum, A. schoenoprasum,* and *A. altaicum* (in Japanese), Abstr. Jpn. Soc. Hortic. Sci. Spring Meet., 1983, 168.
22. **Nishitani, S.,** Cytogenetical studies on the cultivated species and its wild relatives in the genus *Allium* (in Japanese), *Jpn. J. Breed,* 34, app.2, 274, 1984.
23. **Aoba, T.,** Effect of temperature on germination of *Allium* species (in Japanese), *Agric. Hortic.,* 41, 791, 1966.
24. **Ishiguro, K.,** Studies on stabilizing and labour saving procedure of raising Japanese bunching onions, *Allium fistulosum* L., var. Koshizu (in Japanese with English summary), *Bull Aichi Hortic. Exp. Stn.,* 6, 1, 1967.
25. **Yakuwa, T.,** Studies on tillering and bulb division in the genus *Allium* (in Japanese with English summary), *Mem. Fac. Agric. Hokkaido Univ.,* 4, 130, 1963.
26. **Inden, T.,** Studies on Japanese bunching onion cv. Kujyo (in Japanese), Rep. Edu. Pro. Assoc., Tokyo, 1942.
27. **Takashima, S.,** Study on growth of Welsh onion seedlings after drying treatment (in Japanese), *Breed. Agric.,* 4, 384, 1949.
28. **Tachibana, Y.,** Meaning of drying treatment and spring sowing in 'Kujyo' cultivars of Japanese bunching onion (in Japanese), *Breed. Agric.,* 4, 257, 1949.
29. **Inden, H., Itsuji, M., and Asahira, T.,** unpublished data, 1985.
30. **Murai, M., Yoshino, A., Jitsukawa, S., and Uchida, T.,** Tillering factors of blanched pseudostem type Japanese bunching onion (in Japanese), Abstr. Jpn. Soc. Hortic. Sci. Autumn Meet., 1979, 186.
31. **Ohkubo, H., Adaniya, S., Takahashi, K., and Fujieda, K.,** Studies on the bulb formation of *Allium wakegi* Araki (in Japanese with English summary), *J. Jpn. Soc. Hortic. Sci.,* 50, 37, 1981.
32. **Fujieda, K., Adaniya, S., Ohkubo, H., Takahashi, K., and Matsuo, E.,** Studies on the intraspecific differentiation of *Allium wakegi* Araki, (in Japanese with English summary), *J. Jpn. Soc. Hortic. Sci.,* 49, 180, 1980.
33. **Yakuwa, T. and Okimizu, S.,** Studies on the flowering of Welsh onion (in Japanese), *Agric. Hortic.,* 44, 1131, 1969.
34. **Watanabe, H.,** Studies on the flower bud differentiation and bolting of Welsh onion varieties, (in Japanese with English summary). *Stud. Inst. Hortic., Kyoto Univ.,* 7, 101, 1955.
35. **Lin, M. W. and Chang, W. N.,** Interspecific hybridization in genus *Allium.* I. Effect of different temperatures on bolting of Japanese bunching onion (*Allium fistulosum* L.) (in Chinese with English summary), *Chinese Hortic.,* 26, 173, 1980.
36. **Inden, H.,** unpublished data, 1987.
37. **Hirata, K. and Akihama, K.,** Über die Chromosomenzahl bei einigen *Allium* Arten, *Bot. Mag. Tokyo,* 41, 597, 1927.
38. **Ono, Y.,** Chromosome numbers in *Allium* (in Japanese with English summary), *Jpn. J. Genet.,* 11, 238, 1935.
39. **Levan, A.,** Cytological studies in *Allium* VI. The chromosome morphology of some diploid species of *Allium, Hereditas,* 20, 289, 1935.
40. **Kurita, M.,** On the karyotypes of some *Allium* species from Japan, *Mem. Ehime Univ. Sect. II (Sci).,* 1, 179, 1952.
41. **Iwasa, S.,** Studies on viviparous onions, with special reference to the origin of top onion, *Allium cepa* var. *viviparum* (Metzg.) Alef. (in Japanese with English summary), *J. Fac. Agric. Kyushu Univ.,* 25, 55, 1970.
42. **Nishimura, Y. and Shibano, M.,** Male sterility in Japanese bunching onion (in Japanese), Abstr. Jpn. Soc. Hortic. Sci. Spring Meet., 1972, 180.
43. **Moue, T. and Uehara, T.,** Inheritance of cytoplasmic male sterility in *A. fistulosum* L. (Welsh onion) (in Japanese with English summary), *J. Jpn. Soc. Hortic. Sci.,* 53, 432, 1985.
44. **Moue, T. and Uehara, T.,** Heterosis of F_1 hybrids produced by utilizing cytoplasmic male sterility and inbreeding depression of inbred line in Welsh onion (*Allium fistulosum* L.) (in Japanese with English summary), *Jpn. J. Breed.,* 35, 175, 1985.

45. **Aoba, T.,** On Seitaka-yaguranegi (Top-onion), *Allium Aobanum* Araki, (in Japanese with English summary), *Bull. Yamagata Soc. Agric.*, 23, 7, 1966.
46. **Prokhanov, J.,** A contribution to the knowledge of the cultivated Alliums of China and Japan (in Russian with English summary), *Bull. Appl. Bot. Plant Breed*, 24, 123, 1930.
47. **Fiskenjö, G.,** Chromosomal relationships between the species of *Allium* as revealed by C-banding, *Hereditas*, 81, 23, 1975.
48. **Vosa, C. G.,** Heterochromatic patterns in *Allium*, *Heredity*, 36, 383, 1976.
49. **Kurita, M.,** Further note on the karyotypes of *Allium*, *Mem. Ehime Univ. Sect. II (Sci).*, 1, 369, 1953.
50. **McCollum, G. D.,** Onion and allies, in *Evolution of Crop Plants*, Simmonds, N. W., Ed., Longman, London, 1976, chap. 53.
51. **Phillips, G. C. and Hubstenberger, J. F.,** Plant regeneration in vitro of selected *Allium* species and interspecific hybrids, *HortScience*, 22, 124, 1987.
52. **Astley, D., Innes, N. L., and van der Meer, Q. P.,** *Genetic Resources of Allium Species*, IBPGR Secretariat, Rome, 1982, 8.
53. **Brian, F. L.,** *Plant Genetic Resources*, Edward Arnold, London, 1986, 100.
54. **Fujieda, K. and Fujita, S.,** Propagation of *Allium* through tissue culture, in *Seed Production Technology of Vegetable Crops* (in Japanese), S.V.S.P., Eds., Seibundoshinkousya, Tokyo, 1978, 68.
55. **Larkin, P. J. and Scowcroft, W. R.,** Somaclonal variation — A novel source of variability from cell cultures for plant improvement, *Theor. Appl. Genet.*, 60, 197, 1981.
56. **Shahin, E. A. and Kaneko, K.,** Somatic embryogenesis and plant regeneration from callus cultures of nonbulbing onions, *HortScience*, 21, 294, 1986.
57. **Dixon, G. R.,** *Vegetable Crop Diseases*, Macmillan, London, 1981, chap. 10.
58. **Itsuji, M.,** Report of Demonstrative Field, Branch Extension of Uji in Kyoto Pref., (in Japanese), 1984, 26.
59. **Inden, H., Asahira, T., and Okui, H.,** Analysis of factors influencing leaf tip drying of Japanese bunching onion in summer (in Japanese), *Abstr. Jpn. Soc. Hortic. Sci. Spring Meet.*, 1987, 272.
60. **Terami, H. and Tsukamoto, Y.,** Effects of soil conditions on the growth of the Japanese bunching onion (in Japanese), *J. Jpn. Soc. Hortic. Sci.*, 10, 120, 1939.
61. **Noguchi, K.,** Digger for Japanese bunching onion and its utilization, in *Problems and Solution in Vegetable Production* (in Japanese), Vol. 2, Assoc. Farm. Tech. Sub. Tokyo, Eds., Seibundoshinkousya, Tokyo, 1977, 230.
62. **Takamatsu, E.,** Seed production of Japanese bunching onion, in *Seed Production Technology of Vegetable Crops* (in Japanese), S.V.S.P., Eds., Seibundoshinkousya, Tokyo, 1978, 369.
63. **Eguchi, T., Oshika, Y., and Yamada, H.,** Studies on the effect of maturity on longevity in vegetable seeds (in Japanese with English summary), *Bull. Natl. Inst. Agric. Sci. Ser. E.*, 7, 145, 1958.
64. *Standard tables of food composition in Japan* (in Japanese), 4th ed., Resources Council, Science and Technology Agency, Tokyo, 1982, 1.
65. **Mizuno, T. and Kinpyo, T.,** Studies on the carbohydrates of Allium species. I. Kinds of carbohydrates of *Allium fistulosum* L. (in Japanese with English summary), *Nippon Nogei Kagaku Kaishi*, 29, 665, 1955.
66. **Mizuno, T., and Kinpyo, T.,** Studies on the carbohydrates of Allium species. II. On the mucilage of *Allium fistulosum* L. (1) (in Japanese), *Nippon Nogei Kagaku Kaishi*, 31, 200, 1957.

Chapter 9

LEEK (*Allium ampeloprasum*)

Q. P. van der Meer and P. Hanelt

TABLE OF CONTENTS

I. TAXONOMY

A. TAXONOMY AND DISTRIBUTION OF *ALLIUM AMPELOPRASUM*

Allium ampeloprasum is a polyploid complex for which we have to adopt a rather broad species concept. Wild taxa from the Mediterranean region and southwest Asia as well as the cultivated forms, namely, leeks, kurrat, pearl onion, and great-headed garlic, are components of this variable species.[1] The general taxonomy of alliums and the relationship of *A. ampeloprasum* to other members of the genus is discussed elsewhere (see chapter on "Taxonomy, Evolution, and History").

Morphologically it is characterized by its robust habit and flat, keeled leaves which sheath a considerable part of the tall, erect scape. It has a large inflorescence covered by a connate and long-beaked spathe which is shed at anthesis. The flowers are campanulate, rose-violet, or whitish with papillate, keeled tepals and long stamens. The inner whorl of stamens are tricuspidate, typical of all species of the genus *Allium* to which *A. ampeloprasum* belongs. The plants, apart from leeks and kurrats, mostly produce bulbs and many bulblets. These consist of only one (rarely two) storage leaf bases and are surrounded by a protective layer of one or two coats which are sclerified in wild taxa and in great-headed garlic.

Wild plants of the species occur frequently in the Mediterranean area from Portugal and northwest Africa in the west to Turkey, Syria, north Iraq, and west Iran in the east. They sometimes occur as introductions in countries outside this area. Isolated occurrences in western Europe (England, Ireland, Channel Islands) have been interpreted as relics of former cultivation.[2] Wild *A. ampeloprasum* is restricted mostly to man-made habitats such as fields, vineyards, roadsides, and walls. In eastern Mediterranean countries populations also occur in natural habitats such as coastal cliffs, ravines or in garigue-like vegetation.[3,4]

A. ampeloprasum represents a polyploid series ($x = 8$) with tetraploids ($4x = 32$), pentaploids ($5x = 40$), and hexaploids ($6x = 48$). So far no diploid populations have been found. Tetraploids are most common among the wild taxa[3,4] as well as in the cultivated groups,[5] of which only the great-headed garlics are hexaploid.[6,7] Wild plants of different ploidy levels are morphologically indistinguishable.[3] The pentaploids rely exclusively on vegetative reproduction by bulblets. Two different types of satellite chromosomes typically occur in *A. ampeloprasum*. First there are rather small satellites on the shorter arms of submetacentric chromosomes. Second, large satellites occur on the short arms of distinctly asymmetrical chromosomes. Both types always occur, but in varying number, in wild *A. ampeloprasum*, in some other related wild species,[8] and in the cultivated groups[9,5] (see chapter on "Cytogenetics").

The cultivated groups of *A. ampeloprasum* are very diverse and variable. Different selection pressures have produced very different crop types, cultivated either for the bulbs (pearl onion, great-headed garlic), the pseudostem (leek), or the leaves (kurrat and primitive land-races of leek).

There is no data on the phylogenetic relationships between wild and cultivated taxa of the species, although spontaneous forms from the Aegean region and Israel have been studied in detail and much is known about their morphological, ecological, and karyological variability, their population structure, and reproductive behavior.[3,4,8,10] There are also many gaps in our knowledge of the relation to *A. ampeloprasum* of numerous related wild taxa from the Iberian peninsula, southern Russia, and the Caucasus region. Therefore, the proposed concept of *A. ampeloprasum* should be considered as preliminary. It is even questionable if *A. ampeloprasum* is the nomenclaturally correct name or if it should be replaced by *A. porrum* which has also long been used in a broad sense.

B. INFRASPECIFIC CLASSIFICATION

1. Wild Taxa

Wild *A. ampeloprasum* plants are characterized by small, helmet-shaped or spherical

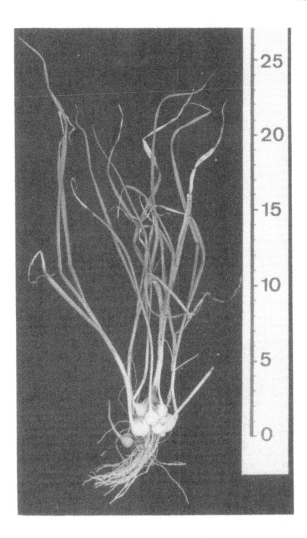

FIGURE 1. Pearl onions.

bulblets appressed to the mother bulb; this character also occurs in great-headed garlic.[3] In some oriental types, bulblets develop on stolons up to 10 cm long and are thus better adapted to vegetative propagation. These forms have been separated as subspecies; ssp. *truncatum* (Feinbr.) Kollm. from Israel and ssp. *iranicum* Wendelbo from west Iran and southeast Anatolia. The typical ssp. *ampeloprasum* includes the rather homogeneous populations from the main Mediterranean area of the species, as described above.[3,10,11] The isolated populations from some sites in western Europe mentioned above have also been placed within ssp. *ampeloprasum*.[2] As far as is known, these are hexaploid. Some of them produce bulbils in the inflorescences, e.g., var. *babingtonii* (Borr.) Syme, from Cornwall and Ireland and var. *bulbiferum* Syme from the Channel Islands.

2. Cultivated Groups of A. ampeloprasum Other Than Leek and Kurrat

In addition to the leek (synonym *A. porrum* L.) and the rather similar kurrat (synonym *A. kurrat* Schweinfurt ex Krause) groups (see Section III below), there are the following cultivated groups.

Pearl Onions (Allium ampeloprasum L. var. sectivum Lued.) — These are grown for small daughter bulblets (Figure 1) in private gardens in the German Democratic Republic[5] and, until 1982, on a commercial scale in the Netherlands. The small bulbs form in a cluster

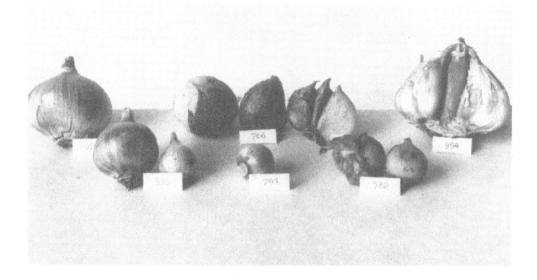

FIGURE 2.   (a) Great-headed garlic bulbs and cloves from different accessions from west Georgia, Transcaucasus U.S.S.R; (b) cloves of garlic, (*A. sativum*), left, and great-headed garlic (*A. ampeloprasum* var. *holmense*), right.

of sessile or short-stalked bulblets with silvery-white skins. The plants are like leeks but smaller and without pseudostems. The flowers are always white. The plants are seed fertile although flower scapes are not regularly developed. They are very winter hardy, like leek, and have a short period of summer dormancy. They are easily propagated by bulblets. They are cultivated for the bulbs which are mostly used for pickling. Pearl onions can easily be crossed with leek, even though the morphology of the chromosome is somewhat different from leek.[5]

Great-Headed Garlic (A. ampeloprasum L. var. ampeloprasum auct.) — This crop (Figure 2) is described by Jones and Mann.[12] Under Dutch conditions it produces floriferous

FIGURE 3. *A. ampeloprasum* subspecies and varieties. From left to right: kurrat, tarée irani, French leek, Dutch leek, Bulgarian leek.

umbels, but seeds are never obtained, even after flowering alongside leek. Under Egyptian conditions a planted clove forms a globe-shaped bulb which, in turn in the following season, gives rise to one flower stalk surrounded by six big cloves.[13] Additional small lateral bulblets appressed to the outside of the main bulb can be present. The plants are robust with a strong garlic-like odor. Flowering scapes form and are much like those of leek; however, seed may not be formed, and if formed, is sterile. It is cultivated for the bulbs and cloves and used like garlic, or the leaves may be eaten as a condiment. Small-scale cultivation has been reported from northwest India, Transcaucasus, south Russia, Greece, and the U.S., but is probably more widely distributed in west Asia and the eastern Mediterranean. It is apparently often confused with garlic (*A. sativum*) from which it differs by its leek-like inflorescence and its larger cloves (Figure 2b). So far as is known great-headed garlic is hexaploid (6x = 48). This was confirmed for accessions from northwest India and Georgia, Transcaucasus.[6,7]

It differs markedly from other crop types of *A. ampeloprasum* but has similarities with the wild populations which have 48 chromosomes. This may indicate a biphyletic evolution of the crop groups of *A. ampeloprasum*, a possibility that needs further investigation.

Tarée Irani — This crop is described by Tabhaz.[14] It is grown near Teheran, Iran for its green leaves, and in this respect it is similar to kurrat (see Section III) (Figure 3). However, under Dutch conditions, kurrat has completely floriferous umbels whereas Tarée Irani, although raised from seeds grown in Iran, has completely bulbiferous umbels (Figure 3).

TABLE 1
Leek Cultivation in European
Community Countries in the Years
1983—1985[a]

Country	Area (ha)		
	1983	**1984**	**1985**
Belgium	2,400	2,600	3,100
Denmark	400	400	400
FRG	1,600	1,700	1,500
Greece	1,900	2,000	1,700
France	10,000	9,600	9,700
Ireland	<100	<100	<100
Italy	1,100	1,200	1,300
Luxemburg	<100	<100	<100
The Netherlands	2,300	2,500	2,900
Portugal	—	—	—
Spain	2,600	2,800	2,800
United Kingdom	2,100	2,300	2,400

[a] Source: Eurostat; via Commodity Board of
Fruit and Vegetables, The Hague, Netherlands.

***Poireau Perpetuél;*[14] *Synonym: des Poireaux non Selectionés*[15]** — The leaves of this
wild leek-type are used in France and Algeria for human consumption. The same type is
collected and used by man in Greece.[16]

Prei anak — A vegetatively propagated type of leeks, called 'prei anak' (leeks with
children), is grown on a commercial scale in Lembang (West Jave, Indonesia) at an altitude
of about 1000 m. After planting a young sprout it will grow out to a moderately sized leek-
like plant producing up to about 10 side shoots. It has a harder texture than the common
European leeks, which are also grown in Lembang, and may originate from France. According to an Indonesian expert (Mr. Herbagiandono) three local strains are available.

II. GENETIC RESOURCES

The center of origin and the natural distribution of *A. ampeloprasum* covers southern
Europe, western Asia, and northern Africa.[2,17] These populations are extremely important
genetic resoues for leek, especially nowadays since the majority of the local strains are
disappearing as a result of intensive breeding and the adoption of more productive and
uniform cultivars. Leek is grown in many European countries (Table 1). Undoubtedly, the
most severe genetic erosion has occurred in Belgium where almost all of the local strains
have been replaced by a limited number of uniform Dutch varieties. This is evident in the
study of Van Parys[18] in which only Dutch strains are compared. Dutch varieties predominate
also in the U.K., but to a lesser extent than in Belgium. Kurrat (see Section III), a leek-
like crop, is very popular in Egypt and several other countries of the Near East. The genetic
resources of this crop are still intact.

Unfortunately, until now large-scale collection and preservation of leek genotypes has
been undertaken only in the Netherlands[19] and Egypt.[20,21] The severe loss of genetic diversity
will inevitably hamper breeding for resistance to pests and diseases.

FIGURE 4. Median longitudinal section of the base of a leek pseudostem, showing adventious roots and leaf sheaths growing from the basal disk of true stem tissue.

III. MORPHOLOGY AND DEVELOPMENT

A. MORPHOLOGY

The following three types of leek can be distinguished by their morphological features (Figure 3):

1. The European type with a relatively short and thick pseudostem
2. The Turkish type, also grown in Bulgaria and northern Egypt, with a relatively long and thin pseudostem
3. Kurrat, grown in the Near East, mainly in Egypt, without a pronounced pseudostem

The third type is grown for its green leaves which are harvested several times a year[12] and is an essential ingredient of ta'amyia, a very famous fried Egyptian dish.

These three types are completely intercrossable and therefore are considered as variants of the same crop.

B. DEVELOPMENT

Leek seeds germinate epigeally. Upon germination the bent cotyledon emerges as in other *Allium* crops. The first leaf emerges through an opening (pore) in the lower part of the cotyledon. The basal plate or disk (Figure 4), a very suppressed stem, is located at the base of the cotyledon. Its apical meristem gives rise to leaves and finally to the inflorescence. Each new leaf is formed from an almost ring-shaped meristem just inside the preceding leaf sheath. Each leaf meristem gives rise to a leaf blade and a tube-shaped leaf sheath. Leaves are initiated oppositely and alternately from the apical meristem. A ligule[22] at the junction

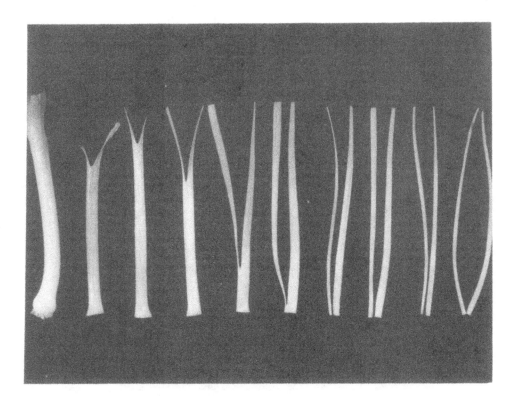

FIGURE 5.   Successive leaf bases dissected from outside to the center of leek pseudostem (outermost on the left). The sheath length and the ratio of sheath length:blade length decrease from outside to inside.

of the leaf and sheath is hardly distinguishable. The pseudostem (shaft or shank) (Figure 5) consists of extended leaf sheaths and also young leaf blades. The ratio leaf-blade length/leaf-sheath length increases from outside to the center (Figure 5).

A flower stalk is formed only by plants above a certain minimum size. It is initiated by the apical meristem and grows up through the center of the pseudostem. Only one umbel per flower stalk is produced, normally bearing hundreds of flowers. The flowers, mostly of a light purple color and a campanulate shape, have six tepals, six tepaloid anthers, and an ovary with three locules, containing two ovules each. Topsets are easily formed in the umbel, especially if flower buds are removed at an early stage of development.[23] Seeds are formed as a result of self- or cross-pollination by insects; the seed weight varies between 2.2 and 3.6 mg. Adventitious roots appear at the stem base after the early decay and loss of the primary root.

IV. PHYSIOLOGY

A. VEGETATIVE DEVELOPMENT

Seeds of leek do not show rest or dormancy. Their germination depends mainly on temperature, 18 to 22°C being the optimum (Table 2). Leeks are rather slow to germinate, the heat sum (in degree days) of 222 for 50% germination[24] is relatively high in comparison with most vegetables (e.g., lettuce requires 71 degree days). Dragland[25] found germination percentages declining from 75 to 97 at 12 to 21°C to 55 to 92 and 2 to 11 at 24 and 27°C, respectively. Priming (osmotic pretreatment) can improve the rate and uniformity of seedling emergence in the field.[26] Mean temperatures of 18 to 22°C seem to be optimal for plant growth[25,27] (Table 3). Brewster found that seedling relative growth rate was maximal at 27°C, but seedlings were more variable in growth rate at 27°C than at 23°C.[27]

TABLE 2
Effect of Temperature on Leek Seed
Germination Rates

	Number of days to 50% emergence			
Genotype	10°C	14°C	18°C	22°C
"Elbeuf improved"	27	19	15	13
Kurrat (Egypt)	19	19	12	9
Tarée irani	15	12	9	9
"Jumbo garlic"	19	12	9	9
"Kajak"	26	16	13	14
"Alaska"	19	13	12	12
"Varna"	19	13	12	12
Mean	21	15	12	11

Note: Experiments were carried out in moist soil in the IVT phytotron in 1987.

TABLE 3
Effects of Temperature on the Development
of Several *A. ampeloprasum* Genotypes

	Fresh weight (g)[a]			
Genotype	10°C	14°C	18°C	22°C
"Elbeuf improved"	0.6	1.3	2.2	1.7
"Géant am. de saulse"	0.5	1.9	2.6	2.5
"Monstrueux d'Elbeuf"	0.5	2.1	2.2	2.1
Kurrat (Egypt)	0.6	1.1	1.1	1.2
Kurrat (Egypt)	0.3	1.1	1.5	1.1
Leek (Egypt)		1.2	2.5	1.3
Tarée irani	0.3	1.1	1.4	1.3
Jumbo garlic	0.2	1.3	1.4	1.2
"Otina"	0.4	2.0	1.8	2.6
"Kajak"	0.2	1.0	1.4	1.6
"Alaska"	0.2	1.4	1.6	1.1
"Varna"	0.3	1.6	1.2	1.2
Mean weight	0.4	1.4	1.7	1.6

[a] Average weight (g) of 15 10-week old plants at different temperatures under glass. Sowing date: 4 April 1987; IVT phytotron, Wageningen.

Although leek is grown as far south as Cuba[28] and as far north as Norway,[29] no information is available on genetic adaptation to the photoperiods and temperatures of different latitudes. The size but not the morphology of the plants is affected by age and environment. Also the considerable heterogenity within cultivars is reflected in differences in size rather than in morphology.[30] It is likely that the high plant-to-plant variability within cultivars results from high rates of self-pollination and the consequent high percentage of inbreds (see Section V).

B. FLOWERING

Only limited quantitative information is available on the effect of temperature and daylength on growth and flowering (bolting). Dragland[25] found stimulation of bulbing by

FIGURE 6.   Leek plants bolting after producing only seven leaves. These plants were grown at 5°C in a glasshouse through the winter and from early April onward outside.

long days. Atanosov and Carrazana[28] reported that bolting is enhanced by long days. Similar results were obtained at constant temperatures between 12 and 18°C and daylengths between 9 and 24 h.[25] In these experiments daylength had a marked effect on the time from sowing to inflorescence appearance, and it was concluded that leek is a quantitative long-day plant without a need for vernalization, although it does show some response to low temperatures.

Flowering can occur only in plants larger than a certain minimum size or over a certain physiological age. At IVT (Netherlands Institute of Horticultural Plant Breeding) plants grown at 5°C through the winter and moved to the field in April bolted after producing approximately seven leaves (Figure 6). In contrast, Dragland[25] stated that at 12°C and a daylength of 19 h bolting occurs only after the formation of a minimum of 13 leaves. For a better understanding of the effect of temperature and daylength on bolting in leek, phytotron experiments similar to those of Dragland[25] should be extended to sequentially varying temperatures and daylengths during development.

FIGURE 7.   Bulbs formed at the base of the leek scape after flowering.

C. SEED DEVELOPMENT

Seed development requires high temperatures and rather a long time. Therefore, the seeds of Dutch varieties are produced under glass in the Netherlands or in warmer areas in the open, e.g., Italy or Idaho (U.S.).

D. VEGETATIVE PROPAGATION

Leek can be propagated vegetatively by the induction of topsets[23] and plantlets in the umbel[31] or in the basal plate.[32] Moreover, leek plants can easily be maintained vegetatively by the basal cloves formed after flower-stalk formation (Figure 7). However, the leek yellow stripe virus can carry over into such vegetatively propagated plants, unlike seeds (see chapter on "Virus Diseases").

V. GENETICS

The members of the ampeloprasum-taxon show different ploidy levels, i.e., 2n = 24, 32, 40, 48, or 56[2], but very probably all commercial leek types are tetraploids.[23,33,34] Schweisguth[23] considers leek to be an autotetraploid whereas Koul and Gohil[35] propose that it is a segmental allotetraploid with the genomic formula AAA'A'. They suggested that tetrasomic inheritance must occur for most genes. They reported that chiasmata localization in leek results in large linkage blocks and is the most important restriction to allelic recombination. They concluded that while giving leek the stability it needs to propagate sexually, the chiasmata localization has limited the variability which is possible with normal allelic recombination.

In spite of this limited recombination, one of the most striking features of leek is its strong sensitivity to inbreeding depression. Even after one generation of selfing the average plant weight is reduced by 26 to 62%.[36] At the same time the uniformity increases considerably.[23] This strong inbreeding depression is evidence for loci occupied by three or four different alleles, as occurs in alfalfa (*Medicago sativa*).[37] The earlier-mentioned heterogeneity within cultivars in size could be the direct consequence of a rather high percentage of self-pollination, e.g., up to 20%.[31,33] The strong sensitivity to inbreeding could be the consequence of excessive selection pressure for uniformity, resulting in a narrow genetic basis. The obvious remedy to counteract the linkage between uniformity and inbreeding seems to be F_1 hybrids, but the breeding of leek hybrids is difficult (see Section VI).

VI. BREEDING

Until recently positive mass selection under heavy stress conditions (namely, winter frost and severe yellow stripe virus attack) was most commonly used. At the same time plants were also selected for maximum uniformity and yield. Nowadays more attention is paid to family selection and recurrent selection, but exact procedures and results are kept confidential by commercial breeders.

The most promising system for improvement of both uniformity and yield seems to be hybrid breeding.[23,38] Nevertheless, the occurrence of tetraallelic loci is likely to hinder the maximization of heterosis in F_1 hybrids generated from inbred lines. This is because increased inbreeding of the parent lines leads to increased diallelic F_1 hybrids, e.g., aacc (here a, b, c, and d signify different alleles of the same gene), whereas the tetraallelic combination (abcd) is desired for maximum heterosis. Consequently, a stronger heterotic effect can be expected from three-way cross hybrids, and the strongest heterosis from double cross hybrids. Unfortunately, practical possibilities for the breeding of hybrids, e.g., a suitable sterile male are still lacking.

A useful tool for the acceleration of breeding procedures is the reduction of the developmental cycle from 2 years to 1. In the Netherlands this is achieved by sowing in August and exposing the plants to artificial light and 10°C in the greenhouse during winter.[39]

An additional goal of breeding should be to select resistance to pests and diseases. However, only a few reports have been published on this. Van der Meer[40] identified in kurrat a complete resistance to yellow stripe virus and started a breeding program to transfer this to leek[41] using the ELISA method for screening[42] (see chapter on "Virus Diseases"). Norman[43] and Burchill et al.[44] reported some differences in varietal resistance to rust (*Puccinia porri*). Both negative or positive line or family selection may increase the resistance of leek varieties to this disease fairly easily. At the IVT recently, a start has been made on breeding for resistance to white tip, *Phytophthora porri*.[45]

VII. CULTIVARS

Many leek cultivars are available in the Netherlands, France, and elsewhere. Many have been evaluated in trials carried out by the Government Institute for Research on Varieties of Cultivated Plants (RIVRO, Wageningen, The Netherlands) and the National Institute of Agricultural Botany (NIAB, Cambridge, U.K.) for yield, shaft length, uniformity, and for adaptation to the different seasons of the year, i.e., earliness and winter hardiness. The results are reported in the following publications; "Rassenberichten" (RIVRO) and "Vegetable Growers Lists" (NIAB). An overview of NIAB results was recently published.[46] Cultivar trials have also been carried out in Norway,[29] Belgium,[18] Switzerland,[47] and the German Democratic Republic.[48] Adaptation to specific geographic latitudes seems to be a matter of winter hardiness at northern latitudes[36,43,46] and tolerance to pests, diseases, and heat in subtropical and tropical regions. Winter hardiness is strongly correlated with a short pseudostem. Individual leek cultivars are adapted to a much wider range of latitudes than are particular onion cultivars.

VIII. PLANT PROTECTION

Sections VIII, IX, and X are based mainly on English and Dutch official advisory literature.[49,50] Leeks suffer severely from many pests and diseases. Onion fly (*Delia antiqua*), *Thrips tabaci*, and leek moth (*Acrolepia asectella*) severely attack leek. The first can be controlled by carbofuran soil treatment, the other two by parathion. Purple blotch *Alternaria porri*, leaf blotch *Cladosporium allii-porri* (synon.; *Heterosporium allii*), *Fusarium culmorum* (soil born), *Leptotrochila porri*, white tip, *Phytophthora porri*, *Puccinia allii*, and damping-off fungi all attack leek. These fungi can be controlled by captafol, benomyl (dipping of transplants), stringent crop rotation, Topaz, and Curater (soil treatment in the seedling nursery), respectively. Sporadic attack by smut, *Urocystis cepulae*, and white rot, *Sclerotium cepivorum*, as well as by nematodes is found in France.[51] Bacterial diseases (*Erwinia* and *Pseudomonas*) may cause severe damage, especially at higher temperatures. Control of these bacterial diseases is difficult. Yellow stripe virus a nonpersistent virus, spreads easily from plant to plant and can cause heavy damage,[50,52] and control is difficult.[54] Interaction with latent shallot virus intensifies this effect.[55] Infection is stimulated by intensive and prolonged, year-round or year-after-year cultivation. Consequently, extensive crop rotation, i.e., disease prevention, is the most effective control measure.

IX. CHEMICAL WEED CONTROL

Recent results and practical instructions are published in advisory literature.[49,50,56,57] Weed control in a transplanted crop in the Netherlands has the following schedule.[50] Shortly after transplanting, weed-free soil can be sprayed with 65% propachlor (7 kg/ha) or with 50% simazin (1 kg/ha). Simazin is cheaper than propachlor but may not be used on a very early crop, on very light sandy soils poor in organic matter, or on organic soils. An alternative method is to spray postemergence a mixture of 20% simazin and 35% prometryn at 1.5 kg/ha or 50% cyanazin at 1.5 kg/ha. This should be done under low light intensity, i.e., during cloudy weather or in the evening. Grasses are only partially controlled by sethoxydim at 1.5 kg/ha applied until 3 weeks before harvest. This herbicide may not be used in combination with others.

X. CULTURAL SYSTEMS AND AGRONOMIC PRACTICES

Leek is grown in practically all soil types, the most important requirement being a loose texture. On peat soil, yields are usually high but quality is bad. Sandy clay soils are the most suitable for leek cultivation. Harvesting is difficult in heavy soils in autumn and winter. Deep ploughing is a prerequisite for the development of long white shafts. In the Netherlands it is found that phreatic water levels of 40 cm or less below the soil surface result in poor yields; optimal levels are 80 to 90 cm.

Ploughing in of 50 to 70 t of farm manure per hectare during autumn or winter is highly advisable, with some additional application of nitrogen, potassium, and phosphorus. Alternatively, 30 to 60 kg P_2O_5, 200 to 250 kg N, and 150 to 300 kg K_2O is recommended under normal soil conditions in the Netherlands.

For early crops leeks are sometimes raised in blocks of compressed peat-based compost, but usually seeds are sown at 1 to 1.5 kg/ha in a nursery, and seedlings are transplanted about 3 months later. Salter[58] obtained high marketable yields after early planting of plants raised in modules (e.g., Speedlings) at high temperature. Because of the high cost of planting labor, direct seeding is increasing in popularity, although this does result in shorter shafts.[49] Direct seeding and mowing instead of digging could be profitable for processing purposes.[59] In Germany sowing in March gave better results than in April or May.[60] For early crops, transplants are grown in a nursery under glass. After sowing, the temperature is kept at 18 to 20°C for 3 to 4 weeks; thereafter the temperature is reduced to 15°C, and later the temperature is lowered to about 12°C. Plants 5 to 6 mm thick are thought to be the optimum size for transplanting. Using older transplants, e.g., 15 weeks instead of 12 weeks old, can result in 25% lower yields. Automatic planting machines are available; also other machines are available for punching planting holes only (depth: 15 to 18 cm; diameter: 3.5 cm).[50]

Increasing the planting depth results in both longer white shafts and later harvesting. Deeper planting of the winter crop (20 cm instead of 15 cm) always gives a lower yield, even with late harvesting. Dutch experiments showed that shortening (cutting) leaf blades and/or roots of transplants result in lower yields. Optimal plant densities are 20 to 40 m², the earlier the crop the higher the density.[50] In the U.K. and the Netherlands leek is grown as a summer, autumn, or winter crop. An overview of the essential points for growing under Dutch conditions is given in Table 4.

Harvesting is done manually or by machine. When harvested by hand the plants are usually lifted mechanically; then the roots are cut, the outer damaged and senescent leaves are removed, the remaining leaves shortened, and the plants are packed into boxes. When mechanical harvesting is used, hand cleaning and machine washing are necessary.

XI. STORAGE

The optimal storage temperature for leek is -1°C at 95% relative humidity (RH). The maximal storage period is 8 weeks. Wilting and fungal disease are important factors in limiting storage life.[61] Controlled atmosphere (10% CO_2 + 1% O_2) can extend the storage period to 5 months.[62] In the Netherlands leeks are stored mainly during May and June in order to close the gap between the late winter and the early summer crop (see Table 4).

XII. COMPOSITION AND NUTRITIONAL ASPECTS

The composition of leek is given in Table 5. The composition of leek does not deviate much from the average composition of 47 of the most common Dutch vegetables.[63] The only noteworthy negative aspect seems to be excessive nitrate content. A nitrate level above 250 mg/100 g fresh weight is considered dangerous, since nitrites and carcinogenic nitrosamines can form after ingestion. The nitrate content increases strongly at high levels of N fertilizer, reaching levels of up to 450 mg NO_3 per 100 g fresh weight. However, at normal levels of N fertilizer (200 to 250 kg N per hectare) the nitrate content does not exceed 50 mg/100 g.

ACKNOWLEDGMENTS

We should like to thank Dr. J. L. Brewster for helpful comments and for providing a number of the references used in preparing this paper.

TABLE 4
Details for Leek Cultivation in the Netherlands

Crop	Sowing time	Sowing place	Planting time	Plants/ha	Harvest time	Yield (ton/ha)
Summer crop[a]	Mid-Dec./mid-Jan	Heated glasshouse	Mid-March/end Apr.	400,000	Mid June/early July	25—30
Summer crop	Mid-Jan./mid-Feb.	Heated glasshouse	Early Apr./end Apr.	350,000	July	30—40
Summer crop[b]	idem	Heated glasshouse	idem	250,000	August	40—50
Autumn crop[c]	March	Low glass cold frame or plastic tunnel	Early May/mid-June	190,000	Sept.—Oct.	40—50
Autumn crop[d]	Mid-May/early Apr.	Low glass cold frame or plastic tunnel	End June	170,000	Nov.—Dec.	45—35
Winter crop[c]	Early Apr./end Apr.	The open	Mid-June/early July	170,000	Early Jan./March	35—30
Winter crop[d]	Mid-Apr./mid-May	The open	Early July/early Aug	170,000	April—May	30—50

[a] Very early.
[b] Normal.
[c] Early.
[d] Late.

TABLE 5
Chemical Composition of
Leek per 100 g Fresh
Weight

Main components	Weight
Water	90 g
Protein	2
Fat	0.3
Carbohydrates	5
Minerals	1.5

Minerals	
Sodium	5 mg
Potassium	250
Calcium	60
Iron	1
Phosphorus	30

Vitamins	
B carotene (provit. A)	600 μg
Thiamin (vit. B_1)	120
Nicotinic acid	500
Pyrodoxine (vit. B_6)	250
Ascorbic acid (vit. C)	25

REFERENCES

1. **Hanelt, P., Fritsch, R., Kruse, J., MaaB, H., and Ohle, H.,** Wild relatives of cultivated *Allium* species, 14th Int. Botanical Congr. — Abstr., Berlin, 1987, 288.
2. **Stearn, W. T.,** European species of *Allium* and allied genera of *Alliaceae*: a synonymic enumeration, *Ann. Musei Goulandris*, 4, 83, 1978.
3. **von Bothmer, R.,** Studies in the Aegean Flora. XXI. Biosystematic studies in the *Allium ampeloprasum* complex, *Opera Bot. Soc. Bot. Lundensis*, 34, 1, 1974.
4. **Kollman, F.,** *Allium ampeloprasum* L. in Israel (Taxonomy), *Isr. J. Bot.*, 20, 263, 1971.
5. **Hanelt, P. and Ohle, H.,** Die Perlzwiebeln der Gartenlebener Sortiments und Bemerkungen zur Systematik und Karyologie dieser Sippe, *Kulturpflanze*, 26, 339, 1978.
6. **Khoshoo, T. N., Atal, C., and Sharma, V. B.,** Cytotaxonomical investigations on the northwest Indian garlics, *Res. Bull. East Punjab Univ.*, 11, 37, 1960.
7. **Fritsch, R.,** unpublished data, 1987.
8. **von Bothmer, R.,** Cytological studies in *Allium*. I. Chromosome numbers and morphology in sect. *Allium* from Greece, *Bot. Not.*, 123, 519, 1970.
9. **Kadry, R. and Kamel, S. A.,** Cytological studies in the two tetraploid species *Allium kurrat* Schweinf. and *A. porrum* L. and their hybrid, *Sven. Bot. Tidskr.*, 49, 314, 1955.
10. **Kollman, F.,** *Allium ampeloprasum* — a polyploid complex. II. Meiosis and relationships between polyploid types, *Caryologia*, 25, 295, 1972.
11. **Wendelbo, P.,** *Alliaceae*, in *Flora Iranica*, Rechinger, K. H., Ed., No. 76, Akad. Druck-u. Verlagsanstalt, Graz, Austria, 1971, 1.
12. **Jones, H. A. and Mann, L. K.,** *Onions and Their Allies*, Interscience, New York, 1963.
13. **Elshami, M. R.,** personal communication, 1987.
14. **Tabhaz, F.,** L'*Allium* Tarée irani du groupe *ampeloprasum* L. cultivé en Iran region de Téhéran, *Bull. Soc. Bot. Fr.*, 118, 753, 1971.
15. **Boscher, J.,** Le patrimoine génétique d'*Allium porrum* L. Observations et caracterésation de Poireaux locaux, *Bull. Soc. Bot. Fr., Lett. Bot.*, 130, 33, 1983.

16. **Apostolos, G.,** personal communication, 1983.
17. **De Wilde-Duyfjes, B. E. E.,** A revision of the Genus *Allium* L. *(Liliaceae)* in Africa, Thesis, Communications Agricultural University, Wageningen, The Netherlands, 76-11, 1976.
18. **Van Parys, L.,** Cultivar onderzoek bij herfst- en winterprei, Mededeling 263; Provinciaal onderzoek- en voorlich tingscentrum voor land-en tuinbouw, Beitem-Roeselare, Belgium, 1986.
19. **Van der Meer, Q. P. and Van Bennekom, J. L.,** Collection of onion- and leek-genotypes in the Netherlands, Inst. Hortic. Plant Breed. Rep. (Wageningen), 1983.
20. **Van der Meer, Q. P.,** Balady strains of Egyptian vegetables. I. Collection from April 21 to May 24, 1985, Inst. Hortic. Plant Breed. Rep. (Wageningen), 1985.
21. **Van der Meer, Q. P.,** Balady strains of Egyptian vegetables. III. Collection from June 17 to July 2, 1986, Inst. Hortic. Plant Breed. Rep. (Wageningen), 1986.
22. **Bonnet, B.,** The leek (*Allium porrum* L.): botanical and agronomic aspects. Review of literature, *Saussurea,* 7, 121, 1976.
23. **Schweisguth, B.,** Etudes préliminaires á l'amélioration du poireau *A. porrum* L. Proposition d'une méthode d'amélioration, *Ann. Amelior. Plant.,* 20, 215, 1970.
24. **Bierhuizen, J. F. and Wagenvoort, W. A.,** Some aspects of seed germination in vegetables. I. The determination and application of heat sums and minimum temperature for germination, *Sci. Hortic.,* 2, 213, 1974.
25. **Dragland, S.,** Effect of temperature and day-length on growth, bulb formation and bolting in leeks, (*Allium porrum* L.), *Meld. Nor. Landbrukshoegsk.,* 51, 21, 1972.
26. **Brocklehurst, P. A., Dearman, J., and Drew, R. L. K.,** Effects of osmotic priming on seed germination and seedling growth in leek, *Sci. Hortic.,* 24, 201, 1984.
27. **Brewster, J. L.,** The response of growth rate to temperature in seedlings of several *Allium* crop species, *Ann. Appl. Biol.,* 93, 351, 1979.
28. **Atanosov, N. and Carrazana, D.,** Formation of reproductive organs in leek in Cuban conditions, *Hortic. Abstr.,* 47, 11345, 1977.
29. **Vik, J.,** Overwintering of leek plants in the field, *Res. Norw. Agric.,* 33, 119, 1982.
30. **Benjamin, L.,** The relative importance of some different sources of plant-weight variation in drilled and transplanted leeks, *J. Agric. Sci.,* 103, 527, 1984.
31. **Currah, L.,** Leek breeding: a review, *J. Hortic. Sci.* 61, 407, 1986.
32. **Debergh, P. and Standaert-Demetsenaere, R.,** Neoformation of bulbils in *Allium porrum* L. cultured in vitro, *Sci. Hortic.,* 5, 11, 1976.
33. **Berninger, E. and Buret, P.,** Etudes des deficients chlorophylliens chez deux espéces cultivees du genre *Allium:* l'oignon *A. cepa* L. et le poireau *A. porrum* L., *Ann. Amelior. Plant.,* 17, 175, 1967.
34. **Gohil, R. N.,** Extent of recombination possible in the cultivated leek, Synopsis 3rd Eucarpia Allium Symp., Wageningen, Netherlands, 4-6 September 1984, 99.
35. **Koul, A. K. and Gohil, R. N.,** Cytology of the tetraploid *Allium ampeloprasum* with chiasma localization, *Chromosoma,* 29, 12, 1970.
36. **Gagnebin, F. and Bonnet, J. C.,** Quelques considérations sur la culture et l'Amélioration du poireau, *Rev. Hortic. Suisse,* 52, 112, 1979.
37. **Busbice, T. H. and Wilsie, C. P.,** Inbreeding depression and heterosis in autotetraploids with application to *Medicago sativa* L., *Euphytica,* 15, 52, 1966.
38. **Kampe, F.,** Untersuchungen zum Ausmasz von Hybrideeffecten bei Porree, *Arch. Zuechtungsforsch.,* 10, 123, 1980.
39. **Van der Meer, Q. P. and van Dam, R.,** Eénjarigheid bij prei, *Zaadbelangen,* February-March 1984, 11.
40. **Van der Meer, Q. P.,** Resistentie tegen preigeelstreep virus in kurrat (*Allium ampeloprasum*), *Zaadbelangen,* June 1980, 173.
41. **Van der Meer, Q. P.,** Breeding for resistance to yellow stripe virus in leeks, (*Allium porrum* L.). A progress report, Synopsis 3rd Eucarpia Allium Symp., Wageningen, Netherlands, 4-6 September 1984, 16.
42. **Schuuring, J.,** Virus testing in ornamentals and other crops by ELISA methods, Synopsis 3rd Eucarpia Allium Symp., Wageningen, Netherlands, 4-6 September 1984, 20.
43. **Norman, B.,** Luddington long shaft leeks are not hardy, *Grower,* March 3, 481, 1977.
44. **Burchill, R. T., et al.,** Annu. Rep. Natl. Veg. Res. Stn. for 1982, Wellesbourne, U.K., 1983, 61.
45. **Van der Meer, Q. P.,** Annu. Rep. 1986, Inst. Hortic. Plant Breed., Wageningen, The Netherlands, 1987.
46. **Anon.,** Leek selection, *Suppl. Grower,* July 2, 20, 1981.
47. **Anon.,** Porree Sorten, *Gemüse,* 23, 166, 1987.
48. **Lindner, U.,** Sortenversuche mit Winterporree 1980/81 und 1981/82, *Rhein. Monatschr.,* 5/83, 233, 1983.
49. **Anon.,** Leeks, Booklet 2069, ADAS, Ministry of Agriculture, Fisheries and Food, U.K., 1982, 8.
50. **Snoek, N. T. and de Jonge, P.,** Teelt van prei, Teel thandleiding nr. 11 CAD Akkerbouw en Groenteteelt Vollegrond, Postbus 369, 8200, AJ Lelystad, The Netherlands, 1985.
51. **Grill, D.,** Les maladies et ravageurs du poireau, *Rev. Hortic.,* 262, Décembre, 39, 1985.
52. **Matthieu, J. L. and Verhoyen, M.,** Leek chlorotic streak, epidemiology and incidence of the disease, *Rev. Plant Pathol.,* 62, 510, 1983.

53. **Anon.**, Streifenvirus in Porree, *Gemüse*, 15, July, 242, 1980.
54. **Bos, L., Huijberts, N., Huttinga, H., and Maat, D.**, Leek yellow stripe virus and its relationships to onion yellow dwarf virus; characterization, ecology and possible control, *Neth. J. Plant Pathol.*, 84, 185, 1978.
55. **Paludan, N.**, Virus attack on leeks: diagnosis, varietal tolerance and overwintering, *Rev. Plant Pathol.*, 60, 2316, 1981.
56. **Pelletier, J.**, Desherbage selectif du poireau par voie chimique. Une mise au point, *Rev. Hortic.*, 275, March 1987, 33.
57. **Frentz, F. W. and Andresen, F.**, Der Einfluss der Hack arbeit auf Auftrag und Qualität von Porree, *Rhein. Monatschrift Gemüse Obst Zierpflanzen*, July 1982, 340.
58. **Salter, P. J.**, Interactive effects of agronomic variables on marketable yield of leeks, *J. Agric. Sci.*, 106, 455, 1986.
59. **Meulendijks, J.**, Prei ter plaatse zaaien, *Groenten Fruit*, 23, January, 1987, 88.
60. **Folster, E. and Kling, M. F.**, Trials with direct sown leeks, *Hortic. Abstr.*, 47, 3536, 1977.
61. **Tahvonen, R.**, *Botrytis porri* Buchw. in leek as an important storage fungus in Finland, *Hortic. Abstr.*, 51, 1146, 1981.
62. **Hoftun, H.**, Storage of leeks, *Meld. Nor. Landbrukshoegsk.*, 57, nr. 39, 1978.
63. **Annon.**, Prei. Produktgegevens Groenten en Fruit, Meded. nr. 30, Sprenger Inst., Wageningen, The Netherlands, 1983.

Chapter 10

RAKKYO *Allium chinense* G. Don

Masao Toyama and Izumi Wakamiya

TABLE OF CONTENTS

I. INTRODUCTION

Rakkyo (*Allium chinense* G. Don, *Allium bakeri* Rgl.) is a perennial herb belonging to the Alliaceae (Figure 1). It is native to the central eastern regions of China, the wild species being found in an area extending from China to the Himalayan region of India.[1] According to ancient documents it has been cultivated in China from early times. In Japan, 10th Century writings indicate that it was used for medicinal purposes, and 17th and 18th Century scripts record its cultivation and use.[2] Today it is widely cultivated in China and Japan, these being the main rakkyo-growing countries. This crop is little known in the western world, although Mann and Stearn[3] have previously written a review about it in English.

II. BOTANICAL ASPECTS

A. MORPHOLOGY

The rakkyo bulb is elliptical in shape, 2 to 4 cm long, with a tapering top where the leaves are easily recognized (Figure 1). The external color is purplish or grayish white, and the bulb is covered by a semitransparent dry membranous skin. The hollow angular leaf blade is 30 to 60 cm long, with 5 ridges on its surface, and somewhat resembles a "D" shape in transverse section. The leaves usually wither around July when the weather in Japan becomes warm and humid. At this stage the bulbs are harvested and are either used as food or kept as planting material. During summer the flower stalk develops inside the bulb (Figure 1), and later in the summer the scape grows out of the old bulb, whose leaves by then have already withered. A new lateral bulb grows rapidly in autumn; thus, the scape appears to stand upright outside the new leaf sheaths. Rakkyo flowers in the fall. The scapes are 40 to 60 cm long and slightly irregular in cross section. Each scape bears between 6 and 30 flowers. The pedicels are about 3 cm long and are much longer than the flowers, such that the flowers are somewhat nodding (Figure 1). The inflorescence is a spherical to bell-shaped umbel, and the flowers are purple in color, tinged with red. The flower has six stamens, with filaments longer than the perianth. Anthers are yellowish brown with a purplish tinge. The style stands upright from the ovary conspicuous in the flower.[4]

B. CYTOLOGY

The chromosome number in the root apex of rakkyo was found by Katayama[5] to be 2n = 16. However, Morinaga and Fukushima[6] and Kurita[7] reported a chromosomal number of 2n = 32. Katayama[8] and Kurita,[7] confirmed it to be an autotetraploid by their observations of cell division in root tips. However, most of the commercial cultivars are tetraploid and infertile. Therefore, they are multiplied vegetatively by offsets.[9]

C. VEGETATIVE GROWTH PATTERN

After planting, in late summer or early fall, the growing point of a bulb divides to form a cluster of sprouting shoots. These multiply further to give in the spring more tiller shoots, each of which develops into a bulb in the early summer (Figure 2). When planting is delayed, the growth is poor, only a few primary dividing bulbs develop in autumn, and few divided bulbs are formed in spring (Table 1). Rakkyo bulbs are usually harvested once a year in August to September following the withering and death of the leaves and after the bulbs become dormant.

The biennial growth habit of rakkyo was studied by Aoba,[11] by Sato and Tanabe[12] and by others.[3] All these authors obtained similar results under natural conditions although they used different cultivars. A summary of the findings is as follows. After planting in August, both leaf number (Figure 3) and leaf weight increases rapidly until October. There is little increase in size during late autumn and winter. Growth renews in spring, and declines in

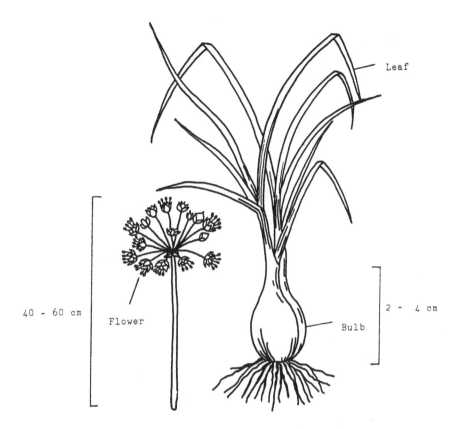

FIGURE 1. Schematic drawing of rakkyo plant showing root, bulb, leaves, and open inflorescence. The expanded leaf length is 30 to 60 cm.

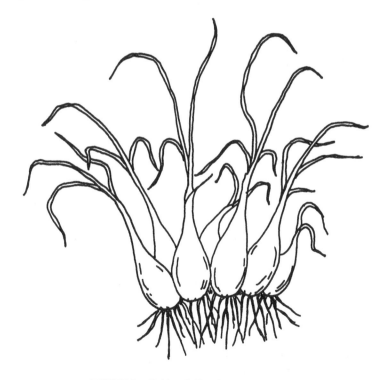

FIGURE 2. Rakkyo bulbs after multiplication.

TABLE 1

Effects of Planting Date and Soil Type on the Growth and the Yield of Rakkyo. Plants were Harvested in August to September.[10]

Planting date	Type of soil	Tiller bud number/plant	Plant weight (g)	Bulb weight (g)	Yield/10 m² (kg)
		cv. Fukui Zairai			
Aug. 30	Clay	10.1	49.8	4.9	18.0
	Sand dune	7.6	41.0	5.4	14.7
Sept. 9	Clay	9.6	44.7	4.8	16.2
	Sand dune	6.0	28.8	4.8	10.5
Sept. 19	Clay	6.8	33.8	5.0	12.3
	Sand dune	5.2	30.0	5.8	10.8
Sept. 29	Clay	5.5	30.6	5.6	11.1
	Sand dune	5.2	33.8	6.5	12.3
		cv. Tama Rakyo			
Aug. 30	Clay	23.3	51.2	2.2	18.3
	Sand dune	23.6	49.8	2.1	18.0
Sept. 9	Clay	22.3	35.6	1.6	12.9
	Sand dune	21.0	35.8	1.7	12.9
Sept. 19	Clay	22.8	45.4	2.0	16.2
	Sand dune	20.7	42.5	2.1	15.3
Sept. 29	Clay	20.0	39.8	2.0	14.4
	Sand dune	14.9	29.7	2.1	10.8

July, when the leaves wither and die. The root number and root weight increase rapidly from August until October, slowly from October to April, and again rapidly in the spring (Figure 3b). Bulb weight decreases from August until October but increases rapidly from April to midsummer when the bulbs become dormant. The number of lateral buds within a bulb increases slowly during the first growing period, is unchanged from February to May, but increases rapidly in June and July. The growth habit of rakkyo in Japan can therefore be divided into the four following periods (Figure 4):

1. The growth of the root system (September until November)
2. The period of no net growth (December until February)
3. The growth of leaf and root (May and June)
4. The period of bulb development and maturation (May to July)

D. BULB FORMATION
1. Morphology and Development

In cross section (Figure 5), the rakkyo bulb resembles a small onion bulb; however, unlike *A. cepa* it contains no bladeless scales. The blades grow to a certain extent while the sheaths are thickening, so that the shape of the rakkyo bulb is not globular but oval. The rakkyo bulb is formed by the thickening of its leaf sheaths, at which time a retardation of leaf-blade growth occurs. At the same time, more leaf buds and more lateral buds differentiate. Following slow winter growth, new leaves grow vigorously in spring, while a succession of inner lateral buds are being produced (Figure 4). Halfway through bulb formation, the mother plant produces lateral buds between the leaf bases. The leaves of each shoot are formed opposite and alternately. Each daughter bulb produces its leaves on an axis perpendicular to the mother shoot (Figure 5). Therefore, the lateral buds do not grow in the direction of the previous phyllotaxis. The phyllotaxis of the mother plant makes a

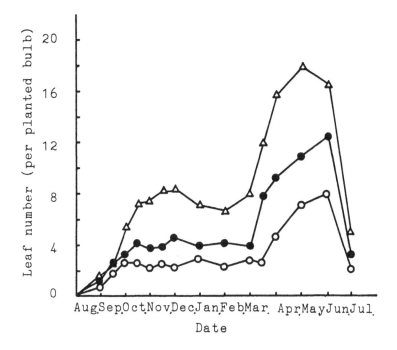

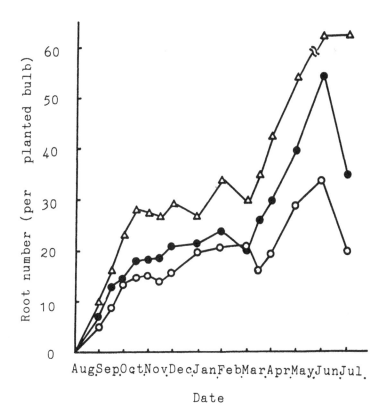

FIGURE 3.   Increase in green leaf number (a) and in root number (b) with time in field-grown rakkyo from planting to maturation.[12] Weight of bulbs at planting: ○ indicates 2 to 3 g; ● 5 to 6 g; △ 10 to 11 g.

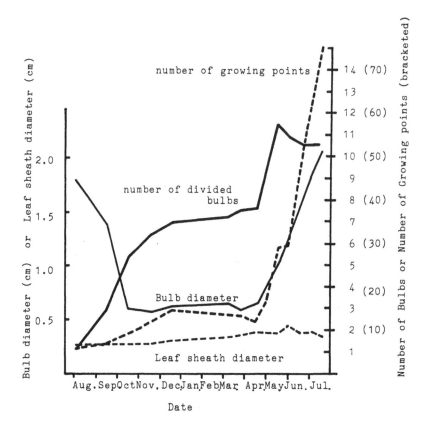

FIGURE 4.   The development of rakkyo bulbs with time. Changes in the number of divided bulbs and in the number of growing points with plant development.[11]

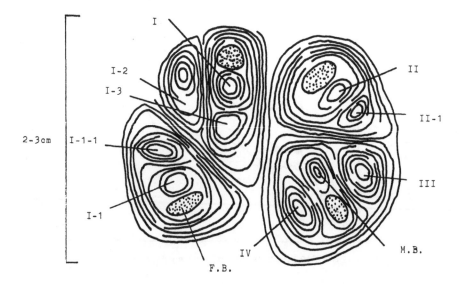

FIGURE 5.   Cross-sectional diagram of rakkyo bulbs illustrating the branching pattern. The bulbs were dissected after 7 months of growth in a glasshouse.[13] I, II, III, IV = first, second, third, and fourth daughter bulbs; I-1, I-2, I-3, and II-1 = granddaughter bulbs; I-1-1 = Great-granddaughter bulb; M.B. = mother bulb; F.B. = inflorescence bud.

TABLE 2
Effects of Photoperiod on the Growth of Rakkyo[11]

Treatment	June 12	June 27	July 13	July 30	Aug. 21	Sept. 1	Sept. 24
			Fresh leaf number				
SD	4.0	3.3	3.0	3.2	1.7	2.0	—
SD-ND	—	—	—	—	2.3	1.5	2.3
ND	4.0	3.3	2.3	2.0	2.5	2.7	8.5
LD	—	4.7	1.3	1.5	—	2.5	—
			Number of growing points				
SD	1.0	1.0	1.4	1.6	1.5	2.0	—
SD-ND	—	—	—	—	2.7	3.0	3.0
ND	2.5	2.3	2.3	2.3	4.0	4.5	7.0
LD	3.8	—	2.3	2.4	—	3.5	—
			Innermost leaf/longest leaf length (cm)				
SD	10/42	19/41	14/33	17/36	9/23	12/29	—
SD-ND	—	—	—	—	16/28	25/31	9/17
ND	29/59	54/63	32/55	14/46	7/25	6/22	18/36
LD	—	35/60	28/44	9/33	—	8/10	—

Note: SD   : short-day treatment from Apr. 15 to Aug. 31.
SD-ND : natural photoperiod from July 26 after short-day treatment.
ND   : natural photoperiod.
LD   : long-day treatment with 16 h from Apr. 15 to June 20.

right-angle turn at the time of initiation of the second or third leaf of each lateral bud. As new leaves differentiate at the growing point, outer older leaves wither. The withering of the leaf sheath which subtends an axillary bud is indicative of the bud's further multiplication.

Rakkyo multiplies by forming lateral buds within the mother bulb, from planting time in August until December. In some cases, however, there is no bulb multiplication until the following spring, probably because winter temperatures are too low, and only then do plants differentiate into daughter bulbs. These bulbs then thicken during the summer maturation period, when the plant height decreases, the leaf blades partly decay, and the weight per bulb increases. During this period, the percentage dry matter and the specific gravity increase to 30% and 1.07, respectively.

2. Environmental Control of Bulb Formation

Photoperiod and temperature are the main environmental factors that affect bulb formation. The effect of photoperiod on bulbing of cv. "Rakuda" was studied by Aoba,[11] who compared the effects of natural photoperiods, long days of 16 h, and short days of 8.5 h. (Table 2). In short photoperiods, the ratio of leaf-blade length to leaf-sheath length is higher in young leaves than is the case in natural photoperiods (Figure 6). Younger leaf blades are longer under short-days than they are under natural days (leaves 5 and 6), while in older leaves (leaves 1 to 4), those exposed to natural photoperiods are longer than those from short days (Figure 6). Retardation of leaf-blade development and bulbing occurred earlier in the long-day plants than in those exposed to short-day (Figure 7). In short-days the thickening of sheaths was less pronounced, scale leaves (the thick leaf bases which form the bulb) were absent, and vegetative growth continued with increased leaf-blade production in July and August. It was therefore inferred that long-days induce the thickening of leaf sheaths and that increase in natural daylength promotes bulb formation from May onward.

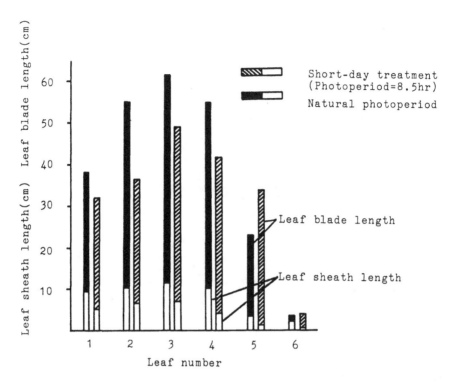

FIGURE 6. Effect of photoperiod on the length of sheaths and blades of successive leaves. 1 = oldest leaf; 6 = youngest leaf.[11]

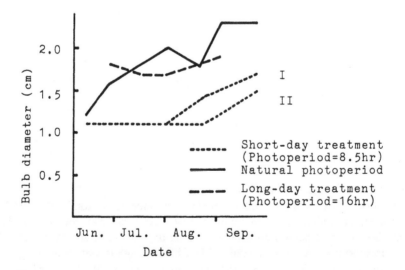

FIGURE 7. Effect of photoperiod on bulb diameter.[15] I had a short-day treatment from April 15 to July 26; II had a short-day treatment.

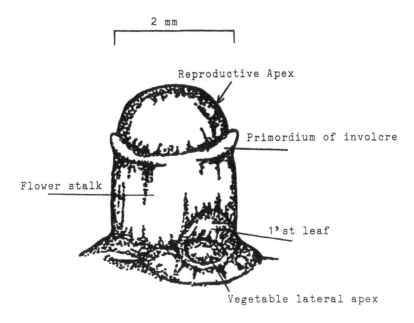

FIGURE 8. Early stage in inflorescence differentiation.[13]

The effect of temperature was studied by Saito.[14] He found that bulbs form and thicken in a temperature range of 15 to 25°C.

E. INFLORESCENCE DIFFERENTIATION
1. The Process
Inflorescence differentiation in rakkyo is shown in Figure 8 and can be summarized as follows. When an inflorescence bud differentiates from late May through June, a new growing point is formed at the base of the flower stalk and develops into a lateral bud. This lateral bud may later form a bulb and become dormant or it may grow and divide to give rise to further daughter plants and form a small cluster. In the initial steps of flower-bud differentiation, the primordium of the involucre is formed as a ridge near the top of the rather columnar growing point. Later the columnar receptacle becomes fully covered with the bracts, and some outgrowths are formed on the surface of the receptacle. These develop into a number of small swellings, each of which is the primordium of one floret. The sequence of floret differentiation begins with the formation of the three outer tepals followed by the three inner ones. Then, the six stamen primordia are initiated within the perianth, and finally the primordium of the pistil is differentiated in the center.

Flower-bud development is similar to that in other alliums except that the leaf blades of the main axis wither and die in the summer before the inflorescence emerges. Newly developed leaves which appear in mid- to late September, at the end of the dormancy period, sprout from lateral buds, but the flower stalk is terminal on the plant's old main axis. The scape therefore appears to be off-centered, i.e., it is not surrounded by the current succession of leaves. In comparison with other cultivated alliums, rakkyo is unique in differentiating its flower primordia during warm season dormancy. The inflorescence opens in the autumn, during mid- to late September. However, no seeds are normally set.[9]

2. Environmental Factors Affecting Inflorescence Differentiation
Aoba[15] investigated the effect of environmental factors on the growth of the inflorescence. He employed the same photoperiodic treatments mentioned earlier for bulb formation (Figure 9). The differentiation and growth of inflorescences and lateral-bud differentiation were promoted by long-days of 16 h, but considerably inhibited by short-days of 8.5 h.

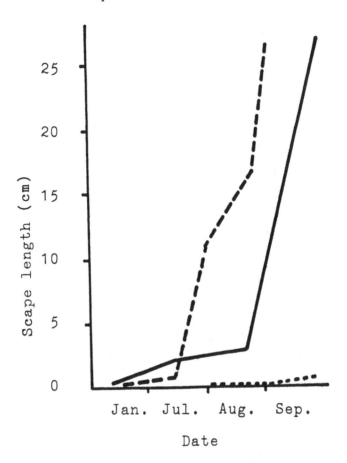

FIGURE 9.   Effect of photoperiod on the elongation of the scape.[15] Long-day treatment of 16 h (— — —) was given from April 15 to June 20. Short-day treatment of 8.5 h (———) was given from April 15 to July 26. Natural photoperiods (———).

3. Relation Between Inflorescence Growth and Bulb Growth

The relationship between flower-stalk development and bulb size was investigated by Sato and Yamane.[16] Small bulbs weighing 3 g or less do not develop flower stalks, while with larger bulbs of about 11 g, approximately 50% of plants flower (Figure 10). Plants which remain vegetative develop more leaves and lateral buds (Figure 11) and also produce heavier bulbs than those which bear flower stalks. Both the number of lateral buds and bulb yields are higher in plants which mature without forming flowers than in those which were disbudded or in which the scape was cut before full flower development. This increase in lateral buds and yield in the absence of flowering is probably due to strong leaf growth (high leaf number) in the spring. This allows increased accumulation of reserves in the bulb since no energy is consumed for the differentiation and growth of the reproductive organs.

III. CULTURAL ASPECTS

A. THE EFFECT OF CLIMATE

The climatic requirements of rakkyo have not yet been well defined. It can be grown in the temperate zones and in the tropics. There are few reports on the effects of environmental conditions on the commercial yield of rakkyo. Toyama et al.[17,18] reported that differences in the environmental conditions between south-sloping and north-sloping fields in the Tottori

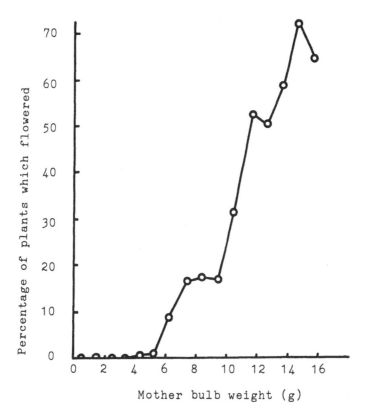

FIGURE 10.   Relationship between the mother bulb size and the percentage of plants developing flowers.[12]

Sand Dunes greatly affected the yield. The south-sloping fields gave markedly higher yields than north-sloping fields. The number of lateral buds per bulb, number of leaves per bulb, and leaf weight were, respectively, 1.4, 1.5, and 2.6 times greater in the south-sloping fields than in the northern ones (Table 3). Environmental factors affecting yield are thought to be air and soil temperatures, relative humidity, and light intensity, all of which were lower in the north-sloping fields. Wind velocity could also adversely affect yields, since in the north-sloping fields it was higher than that in the south-sloping fields throughout the year, and especially so in winter when wind speeds are high and plants can be physically damaged. Toyama et al.[19] built windbreak nets in north-sloping fields and found weight increases in bulbs of 1.5 and in leaves of 2.6 times of those in the control (Table 4). Toyama et al.[20,21] examined the effects of temperature and light intensity on photosynthesis and transpiration of rakkyo in controlled environments. Photosynthesis decreases as the temperature increases from 15 to 35°C whereas transpiration increases. Consequently, there was a decrease in water use efficiency (Figures 12 to 14).

B. SOIL CONDITION
1. Soil Type

Because of its vigorous growth habit and strong ability to absorb nutrients, rakkyo produces high yields of small high-quality bulbs even in poor soils. These characteristics make rakkyo a suitable crop for reclaimed soils and sand dunes. Rakkyo requires well-drained soil. In fertile soils such as clay loam or volcanic ash, the yield increases (Table 1), but the bulbs become too large and therefore have a lower market value. Moreover, these bulbs are too soft and are much inferior in quality to those grown on sand dunes.

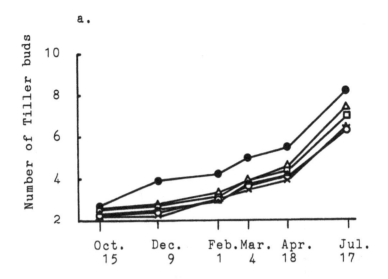

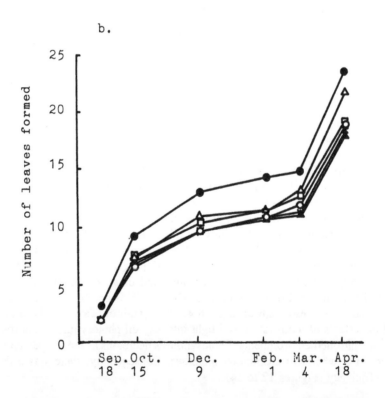

FIGURE 11. Changes of (a) tiller bud number,[12] and of (b) leaf[16] number, with the growth of rakkyo as influenced by the presence or absence of inflorescence. •——• Flowers naturally not present; ○——○ flower stalks present; △——△ flower buds removed; X----X scapes removed; □——□ covering with transparent polyethylene; ▲——▲ covering with black polyethylene.

TABLE 3
Comparison Between the Yield of Rakkyo in a Field Sloping to the South and that in a Field sloping to the North[18]

Direction of slope	Seed bulb number	Leaf number		Leaf weight (g)		Bulb number/ seed bulb	Bulb weight (g)/ seed bulb (g)	Bulb weight (g)
		/Seed bulb	/Bulb (g)	/Seed bulb	/Bulb (g)			
North	83	10.6	2.4	11.8	2.6	4.4	21.0	4.7
South	79	16.1	2.7	30.9	5.2	6.2	32.3	5.3

TABLE 4
**Effect of Windbreaks on the Growth of Rakkyo
on North-Sloping Sand Dune Fields[17]**

Growth measurements	Distance from Windbreak (m)		
	1	5	10
Total plant weight (g)	71	60	49
Plant height (cm)	60	54	51
Bulb cluster weight (g)	32	26	20
Individual bulb weight (g)	6	6	5
Leaf number/plant	25	23	20
Leaf number/bulb	5	6	6
Leaf weight/plant (g)	36	24	19
Leaf weight/bulb	8	6	5
Number of tiller buds	5	4	4

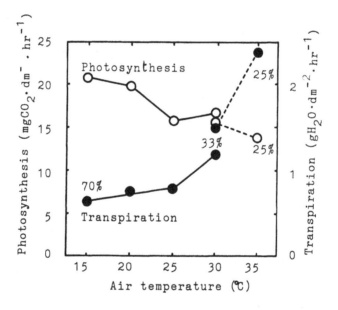

FIGURE 12.   Effect of temperature on photosynthesis ○ and transpiration ● per unit leaf area of rakkyo leaves under a constant light intensity. The experiments were carried under constant light intensity of 22 klx. The percentage numbers in the figure represent the relative humidity.[21]

2. Mineral Uptake

Sato and Tanabe[22] investigated the change in the mineral content of rakkyo with growth. Nitrogen, potassium, magnesium, and phosphorous uptake rates of shoots are high in the spring and early summer, especially in May and June. Nitrogen, magnesium, and particularly potassium are accumulated in the roots in the autumn and winter, but at slower rates (Figure 15). Although this mineral intake is not accompanied by an increase in dry matter or leaf growth (Figures 3 and 4), it probably has an important role, since the root system is developed in these cold months and thereafter it supplies the vegetative growth in the spring. In spring the mineral content of the roots declines sharply (except for magnesium), but the mineral content of the bulb increases as it swells (Figure 15). The high potassium content in winter

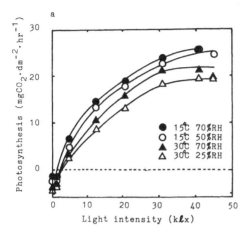
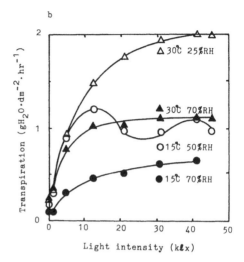

FIGURE 13. Effect of light intensity on photosynthesis (a) and transpiration (b) per unit leaf area under two temperature and two relative humidity regimes.[21]

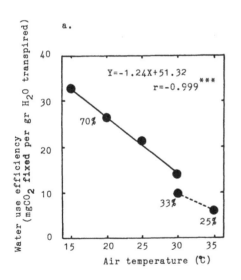
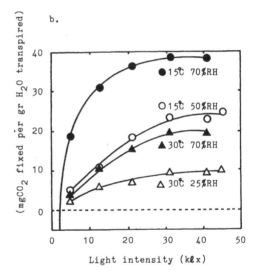

FIGURE 14. Effect of temperature (a) or light intensity (b) on water use efficiency of rakkyo as measure by CO_2 fixation per unit water transpired. The experiments in (a) were carried out at constant light intensity of 22 Klx. Percentage numbers in the figure refer to the relative humidity.[21]

(Figure 15a) suggests a close relationship between this and the winter hardiness of the rakkyo, similar to many other crops.

3. Leaf Chlorosis Caused by Zinc Deficiency

Chlorosis is one of the major physiological diseases of rakkyo and was described in detail by Yanagawa et al.[23-26] The symptoms are as follows: the plant turns yellow, becomes shorter, and the leaves thicken. The new leaves cluster, curl, and become malformed. Following severe chlorosis at the tips, the outer leaves wither and drop off, except for those with mild symptoms. As a result, bulb yields are low and the roots become very weak. Chlorosis symptoms are especially strong in the spring between mid-April and mid-May.

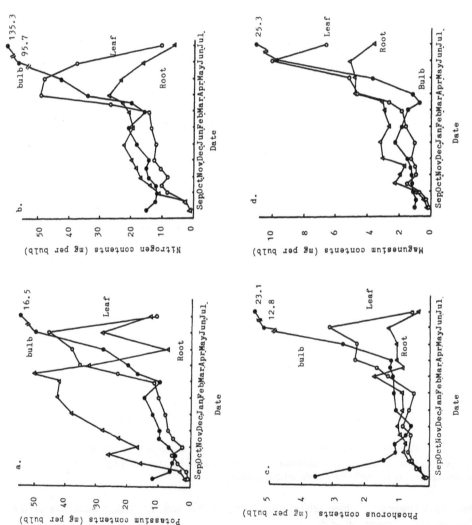

FIGURE 15. Change of content of potassium (a), nitrogen (b), phosphorous (c), and magnesium (d) in rakkyo bulbs, leaves, and roots during field growth.[22]

The phenomenon is sporadic and is initially limited to small areas, but spreads slowly to the rest of the field. A high incidence of chlorosis can occur in the first seasons after the establishment of a rakkyo field. Year by year the number of outbreaks decreases and the severity of symptoms declines. Chlorosis is probably caused by a shortage of zinc to the roots, as a result of a decrease in zinc supply in the soil. This may be caused by increased pH and by the antagonism between phosphate and zinc. Plants grown on shallow (10 cm) or on deep soil (50 cm) show a high frequency of severe chlorosis symptoms. These are attributed to both above-mentioned reasons: low Zn content in the humus and high pH, since zinc can easily form insoluble compounds in these conditions. Three foliar applications at 5-d intervals of 0.3% zinc sulfate at 10 l/100 m² plus a surfactant drastically reduced both, the symptoms and the damage (Table 5).

C. IRRIGATION

Rakkyo is resistant to drought and can be grown with no irrigation. However, it is almost impossible to obtain good growth in the dry months from May to September in Japan, especially in sand dune fields.

Sato and Tanabe[27] showed that spring irrigation of sand dune fields results in an increase of both bulb weight and yield. Autumn irrigation promotes mainly lateral bud formation and growth (Table 6). Ueda[28] repeated these experiments and showed a 20 to 30% increase in bulb yield following spring and autumn irrigations of sand dune fields.

D. CULTIVARS

There are a great number of rakkyo cultivars originating in both China and Japan. Most of them have not been fully characterized or evaluated. Recognized cultivars can be broadly divided into two groups, according to the size of their bulbs. The majority of the large-bulbed cultivars originated in Japan, and most of them are quite similar. Descriptions of some small- and large-bulbed cultivars are given below.

"**Tama Rakkyo**" — This is representative of the small-bulbed group which was introduced to Japan from Taiwan by the horticulturalist Kumazawa. The plant is short; the leaves are thin and curled. The number of side bulbs produced from one mother bulb is very high, namely, 10 to 25, but they are small (1.5 to 2.9 g per bulb). The bulbs are white, soft, almost odorless, have a firm neck, and a high proportion of the bulbs are suitable for processing. However, the plants are highly susceptible to the *Rhizoglyphus* mite.

"**Rakuda**" — This is a large-bulbed group which is cultivated all over Japan. The high yielding plants are large (4 to 10 g per bulb in the sand dune fields) and taller than cv. "Tama Rakkyo". The plants are strong and develop large foliage, but have fewer side bulbs (6 to 9 bulbs per year). The large bulbs are elliptical with a long neck. "Rakuda" has a high yield potential, but needs some selection for more uniformity since there is much variation between individual bulb weights.

"**Yatsufusa**" — These produce more side bulbs with firmer neck than cv. "Rakuda" but of lighter weight. They are very suitable for processing. Both yield and quality are low. In the clay soils of the Kanto districts (Japan) it matures earlier than "Rakuda".

IV. CULINARY ASPECTS

A. COMPOSITION

The general chemical ingredients of the parts of Rakkyo are given in Table 7, and the food values in the edible parts of rakkyo are shown in Table 8. The content of crude protein (protein conjugated with other compounds) in all parts of the plant is similar, ranging from 15 to 18%, but the content of true protein in the bulbs is low compared to the other parts. The highest values of crude fiber (40%) are found in the roots, and lowest (3%) in the bulbs.

TABLE 5
Effects of Spraying Zinc Sulfate Solution on the Yield of Rakkyo Showing Chlorosis Symptoms[13]

Chemical	Concentration (%)	Times of spraying	Leaf weight (kg/5 m²)	Bulb number (kg/5 m²)	Bulb yield (kg/5 m²)	Yield as % of healthy plant yield	Mean weight bulb (g)
Zinc sulfate	0.1	1	4.31	1878	2.40	60	1.3
	0.1	3	5.02	1707	2.97	74	1.7
	0.3	1	5.16	1650	2.88	72	1.7
	0.3	3	5.49	1768	3.56	89	2.0
	1.0	1	6.14	1800	3.54	89	1.9
	1.0	3	5.80	1690	3.38	85	2.0
Control	—	—	1.55	1470	1.21	30	0.8
Healthy plants	—	—	5.00	1494	3.99	100	2.7

TABLE 6
Effects of Irrigation of Sand Dune Fields on the Growth and Yield of Rakkyo[27]

Irrigation season	Date of irrigation			Water applied (mm)			Tiller bud number	Bulb weight /plant (g)	Bulb weight (g)	Yield (kg/m²)
	Aug.	Sept.	May	Aug.	Sept.	May				
Spring and autumn	15	11	10	150	110	100	8.6	64.2	7.47	4.11
Autumn	15	11	—	150	110	—	8.5	57.9	6.80	3.71
Spring	—	—	10	—	—	100	6.8	63.5	9.30	4.06
Without irrigation	—	—	—	—	—	—	7.2	52.7	7.27	3.37

TABLE 7
Chemical Components of Each Part of *Allium bakeri* Regal (Rakkyo) Percent of Dry Matter.[9] At the Sampling Times 37, 22, 35, and 6%, Respectively, of the Dry Weight was in Green Leaves, Middle Leaves, Bulb, and Root

	Green leaves	Middle leaves	Bulb	Root
Crude ash	11.67	9.22	4.30	18.95
Alkalinity of ash[a]	8.18	5.56	1.67	4.28
Crude protein	17.57	15.38	18.38	16.29
True protein	8.80	8.13	4.71	8.32
Nonprotein nutrient	1.40	1.16	2.19	1.27
Crude fat	5.09	4.75	3.20	3.08
Crude fiber	29.82	20.04	3.27	40.28
N-free extract	35.83	50.61	70.85	21.40
Pentosan	3.71	5.83	7.16	4.45
(Moisture of fresh sample)	91.67	87.35	76.70	77.35

[a] The alkalinity of the ash was measured by titration with N/10 HCl. The figures are milliliters of HCl required to neutralize 1 mg of ash.

The bulbs have the highest content both of N-free extract (71%) and of pentosan (a type of hemicellulose). Soluble sugars are high both in bulbs and in the middle leaves. They are divided equally between reducing and nonreducing sugars. Most of the reducing sugars are glucose and fructose. The nonreducing sugars are fructo-oligosaccharides and fructans, all of which are characteristics of *Allium* species.[30]

Allyl sulfides are the source of the characteristic odor of rakkyo much as in other *Allium* species. (See chapter on ''Flavor Biochemistry''.) The steam vapor volatiles of rakkyo include thiolanes, alcohols, ketones, and the oily compound 2,3-dihydro-2-hexyl-5-methyl-furan-3-one (approximately 20%) and others.[31]

B. PROCESSING

Most rakkyo is consumed as pickles. The preferred quality pickles are made from small bulbs. The bulbs are usually first steeped in brine for several days before being transferred to vinegar, sake, and sugar to produce sweet pickle or vinegar and salt to produce a sour pickle. Soy sauce may be added to produce an amber-colored pickle. The pickles are transported in bulk in barrels, and they may then be bottled or canned and heat sterilized. Pickled rakkyo is exported from Japan to many countries, particularly to the U.S. and to Southeast Asian countries.[9]

TABLE 8
Ingredient of Rakkyo as Food (Values Per 100 g of Eatable Fresh Weight)[29]

	Energy (kcal)	Moisture (g)	Protein (g)	Carbohydrate (g)		Ash (g)	Mineral (mg)				Vitamin				
				Saccharides	Fiber		Calcium	Phosphorous	Iron	Sodium	A (µg)	B_1 (mg)	B_2 (mg)	Niacin (mg)	C (mg)
Fresh	52	86.2	0.6	12.6	0.3	0.2	6	15	0.2	1	0	0.03	0.02	0.9	10
Sweet-and-sour pickles	126	64.9	0.8	31.1	0.6	2.4	16	23	2.4	940	0	0.01	0.01	0.2	0

REFERENCES

1. **Iwasa, S.,** Greens, in *Tropical Vegetables,* ed., Tropical Agricultural Center, Yokendo, Tokyo, 1980, 365.
2. **Ageta, H., Ando, T., Ikuse, M., et al.** Rakkyo, in *Hirokawa Encyclopedia on Medicinal Plants,* Kijima, M., Shibata, S., Shimomura, T., and Higashi, T., Eds., Hirokawa Bookstore, Tokyo, 1980, 380.
3. **Mann, L. K. and Stearn, W. T.,** Rakkyo or ch'iao t'on (*Allium chinense* G. Don, syn, *A. bakeri* Regal), a little known vegetable crop, *Econ. Bot.,* 14, 69, 1960.
4. **Namikawa, I.,** *System of Agriculture, Horticulture. Leaf Vegetables,* Yokendo, Tokyo, 1952, 212.
5. **Katayama, Y.,** The chromosome number in *Phaseolus* and *Allium,* and an observation on the size of stomata in different species of *Triticum, J. Sci. Agric. Soc.,* 303, 562, 1928.
6. **Morinaga, T. and Fukushima, E.,** Chromosome number of cultivated plants. III, *Bot. Mag. (Tokyo),* 45, 140, 1931.
7. **Kurita, M.,** On the karyotypes of some *Allium* species from Japan, *Mem. Ehime Univ. Sec. II,* 1(3), 179, 1952.
8. **Katayama, Y.,** Chromosome studies in some *Alliums, J. Coll. Agric. Tokyo Imp. Univ.,* 13, 431, 1936.
9. **Jones, H. A. and Mann, L. K.,** *Onions, Their Allies,* Interscience, New York, 1963.
10. **Kunitomi, S. and Kosuga, M.,** Cultivation of rakkyo in Fukui District, *Agric. Hortic.,* 35, 523, 1960.
11. **Aoba, T.,** Process of bulb formation and effects of day-length on bulb formation and inflorescence initiation in rakkyo (*Allium chinense* G. Don), *Bull. Fac. Agric. Yamagata Univ.,* 5, 2, 1967.
12. **Sato, I. and Tanabe, K.,** Studies on growing Baker's garlic (*Allium Bakeri* Regal) in the sand dune field. I. On the growth habit, *Bull. Sand Dune Res. Inst. Tottori Univ.,* 9, 1, 1970.
13. **Yamada, Y.,** Studies on the breeding of Allium, in *Bull. Fac. Agric. Saga Univ.,* 6, 35, 1957.
14. **Saito, T.,** *Vegetable Crop Science, Part of Pulse, Root Vegetables,* Nosangyoson Bunka Kyokai, Tokyo, 1983, 282.
15. **Aoba, T.,** Studies on bulb formation and dormancy in onion, *Bull. Fac. Agric. Yamagata Univ. Agric. Sci.,* 4(3), 1, 1964.
16. **Sato, I. and Yamane, M.,** On the relation between yield and flowering in Baker's garlic (*Allium Bakeri* Regal), *Bull. Sand Dune Res. Inst. Tottori Univ.,* 22(2), 1, 1976.
17. **Toyama, M., Takeuchi, Y., and Nishiyama, O.,** Meteorological conditions which affect to the yield of rakkyo (Baker's garlic, *Allium Bakeri* Regal) cultivated in a sandy field, *Bull. Sand Dune Res. Inst. Tottori Univ.,* 22(2), 47, 1983.
18. **Toyama, M., Takeuchi, Y., Nakade, Y., and Sugimoto, K.,** Relationship between meteorologic conditions and yield of rakkyo (Baker's garlic, *Allium Bakeri* Regal) in sandy fields, *Bull. Sand Dune Res. Inst. Tottori Univ.,* 30(2), 283, 1983.
19. **Toyama, M., Takeuchi, Y., Kuroyanagi, N., and Sugimoto, K.,** Water use efficiency and leaf conductance of rakkyo (Baker's garlic, *Allium Bakeri* Regal) under high air temperature and low relative humidity conditions by change of light intensity, *Bull. Sand Dune Res. Inst. Tottori Univ.,* 23, 11, 1984.
20. **Toyama, M., Takeuchi, Y., Nishiyama, O., and Sugimoto, K.,** Effects of wind breaking net to yield of rakkyo (Baker's garlic, *Allium Bakeri* Regal), *Bull. Sand Dune Res. Inst. Tottori Univ.,* 30(2), 276, 1983.
21. **Toyama, M., Takeuchi, Y., Kuroyanagi, N., and Sugimoto, K.,** Relationship between photosynthesis, transpiration and water use efficiency and air temperature, relative humidity and light intensity, *J. Jpn. Soc. Hortic. Sci.,* 53(4), 444, 1985.
22. **Sato, I. and Tanabe, K.,** Studies on growing Baker's garlic (*Allium Bakeri* Regal) in the sand dune field. II. Changes of mineral contents following the growth, *Bull. Sand Dune Res. Inst. Tottori Univ.,* 9, 9, 1970.
23. **Yanagawa, T. and Nakamura, K.,** Studies on the chlorosis of Baker's garlic in sandy field. I. Research on the actual condition of its occurrence, *Bull. Tottori Agric. Exp. Stn.,* 10, 66, 1970.
24. **Yanagawa, T., Ueda, H., and Tanaka, A.,** Studies on the chlorosis of leaves in Baker's garlic grown in sandy soils. II. Recovery response from the chlolorosis by zinc applications, *J. Jpn. Soc. Hortic. Sci.,* 40(2), 57, 1971.
25. **Yanagawa, T., Ueda, H., and Tanaka, A.,** Studies on chlorosis of leaves in Baker's garlic grown on sandy soils. III. Investigation on soil zinc status in the field, *J. Jpn. Soc. Hortic. Sci.,* 40(2), 63, 1971.
26. **Yanagawa, T. and Fujii, S.,** Studies on the chlorosis of leaves in Baker's garlic grown on sandy soils. IV. Foliar sprays for correcting zinc deficiencies in garlic plants, *J. Jpn. Soc. Hortic. Sci.,* 41(1), 61, 1972.
27. **Sato, I. and Tanabe, K.,** Studies on growing Baker's garlic (*Allium Bakeri* Regal) in a sand dune field. IV. Effects of irrigation in the autumn and in the spring on the growth and yield, *Bull. Sand Dune Res. Inst. Tottori Univ.,* 11, 11, 1972.
28. **Ueda, H.,** Irrigation and soil management in rakkyo fields, preparation of ground basis and introduction of irrigation to the fields in Fukube Sand Dune, in *Fields Soil,* 13(10), 103, 1981.
29. **Yamaguchi, F., Yoshida, K., and Takiguchi, N.,** *Guide Book on Vegetables,* Public Department, Kagawa Nutrition College Tokyo, 1982, 64.

30. **Mizuno, T., Yokoyama, M., and Kinpyo, T.,** Studies on the carbohydrates of *Allium* species. VI. Free sugars and polysaccharides of *Allium bakeri* Regal, *Bull. Fac. Agric. Shizuoka Univ.,* 11, 117, 1961.
31. **Kameoka, H., Iida, H., Hashimoto, S., and Miyazawa, M.,** Sulphides and furanones from steam volatile oils of *Allium fistulosum* and *Allium chinense, Phytochemistry,* 23(1), 155, 1984.

Chapter 11

CHINESE CHIVES *Allium tuberosum* Rottl.

Susumu Saito

TABLE OF CONTENTS

I. ORIGIN

The scientific name of Chinese chives is *Allium tuberosum* Rottl. It is known as "Kau tsai" in China and Nira in Japan.[1,2] It is believed to have originated in China. It grows naturally in central and northern parts of Asia, and is cultured in Japan, China, Korea, India, Nepal, Thailand, and the Philippines. It is a perennial plant and both the leaves and the inflorescences are eaten. It has also long been used as a herbal medicine, and it is considered effective for recovery from fatigue.

II. MORPHOLOGY

Chinese chives spreads via rhizomes. These bear the long, approximately 5-mm-wide, flattish, and slightly keeled leaves. Spread by rhizomes results in the formation of dense clumps of leaves. The rhizomes may branch and are covered by a pale, brown, fibrous reticulum which is formed from the remains of old foilage leaf bases. The rhizome resembles that of the common bearded iris. Thick roots emerge from the underside of the rhizome. These may persist for much longer than is the case with garlic or onion. The leaves typically bend downward at the tip, are solid in transverse section, and do not have a ligule. In summer lateral vegetative bulbs form at the same time as flowering. The extent of bulbing varies with genotype. Some bulbs are large enough to perrenate plants from season to season, but in others the bulbs are too poorly developed for this. The rhizome rather than bulbs is the main storage tissue.

The inflorescence is carried on a tall, very erect scape which is solid in cross section and has two or more sharp angles. The inflorescence is a flat-topped umbel typically containing two cymes of approximately 20 flowers each. A short spathe remains at the base of the umbel. The flowers are white with a green line on the outside. They open into wide star-shaped flowers and are fragrant. Inflorescences initiate from the terminal bud. A lateral bud in the axil of the leaf immediately below the inflorescence continues the growth of the vegetative axis. Every two to four leaves the rhizome produces another inflorescence, and, typically, each rhizome produces two to four inflorescences per year.[1]

III. PHYSIOLOGY

Short photoperiods may induce dormancy in Japanese cvs. When plants are transferred in autumn to a greenhouse and exposed to 6-, 12-, 14-, and 16-h photoperiods, only those exposed to 16 h continue leaf growth, and after about 30 d leaf elongation ceases in the other treatments.[3] When plants are transferred to a greenhouse on successive dates during the winter, the leaf growth of plants after sprouting is faster the longer the previous exposure to low temperatures.[4] This indicates that low temperatures favor the breaking of dormancy. In this experiment Japanese cultivars sprouted 7 to 13 d after transfer to the greenhouse, but south Asian varieties sprouted after only 1 d.

Chinese chives differentiate inflorescences in long photoperiods. When exposed to 8.5-h days from early May to late July, Japanese cvs. did not flower, but some South Asian cvs. did flower. When the Japanese cvs. were transferred to 16-h photoperiods after this treatment, they began to develop flower stalks after 20 d.[5] The critical photoperiod below which flowering cannot occur is estimated as between 8.5 and 12 h for some cvs.[5] and between 12 and 14 h for others.[6] Many cvs. do not flower in response to long photoperiods in the first year after sowing; however, cvs. from south Asia will flower in their first year when they have only about six leaves.[5]

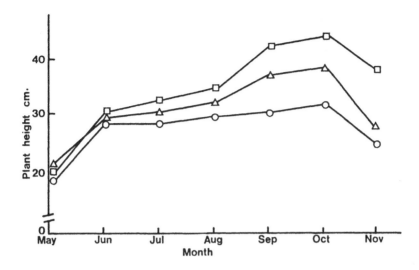

FIGURE 1. (A) Changes in Chinese chives height with time in the open field and under protection. ○ = control (outdoors); △, 0.06-mm vinyl cover; □, 0.20-mm vinyl cover.[7] (B) Chinese chives growing in a vinyl greenhouse.

IV. GROWTH AND PRODUCTION

The optimum temperature for Chinese chives culture is 20°C. It grows well in rather cool climates. In August and September, the scape grows and it blooms. Assimilates are then transferred both to flowers and seeds, and the vegetative growth declines slightly. Consequently, both yield and quality decrease in these months. Although it can be cultured throughout the year. harvestable quantities are low in the winter, as growth is slowed by the low temperatures.

Chinese chives grow well in most soils, but rich, well-fertilized fields and strains which are well adapted to the climatic zone are essential for production of commercial yields. The outdoor season is quite long. In some places, protected cultivation is practiced, thus supplying the market demand in the winter months (Figure 1). However, the quality and chemical composition of the winter crop raised under vinyl tunnels at low temperatures and light intensities is inferior to the one grown in the spring and autumn in the open. Also, the structure is somewhat different, the plants are longer, and the shape of the stomata is different in the protected crop as compared to the plants grown in the open fields (Figure 2). Yields and plant size vary with cultural techniques (Figure 3). Hence, hydroponic culture considerably promotes root growth and also results in taller plants.

Cultural practices in China have recently been described based on Chinese advisory literature from Liao Ning province in northern China.[8] In Liao Ning, Chinese chives are grown as a perennial crop with a life ranging from 7 years up to 20 or 30 years. Traditionally, seeds are sown 1 cm deep in spring into four, flat-bottomed, 10-cm-deep drills in 1-m-wide beds. During the first season, watering is minimal in order to encourage a deep root system, and leaves are not cut so as to maximize the built-up reserves. Early in the second season, decayed leaves are removed and soil is pulled around the base of the plants, and a further 4 to 6 cm of sandy soil may be put into the furrows in which the plants are growing to blanch the lower parts of the leaves. Three cuts of leaves are made at 20 to 30-d intervals when the leaves reach a height of approximately 20 cm. To allow root reserves to build up, leaves are not cut during the autumn. In later seasons the beds must be raked to remove old and dead roots at the start of growth in the spring. The beds must also be top-dressed with 1 to 5 cm of sandy soil, since the roots have a tendency to work themselves up out of the soil. Fertilizer must be applied between leaf cuts. In the third season and thereafter, a fourth cut of flower stems is also made. More recently, chives have been raised in seedbeds and transplanted to beds in the second season. Clumps of 20 seedlings are transplanted 16 cm apart in rows that are 35 cm apart.

For winter production in Liao Ning province, beds of Chinese chives are covered by glass or plastic structures in winter after the ground has frozen to a depth of 10 to 15 cm and the leaves have died back. The temperature is gradually increased to 20°C in the day and 10°C at night. Beds are raked and soil is earthed up round the plants as they grow. Three harvests are made, starting 40 to 50 d after applying the glass, after which the roots are discarded.

Bleached chinese chives are widely produced in China. In the field light may be excluded and blanched chives produced using "tents" of straw matting, roofing tiles, or dark paper. In Guangzhou province of southern China alternate crops of blanched and green leaves are taken. In the warm climate of this region nine crops a year are harvested. Here, blanching is achieved using clay pots. Blanched chives can be produced in winter in colder areas by excluding light from plants growing in heated greenhouses. Alternatively, clumps of roots may be lifted just before flowering, the flower stalks removed, and the roots grown to sprout in darkened cellars. Three cuts can be made from such roots.

In China seed is saved from plants flowering in their third or fourth season. Good plants which flower in midseason are selected. Seed heads are cut with 30 cm of stem attached, and they are air or sun dried, threshed, and cleaned. The seed is short-lived and is sown in the following spring.

V. COMPOSITION AND NUTRITIONAL VALUES

The main components of the Chinese chives are shown in Tables 1 and 2. Plants grown under protection produce less volatile flavor compounds as measured by gas chromatography using the head-space method (Figure 4). This coincides with results obtained from olfactory sensory testing. Chinese chives are relatively rich in minerals, especially in calcium and iron (Table 2).

Alliaceous vegetables all have a high content of sugars, and Chinese chives are no exception. The leaves contain 3% sugars on a fresh-weight basis and taste very sweet. The main sugars are fructose, sucrose, and glucose. Sugar content is markedly influenced by

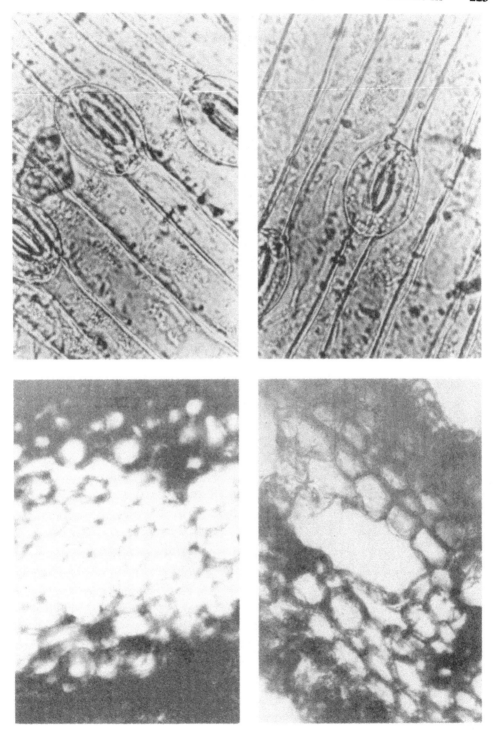

FIGURE 2.   Microscope observations of Chinese chives (A) epidermis and (B) messophyl tissues from plant grown with (right) and without (left) protection. Protected plants were cultivated in vinyl tunnels. (Epidermis magnification × 600; mesophyl magnification × 125.)

A

B

FIGURE 3.   Hydroponic and soil cultures of Chinese chives. (A) Individual plants are planted in a supportive layer; (B) comparative length of plants of the same chronological age developed in hydroponics (left) or in the soil.

TABLE 1
Contents of Moisture and Crude Fiber Per 100 g Fresh Weight of Chinese Chives, Grown with and without Vinyl Cover[7]

	Composition			
	Crude fiber (g)		Moisture (g)	
	Harvest time			
Cultural practice	July	September	July	September
Open field	1.53	1.61	86.2	85.9
Protected cultivation 0.06-mm vinyl	1.47	1.17	86.7	86.6
Protected cultivation 0.20-mm vinyl	1.14	0.91	86.8	88.1

TABLE 2
Chemical Composition and Nutritional Properties of Chinese Chives Per 100 g of Edible Fresh Weight[9]

	Fresh	Boiled
Macronutrients (g)		
Proteins	2.1	2.3
Fat	0.1	0.0
Sugars	2.8	5.2
Vitamins		
A (IU)	1800	2200
Thiamin (μg)	60	40
Riboflavin (μg)	190	110
Niacin (μg)	600	300
C (mg)	25	10
Minerals (mg)		
Ca	50	46
Fe	0.6	0.6
P	32	23
Others		
Energy (kcal)	19	28
Water (g)	93.1	90.8
Fiber (g)	0.9	1.1
Ash (g)	1.0	0.6

cultural conditions and, in the winter, depends on the thickness of the vinyl-film protecting cover. The thicker the film, the lower the light intensity at the plant surface and the lower the chlorophyll and the sugar content (Figure 5). Sugar content tends to be lower in plants raised hydroponically than those grown in the soil (Table 3).

Chinese chives contain much vitamin C and carotene, both of which may have anti-carcinogenic effects. It also contains vitamin B_1 and B_2. The vitamin content again depends on the cultural conditions, and protected culture results also in lower vitamin levels (Table 4, Figure 6), as does hydroponic culture.

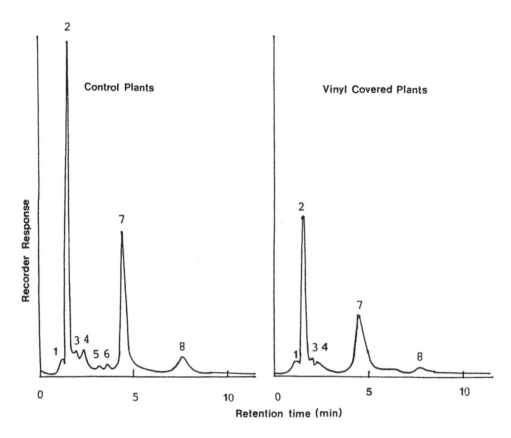

FIGURE 4.   Effect of culture conditions on sulfur volatiles of Chinese chives. Gas chromatogram output graph showing volatiles accumulated in closed vessels after heating (head-space technique). The column 3.0 mm by 2 m was composed of shimalite W, 60 or 80 mesh, and was heated to 100°C.[10]

VI. QUALITY AND GRADE

Thick dark-green leaves are rated as high quality. In Japan, quality is graded by leaf length, flavor, and tenderness. Leaves 23 to 28 cm long are regarded as top quality, 18 to 22 cm long as good, and 15 to 17 as medium quality.

All aspects of quality are affected by the conditions of culture. Winter-produced leaves are better tasting, deeper in color, and thicker the thinner the vinyl cover (Table 5, Figure 7). Plants grown under vinyl tunnels are more tender as measured by a texture meter than those grown in the open (Table 6). Probably because they have larger and thinner-walled cells (Figure 2B).

Chinese chives are used fresh and are not processed. Typically, they are fried together with other vegetables and meat or, are used as seasoning with meat. Vitamin C content is reduced to approximately one third of that in the fresh leaf by boiling and to about 80% of that in the fresh leaf by frying.

VII. STORAGE

The leaves of Chinese chives are very soft and tend to lose their freshness quite rapidly. They are impossible to store for longer than 2 to 3 d, and even then must be stored at 0 to 2°C if quality is to be retained. Carotene content falls to less than half its original value during these few days of cold storage. Nevertheless, such a storage is much superior to

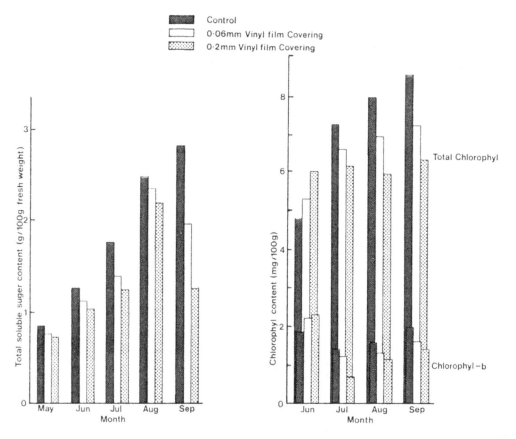

FIGURE 5.   Effect of cultural practice on (left) total soluble sugars of Chinese chives, based on fresh weight, and (right) chlorophyll content, based on fresh weight.

TABLE 3
Vitamin C, Carotene, and Sugar Content Per
100 g of Fresh Weight Chinese Chives Grown
Hydroponically or in the Soil

	Culture	
Composition	**Soil**	**Hydroponics**
Sugars (g)		
Total	3.39	3.00
Reducing	2.66	2.17
Vitamin C (mg)		
Reduced	4.73	5.54
Oxidized	20.09	14.54
Carotene (mg)	4.44	3.11

ambient conditions for maintaining turgidity and carotene content (Table 7). Therefore, a cold chain is required for harvested Chinese chives, i.e., the leaf temperature is dropped immediately after harvest, and the produce is transferred to the market in refrigerated transport.

TABLE 4
Vitamins B₁, B₂, C, and Moisture Content
Per 100 g Fresh Weight of Chinese Chives,
Grown with and without Protection[11]

	Cultural conditions	
Composition	Open field	Vinyl tunnel
Vitamin B_1 (μg)	50	40
Vitamin B_2 (μg)	140	100
Vitamin C (mg)		
Reduced	2.30	1.40
Oxidized	46.3	29.2
Water (g)	91.3	92.6

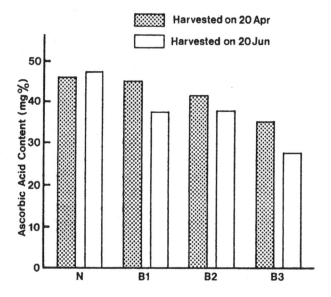

FIGURE 6.   Effect of cultural practice on Chinese chives ascorbic acid content, based on fresh weight.[11] N — control; B_1 — vinyl cover at 0.06 mm; B_2 — vinyl cover at 0.10 mm; B_3 — vinyl cover at 0.20 mm.

TABLE 5
Sensory Test[a] for Quality Evaluation of Chinese Chives, Grown
with and without Protection[11]

	Cultural practice			
		Vinyl tunnel		
Quality component	Open field	0.06 mm	0.10 mm	0.20 mm
Color	4.5 ± 0.22[b]	4.3 ± 0.13	4.0 ± 0.18	1.5 ± 0.19
Flavor	4.0 ± 0.30	2.8 ± 0.35	3.0 ± 0.32	1.2 ± 0.27
Taste	4.0 ± 0.15	3.3 ± 0.2	3.5 ± 0.18	1.4 ± 0.24
Texture	4.8 ± 0.39	4.2 ± 0.41	3.0 ± 0.52	1.0 ± 0.42

[a]   Scale: 5 — excellent; 4 — good; 3 — medium; 2 — rather poor; 1 — very poor.
[b]   Mean ± SE.

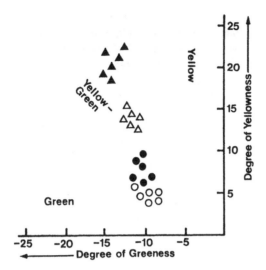

FIGURE 7.   Effect of cultural practice on Chinese chives leaf pigmentation. Refrective values were read using a color-difference-meter. ○ — control; ● — vinyl cover at 0.06 mm; △ — vinyl cover at 0.10 mm; ▲ — vinyl cover at 0.20 mm.

TABLE 6
Firmness Values for Chinese Chives Issues Grown with and without Protection[12]

Cultural practice	Tissue		
	Leaf	Base	Stem
Open field	0.60	0.79	0.40
Vinyl tunnel	0.40	0.43	0.29

Note: Firmness was measured with texturometer. Cylindrical plunger cutter blade = 3 mm diameter; units, kg/cm.² Mean values of five replication are given.

TABLE 7
Changes in Fresh Weight and Carotene Contents of Chinese Chives During Storate[13]

Storage day	Storage mean	
	Cold room	Room temperature
Fresh weight (g)		
0	100	100
1	99.6	84.4
2	99.5	54.9
Carotene (mg/100 g fresh weight)		
0	13.9	13.9
1	10.6	10.2
2	10.0	5.4
3	6.6	6.7
4	5.2	—

REFERENCES

1. **Jones, H. H. and Mann, L. K.,** *Onions and Their Allies: Botany, Cultivation, and Utilization,* Interscience, New York, 1963.
2. **Yamaguchi, M.,** *World Vegetables: Principles, Production and Nutritive Values,* AVI Publishing, Westport, CT, 1983, 203.
3. **Yukawa, T. and Tagai, S.,** Studies on the flowering of genus *Allium.* II. Effects of daylength on flowering and dormancy of Chinese chive, *Agric. Hortic.,* 46, 369, 1972.
4. **Nakamura, E.,** *Allium* — minor vegetables, in *CRC Handbook of Flowering,* Halevy, A. H., Ed., CRC Press, Boca Raton, FL, 1985, 410.
5. **Aoba, T. and Iwasaki, T.,** Studies on the ecological characteristics of Chinese chive. II. Differentiation and development of inflorescences, *Agric. Hortic.,* 45, 845, 1970.
6. **Watanabe, H.,** Studies on the flower bud differentiation and bolting of Welsh onion varieties, *Stud. Inst. Hortic. Kyoto Univ.,* 7, 101, 1955.
7. **Saito, S. and Takahashi, Y.,** Effect of vinyl covering on the growth, quality and chemical composition in vegetables. V. Effect on growth, sugar and chlorophyll contents of the Chinese chives. *J. Agric. Sci.,* 29, 122, 1984.
8. **Larkcom, J.,** Chinese chives, *Garden (J. R. Hortic. Soc.),* 112, 432, 1987.
9. **Anon.,** *Standard Tables of Food Composition for Japan,* Japan Scientific Agency, 1984.
10. **Saito, S. and Takama, F.,** Effect of vinyl covering on the growth, quality and chemical composition in vegetables. IV. Effect on the quality and volatile aroma component of the Chinese chive, *J. Agric. Sci.,* 21, 177, 1976.
11. **Saito, S., Takama, F., and Mayama, T.,** Effect on the vitamin content and quality of Chinese chive, *Jpn. J. Nutr.,* 34, 103, 1976.
12. **Saito, S., Takama, F., and Fukuda, S.,** Effect of vinyl covering on the growth, quality and chemical compositions in vegetables. III. Effect on the texture of the sweet pepper fruit and Chinese chive, *J. Agric. Sci.,* 20, 231, 1976.
13. **Takama, F. and Saito, S.,** Studies on the storage of the vegetables and fruits. II. Total carotene contents of sweet pepper, leek and parsley during storage, *J. Agric. Sci.,* 19, 11, 1974.

Chapter 12

CHIVES *Allium schoenoprasum* L.

N. Poulsen

TABLE OF CONTENTS

I. INTRODUCTION

Chives (*Allium schoenoprasum* L.) are used in many dishes. Only the leaves are eaten, the bulbs being too small. In many domestic gardens, perennial plants are kept for food or as ornamentals. Small bundles of chives, harvested from perennial fields or forced in greenhouses, provide a year-round supply. The processing industry freezes fresh leaves or dries them with hot air or by freeze drying. Processed chives are mostly used by the food industry and by caterers, and few are sold retail. Chives have never been of great economic importance. For many years their cultivation was confined to domestic gardens. In more recent times, they have been grown and processed commercially. The world area of commercial chive production is approximately 1000 ha. Little effort has been made in breeding; therefore, the plant material grown domestically and commercially is not uniform.[1]

II. BOTANICAL ASPECTS

A. TAXONOMY AND MORPHOLOGY
1. Taxonomy

Recent taxonomic literature[2] places *A. schoenoprasum* L. in the family Alliaceae, order Asparagales. In older literature, however, it was placed in the Liliaceae[3] Chemical analysis has shown that the Alliaceae contain steroid saponins, but no alkaloids.[2] *Allium* species also contain essential oils, of which the main components are sulfur compounds.

Allium schoenoprasum L. grows in a variety of habitats. Formerly, it was thought that different ecotypes were separate species. However, Levan[4] synchronized flowering and showed that all forms were interfertile and were therefore members of the same species. He divided the different forms of the species into three morphological types. These are distinguished by both the form of the plant and the size of the pollen. The three types are (1) diploids which include numerous biotypes of *A. schoenoprasum* L., e.g., var. *sibiricum* L., var. *ledebourianum* Roem et Schult. and var. *alpinum* Heg. These have a pollen diameter of 28 to 30 μm; (2) the autotetraploid gigas type, which includes *A. schoenoprasum* var. *sibiricum* Garke. The pollen diameter is 36 to 40 μm. The plants are usually larger than the diploids. However, one gigas diploid was also found; therefore, the best distinguishing feature is pollen size; (3) the allotetraploid types which have pollen diameter 36 to 37 μm, form very distinct bulbs, and are late flowering. No name examples were given. Later, it was found[5] that viable diploid pollen had an average length of 33.1 μm and width of 24.4 μm.

In addition, many local ecotypes have arisen. For example, *A. schoenoprasum* var. *alvarense* was found on the Swedish islands Öland and Gotland.[6]

Taxonomic relationships of chives have been studied using cytological and molecular biological methods as well as chemical composition[7-12] and morphological characteristics. The *Allium* subgenus *rhizirideum* has been divided into 15 taxa, which were termed "operational taxonomic units" (OTUs).[13] These were based on 49 morphological and cytological characters and 38 chemical properties.

A strong relationship between the four *A. schoenoprasum* OTUs is clear (Figure 1). The nearest relatives to this group are *A. cepa* (common onion), *A. cepa* var. *aggregatum* (shallot), *A. fistulosum* (Japanese bunching onion), and *A. galanthum*.

Some of the relationships between species within the genus *Allium* have been elucidated by chromosome mapping.[13-19] A close relationship between the species in the *cepa*-group and *A. schoenoprasum* L. as well as with *A. ledebourianum* Roem and Schult. was shown.[16] However, seed protein analyses gave no indication of a relationship between *A. schoenoprasum* L. and the *cepa*-group.[8-10]

The content of volatile sulfur compounds has also been used to study taxonomic rela-

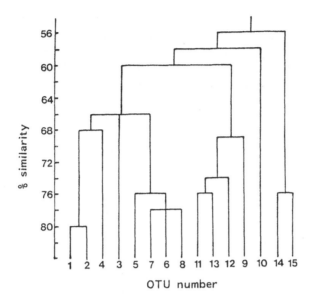

FIGURE 1. Dendrogram showing percentage similarity of Operational Taxonomic Units (OTUs) from the hierarchical analysis of 49 morphological and cytological characteristics, and 38 chemical characteristics. Key to OTUs: 1. *Allium cepa* L., 2. *A. cepa* var. *ascalonicum* shallot, 3. *A. fistulosum* L., 4. *A. galanthum* Kar. and Kir., 5. *A. schoenoprasum* L. (2n = 16 + 4B)., 6. *A. schoenoprasum* L. (2n = 16), 7. *A. schoenoprasum* L. (2n = 16), 8. *A. schoenoprasum* L. (2n = 32), 9. *A. albidum* Fisch. ex Bess, 10. *A. farreri* Stearn, 11. *A. senescens* L., 12. *A. senescens* var. *glaucum* (Schrader) Rgl., 13. *A. montanum* Schmidt, 14. *A. tuberosum* Rottler ex Spreng., 15. *A. ramosum* L. (From El-Gadi, A. and Ellkington, T. T., *New Phytol.l*, 79, 183, 1977. With permission.)

tionships in the genus *Allium*. The content of these chemicals in chives leaves was compared with that in bulbs of other species.[7] The chromatographic pattern of chives was dominated by a large di-*n*-propyl disulfide peak and the rather substantial peak "18". However, there could be some problems in such comparisons, and it is best if same plant organs are compared in such analyses.[20]

2. Morphology

The species is highly polymorphic.[4] The bulbs are clustered on a very small rhizome and are covered by a membranous skin. Leaves are hollow, cylindrical, or semicylindrical, as long as the scape, or even longer. The umbel is many flowered, pedicels vary in length, but are shorter than the flowers. Tepals are rose to rose violet in color, 6 to 15 mm long, and approximately five times as long as wide. Stamens are one third or one half the length of the tepals and are united to the perianth and to each other for one quarter or one third of their length. The ovaries have nectaries. The scape can be up to 60 to 70 cm long.

In *A. schoenoprasum* var. *orientale* Rgl., the umbels are loose, the tepals are lanceolate, and filaments are nearly as long as the tepals. *A. schoenoprasum* var. *sibiricum* (L.) Hartm. differs mainly in having leaves shorter than the scape.[21] In addition, there are white-flowered forms.[22]

According to Weber[23] bulbils do not form in the inflorescence. The flowers open first at the top of the umbel and then successively toward the base.[24]

B. EVOLUTION, WORLD DISTRIBUTION, AND ECOLOGY

Chives prefer moist soil conditions. The typical biotopes are wet meadows, small islands in rivers, riversides, and seashores.[4,6,25-31] The world distribution of the species is circumpolar and extends as far south as southern Europe, Iran, India, and China. In North America, it ranges from New York to northern Colorado, and it is also found in Japan.[6,22,25,32,33] Because chives are very cold hardy, they range into the arctic.[34] The northern limit of natural distribution is about 70° of latitude.[35] In northern Europe, the plant is usually found wild in the lowlands, but in southern Europe, it grows on mountains, e.g., up to 2650 m in the Alps. It grows beside streams, in wet meadows, among rocks, and in rocky pastures where there is little competition from other plants.[36]

Allium fossils have not been found,[34] so understanding of the evolution of *Allium* species must be based on their present distribution and our knowledge of geological history. The geographic origin of the plant is unknown, but was probably in the U.S.S.R. Turesson[6] described two forms which he asserted arose in East-Altai.

During the late Cenozoic Era, between 25- and 1-million years ago, the Bering Strait was bridged by a mountain range. This probably facilitated migration of plants and animals from Siberia to Alaska and back. Consequently, the distribution of flora encircled the Northern Hemisphere. The present natural distribution of the chive indicates that this was true for it.

C. HISTORY

The Latin name for chives was first given by Carl Linnaeus (1707 to 1778) in *Species Plantarum* (1753). The name of the genus, *Allium*, can be traced back to Plautus (184 B.C.) when it was the name for garlic.[37] The species name *schoenoprasum*, is formed from the Greek word *skhoinos* (meaning ''rush'') and *prason* (meaning ''leek''), and it refers to the rush-like leaves of the plant.[38] The English name ''chives'' is derived from the Latin *cepa*.[24] Synonyms are *Cepa schoenoprasa* Moench., *A. tenuifolium* Salisb., *Porrum schoenoprasum* Schur, *Schoenoprasum vulgare* Fourr.[39]

Local names are (England) cives, chives, civegarlic, civet; (Czechoslovakia) pazitka; (France) ciboulette, (ail)civette, petit porreau, fausse échalote, appétit; (Germany) Schnittlauch, Schnittling, Pries(e)lauch, Graslauch, Binsenlauch, Suppenlauch, Suppenkraut, Pankokenkraut, Schnittzwiebel; (Italy) Cipoletta, Cipollina, (Poland) luczer-lupny; (Russia) chnittlauk; (Spain) cebollino;[39] (Bulgaria) resanez luk; (Denmark) purløg; (Holland) bieslook; (Hungary) metélähaggma; (Israel) irit; (Portugal) cebolinha; (Sweden) gräslök.[26]

Chives have been cultivated in Europe at least since the 16th Century. Cultivation is thought to have originated in Italy and spread first to Germany and then to other European countries.[39]

D. CHROMOSOMES AND NUCLEIC ACIDS

The cytology of chives has been studied extensively.[4,6,14-16,18,19,28-31,33,40-44] Levan[4] investigated 150 different varieties of different ploidy levels and intercrossed them. The most frequent chromosome number is 2n = 16. Tetraploids, however, with chromosome numbers near 30 also occur as do triploids.[6] The tetraploids are divided into the autotetraploid gigastype and the allotetraploids. By intercrossing these types, Levan[4] obtained 21 different chromosome numbers; namely, 2n = 16—19; 2n = 21—34; 2n = 38—39 and 2n = 45. Diploid gametes occur spontaneously in the diploid 2n = 16 form of the species.[41] Kurosawa[33] studied *A. schoenoprasum* L. var. *caespitans* Ohwi. from Tochigi and Nagano (Japan) and found 2n = 24 and 2n = 16, respectively. One or more B-chromosome have been found.[15,28-31] The length of the chromosomes ranges between 3 and 7 μm.[4]

Jones and Rees[45] measured the DNA content of diploid chives to be 16.9×10^{-12} g per nucleus. A similar value of 15.6×10^{-12} g per nucleus was found later.[40] The relative

TABLE 1
Mineral Composition of Chives (% Dry-Matter)

Element	Germany[26a]	Sweden[47]	Italy[48a]
N	—	3.6	—
K	3.3	2.6	2.5
Ca	2.1	1.2	0.8
P	1.5	0.44	0.5
Mg	0.5	0.17	0.5
Na	6.4	0.03	0.06
Mn (ppm)	—	83	—

[a] Calculated from values in fresh material assuming a dry-matter content of 10%.[26,48]

amount of DNA in two- and three-nucleated pollen was found to be 9.6 in the telophase of the first and second meiotic divisions and 22.95 in the nucleus of the generative cell.[46]

E. CHEMICAL COMPOSITION AND BIOCHEMISTRY
1. Mineral Content
The mineral composition of chives is similar to that of other leafy vegetables (Table 1). The high sodium value found in Germany is probably experimental error (see chapter on "Chemical Composition").

2. Vitamins
Different investigators have reported different levels of vitamin C per 100 g fresh weight; namely, 60 mg,[49] 78 mg,[48] 80 mg,[50] and 140 mg.[51] The content of vitamin C depends on climatic conditions. The content was measured in plants grown at 12, 15, 18, 21, and 24°C. The highest content, 74 mg/100 g fresh weight, was found at 12°C, and the lowest content, 53 mg, at 21°C.[49] Thiamin (vitamin B_1)-levels range from 71.5 to 100 μg/100 g fresh weight.[48,51] Vitamin B_2 was 180 μg/100 g fresh weight, and vitamin A was 6400 IU (international units)/100 g fresh weight.[48] Fumaric acid content was found to be 6.8 mg/100 g fresh weight.[52]

3. SULFUR-CONTAINING COMPOUNDS
A vitamin B_1-derivative, "allithiamine", has been found in chives.[53] The two most frequent aliins are propyl-aliin and methyl-aliin. These two compounds are present also in other *Allium* species (see chapter on "Flavor Biochemistry"). The lacrimatory substance propenyl sulfenic acid[54] is produced from propenyl cysteine sulfoxide when leaves are crushed. The latter is the most important S-compound in chives.[55] Using gas liquid chromatography, the volatiles 2-methyl-2-butenal, 2-methyl-2-pentenal, methyl-propyldisulfide, and the *cis* and *trans* forms of propenyl-propyldisulfide were identified, but the allyl derivatives were not found.[7,56-58]

The study of organic sulfur compounds in chives and other alliums led to the isolation of a number of new cysteine derivatives.[59-61] It was possible to follow the biosynthetic pathways for a number of biologically active substances. In the seeds of chives were found the novel gamma-glutamyl peptides glutamyl-*S*-(prop-1-enyl)-cysteine, *N,N′*-*bis*-(glutamyl)-cystine, *N,N′*-*bis*-(glutamyl)-3,3′-(2-methylethylene-1,2-dithio)-dialanine, and glutamyl-*S*-propyl cysteine.

Due to the exceedingly low concentrations of cysteine derivatives, a sensitive analytical method was required for their identification and quantification. This was developed based on two-dimensional chromatography[62] and was used to study the biosynthesis and decomposition of these compounds[63] (see chapter on "Flavor Biochemistry"). Two new sulfur compounds, methyl pentyl disulfide and pentyl hydrodisulfide have recently been detected.[64,65]

The sulfur compounds in chives have relatively little antibiotic effect compared to those of garlic.[20,55]

In coastal areas of Finland, where chives are a common weed, cows often eat them, and this causes a bad flavor in milk. A vacuum treatment can remove these sulfur compounds and the bad taste.[63]

4. Other Constituents

The three glycosides, kaempferol-3-glycoside, quercetine-3-glycoside, and isorhamnetine-3-glycoside were found in chives.[66] The kaempferol glycosides and di- and triglycosides are dominant. The associated sugars are glucose and galactose. Quercitin and isorhamnetin occur as diglycosides.[67]

Ferulic acid and *para*-cumaric acid were found.[68] Some compounds in chives which are not essential as foods have recently been reviewed.[69]

The enzyme phosphodiesterase I has been found in the leaves.[70] In another study[71] two leaf fractions with different phosphofructokinase activity were examined. This enzyme which plays a very important role in the regulation of glycolysis, but is also necessary for the net conversion of starch to triose-P, was not found in the chloroplasts from chives. Chloroplast phosphofructokinase activity may be connected with the potential starch content of chloroplasts. Alliums do not contain starch in their chloroplasts.

The gel isoelectric focusing method for seed protein analysis, as well as immunological methods, is useful in determining interspecific relationships in *Allium*.[8-11]

Cholinesterase activity has been found in some species belonging to the subgenera *Allium*, *Rhizirideum*, and *Phyllodolon*. In chives, however, only a very weak activity was found.[12]

A content of malic acid, 240 mg/100 g, and of citric acid, 19 mg/100 g fresh weight, has been found. In comparison with many other species the citric acid content was very low, and the ratio malic acid/citric acid was relatively high.[52]

F. VEGETATIVE GROWTH
1. Growth Habit

The plant grows in clusters similarly to Chinese chives (*A. tuberosum*).[32] The clusters arise from two axillary buds to the main axis which develop after the primary bulb has reached a diameter of about 8 mm. This normally occurs 2 to 3 months after sowing in the spring. These two buds develop into new plants which compress the levels of the original axis from both sides. At this time the original plant has three to six leaves. When a further two to three leaves are formed in all plants, the cluster divides. Hence, the number of leaves is doubled by each division.[72] In the first year, five to seven such divisions take place and more than 80 leaves are formed per cluster. The plants in a cluster remain attached by a linking rhizome just below the bulb.[35]

Growth analysis of field grown plants indicates that the initiation of leaves that will emerge in the following year takes place in late July. These are also the leaves which emerge upon forcing[73] (see Section III.B).

2. Temperature and Light Effects

Chives are resistant to extremes of cold and heat. In controlled environments, under fluorescent lamps giving a total irradiation of 120 W/m^{-2} for 12 h/d, the optimum temperature for growth was between 19 and 23°C.[74] Relative growth rate, leaf-area ratio, and net-assimilation rate have maximum values at 19, 23, and 19°C, respectively.[74] The leaf-weight

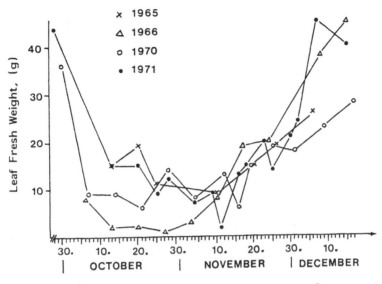

× 1965
△ 1966
○ 1970
● 1971

Date of Transfer to Glasshouse at 20°C

FIGURE 2.   Changes in the intensity of dormancy of chives with time. Plants were lifted from the field at various dates and transplanted to a greenhouse for 20 d at 20°C, after which the shoot fresh weight was recorded. (From Fölster, E. and Krug, H., *Sci. Hortic.*, 7, 213, 1977. With permission.)

ratio increased with temperature throughout the range from 10 to 31°C.[74] The optimum temperature for leaf growth was later confirmed to be in the range 17 to 25°C.[75] The optimum for root growth was found to be in the range 21 to 25°C.[75]

The effect of light intensity or light quality on growth has been little studied. All studies on these factors seem to concern the induction of dormancy. When studying the effects of environmental factors on growth rate, it is important to ensure that no photoperiodic induction of dormancy has taken place. Photoperiod extensions, using light intensities higher than 50 lx, can be used to maintain growth and prevent the induction of dormancy as natural photoperiods decrease.[76]

G. THE INDUCTION OF DORMANCY

Growth is rapid during spring and summer, but during the autumn the plants naturally enter a phase of dormancy. Factors involved in the induction of dormancy have been investigated in Germany, both in the field and in controlled environments.[76] The following is a summary of the findings. In the field, chives rapidly become dormant in early October (Figure 2). Dormant plants cannot be induced to grow even if they are transferred to optimal conditions for growth. Dormancy begins to disappear in mid-November and can be completely dissipated by mid- to late December (Figure 2).

1. Factors Inducing Dormancy

a. Effect of Temperature

Under short photoperiods, approximately 11 h, a constant induction temperature of about 14°C was most efficient in inducing dormancy (Figure 3). After 4 weeks at this temperature, a partial induction was found, and after 6 to 8 weeks induction was completed. Low temperatures of 2 to 6°C did not induce dormancy (Figure 3).[76]

Plants exposed to short-day conditions and with fluctuating day/night temperatures, with a mean of 13 to 14°C, were induced into dormancy when the temperature amplitude was

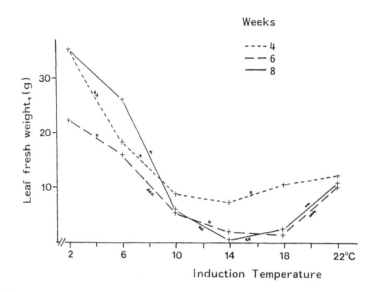

FIGURE 3.   Effect of temperature on the induction of dormancy in chives under short photoperiods. Experimental conditions weere 11 h light of 8 lx for 4, 6, or 8 weeks. Plants were then exposed to 20°C for 20 d, after which leaf fresh weight was measured. (From Krug, H. and Fölster, E., *Sci. Hortic.*, 4, 211, 1976. With permission.)

8°C (18/10°C). With an amplitude of 16°C (22/6°C), the dormancy-inducing effect was amplified, even though neither the day nor the night temperature alone was inductive.[76] During the summer months the temperature in some areas becomes so high that it induces dormancy. In Japan such induction takes place in early summer.[75]

b. Effect of Photoperiod and Light Intensity

The effect of photoperiod on the induction of dormancy was tested at a constant temperature. Photoperiods of 15 h or less induced dormancy. This was evident after only 4 weeks, and dormancy was complete after 6 weeks at a photoperiod of 13 h or less (Figure 4).[76] In controlled environments, short photoperiods (12 h as compared to 18 h) at 14°C were the most efficient for induction of dormancy.[79]

A field experiment was made in which the natural photoperiod in Germany between September and November (13 to 11 h) was extended using incandescent lamps suspended above the crop. It was shown that a minimum light intensity of 50 lx was needed for plants to respond to the extended photoperiods and thereby fail to enter dormancy.[76]

2. Carbohydrate Translocation in Relation to Dormancy

In field-grown plants, accumulation of starch and sugar reserves in the roots and bulbs continues until the beginning of October, when growth naturally decreases. The maximum carbohydrate content occurs before rest is induced.[77]

3. Hormonal Control of Dormancy

A growth inhibitor resembling abcisic acid (ABA) can be extracted from intact, dormant chive bulbs. Furthermore, ABA is capable of inducing growth retardation akin to dormancy in actively growing plants.[78] When breaking dormancy by hot-water treatment (see Section III.B), it is important to use freshwater, probably because ABA or other growth inhibitors accumulate in the water. During the induction of dormancy, levels of both auxin and giberellin (GA) decrease and growth inhibitors increase, particularly ABA. During the dormant period,

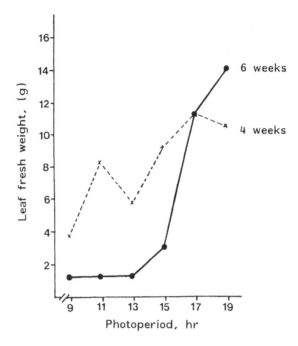

FIGURE 4. Effect of photoperiod on the induction of dormancy in chives. Experimental conditions were 14°C, 8 lx for 4 or 6 weeks. Plants were then exposed to 20°C for 20 d, after which leaf fresh weight was measured. (From Krug, H. and Fölster, E., *Sci. Hortic.*, 4, 211, 1976. With permission.)

a low concentration of both promoters and inhibitors is found. At the end of dormancy, the level of growth promoters increases rapidly.[79]

H. FLOWERING

Even though unwanted flowering is a great problem in forcing as well as in perrenial field production, the topic has been little investigated in chives. Most of what we know comes from investigations on other topics and from what can be observed during agronomic practice.

It is possible that the induction of flowering depends on photoperiod as well as temperature. Also, some minimum of reserve material must be present for the flowering process.

In the field, spring-sown chives do not flower in the year of sowing. However, if sowing takes place in January in a glasshouse, flowering can take place in the first season.[35] Plants sown in the field in the late autumn do not flower in the next year. It seems that plants must reach a certain minimum growth stage before short photoperiods can induce flowering. In forcing plants (see Section III.B) flowering can be eliminated by keeping temperatures above 18°C. Flowering takes place at the end of May when two or three leaves have developed.[35] Frequently, weak flowering of some plants occurs in the early autumn also. The mechanisms and control of flower induction in chives need further investigation.

I. *IN VITRO* CULTURE

In a tissue culture experiment the response of chives to gibberellin (GA) was carried out both with and without light, and a range of sucrose concentrations was used. The temperature ranged from 16 to 25°C. Irrespective of the sucrose concentration, GA always inhibited the growth of chives while enhancing the growth of the dicots.[80] Another interesting observation was the ability of chives to grow at high sucrose concentration under the consequent low water potential.[80] This seems consistent with its ability to grow on seashores (i.e., in saline conditions). Chive calluses ($2n = 16$) can be induced to differentiate *in vitro* when grown on a medium containing 1 mg IAA per liter and 5 mg kinetin per liter.[81]

III. CROP PRODUCTION

Chives are grown commercially for forcing (selling as fresh leaves), for processing (air dried, freeze-dried, and frozen), and for seed production. Fresh leaves and those for processing are harvested from perennial fields. Plants for all forcing are sown every spring. The forcing of bulbs is mostly practiced in industrialized countries, e.g., Germany, since it is expensive. The world area of commercial chive production is approximately 1000 ha.

A. PLANT BREEDING

Recent breeding of chives has concentrated on the forcing crop (see Section III.B) and not on the outdoor crop. The plentiful information on the cytology of chives (see Section II.D) provides a valuable basis for plant breeding. However, the breeding of chives began only a few decades ago. Consequently, cultivars are not uniform, being commercially maintained and probably quite variable populations.

In Europe, serious breeding work began in the early 1960s to produce genotypes suitable for forcing (see Section III.B).[82] In addition, breeding aims to produce plants with erect leaves and of uniform curvature. The leaf diameter should be at least of medium size. Many new bulbs should be formed every year, flowering should be late, and the yield should be high.

German-bred hybrids produced higher yields than open-pollinated cultivars.[83] There was a significant positive correlation between plant height, large leaf diameter (large diameter is more than 3.5 mm and small diameter is 2.6 mm), and yield. Consequently, plants were selected for these characteristics.

Despite the fact that chives are protandric,[84] there is a high degree of self-pollination.[82] Consequently, some plants show inbreeding depression and grow and produce poorly. The use of male sterile plants in hybrid seed production can overcome this problem.[5,83,85-87] Kobabe[82] found 1 to 2% male sterile plants in chive populations and suggested that F_1 hybrids be produced. Male sterile flowers have relatively short anthers which are shrunken and yellowish, and the filaments are shorter than normal.[88] This makes it easy to recognize male steriles. Within a few years, a number of F_1 hybrids were produced, e.g., cvs. "Friilau" and "Treilau" by the German seed company Sperling. Thus, much more uniform plant material became available for use in forcing. Unfortunately, the F_1 seed is more expensive than the open-pollinated, and despite its higher productivity the extra cost is not always economically worthwhile. Therefore, most seed sown for forcing is still produced in the traditional way (see Section III.F).

It was found that a male sterility factor is carried by the mitochondria. This male sterility can be total or partial and could be of great value to plant breeders in the future.[5,86,87] Some plants were sterile under 17/17°C or 20/17°C conditions, but fertile at a constant temperature of 24°C. Sterility recurred if the plants were exposed to low temperatures again in the following year. Hence, temperature-induced sterility is reversible.[86] The effect of some protein synthesis inhibitors, especially antibiotics, on male sterility was also studied.[87] Fertility was restored, in part or in full, after treatment with tetracycline-HCl or basic-tetracycline. The plants made fertile by tetracycline became male sterile again the following year, so tetracycline induced fertility was also reversible. Possible interrelationships between the temperature and tetracycline effects need future investigation.

A form of apomixis in which plants appear to show polyembryony has been found in chives.[89]

B. FORCING

McCollum, in 1935, first showed that dormant chives could be induced to grow following a heat treatment.[90] Since then, procedures have been investigated in order to develop optimal treatments for breaking dormancy, so that chive leaves can be produced year-round utilizing glasshouses for winter production. Such winter production is termed "forcing".

Early work examined the effect on subsequent growth of immersion of dormant plants in cold water (14 to 16°C) and warm water (between 40 and 45°C, depending on experiment).[91-93] The influence of plant age was examined by comparing plants sown the previous year, transplanted seedlings a few weeks old, and newly sown plants. It was concluded that no plants can be forced successfully until the middle of October. A heat treatment in September did not result in better growth, and the growth of young plants was even inhibited by heat treatment then. Therefore, actively growing plants are sensitive to heat treatment and can be damaged by it. However, by mid-October the plants can be forced successfully. In general, the older the plants, the earlier they were suitable for forcing and the higher the yields in all treatments.[77]

Immersion of plants in warm water (approximately 40°C) is now used commercially to break dormancy. Thereafter, plants are transferred to a glasshouse at approximately 20°C where leaf growth occurs rapidly. Fölster and Krug[77] showed that such an immersion for 3 h breaks dormancy effectively. Immersion at a temperature of 44°C for more than 1 h damages the plants and reduces subsequent growth. Temperatures above 44°C are lethal. In December, when dormancy is naturally decreasing (see Section II.G), warm-water treatments become progressively harmful,[77] and this is exacerbated the longer their duration. Nevertheless, in both November and December, good growth results after immersion of plants in water, initially at 40°C, which is then allowed to cool to 25°C over a 16-h period. Results from this treatment are so good and consistent that it is termed the "standard warm-water treatment".[77] Cooling from 40 to 25°C over an 8-h period is less effective (Figure 5). Treatment of plants for 1 or 2 weeks at −2 or 2°C after lifting, at times between mid-September and early October, does not improve the results from subsequent warm-water treatments. In a controlled environment experiment, exposure of plants to an air temperature of −5°C for 3 d did have a significant effect on the breaking of dormancy, but it was less effective than the above warm-water treatment.[77]

As an alternative to warm-water treatment, plants may be lifted after removal of leaves and transplanted to boxes. Good yields are obtained after exposure to air temperatures of 38 to 44°C for between 16 and 32 h (Figure 6). Otherwise, plants may be lifted intact and simply spread on the floor of a glasshouse. Thereafter, the best yields are obtained from treatment at 33 and 36°C for 96 or 72 h, respectively (Figure 6). In this procedure, the duration of treatment is more important than the precise temperature used.[77]

When using warm-air treatments, it is most important to maintain a high relative humidity. For example, at 35°C, air humidities below 80% result in lower yields (Figure 7). The roots are important as a source of reserves for leaf growth, and as many as possible should be kept intact and healthy after harvest until plants are subjected to dormancy-breaking treatments.[77]

The effect of temperature and photoperiod on the growth of plants after heat treatment was investigated in controlled environments. Highest yields were obtained at 24 to 30°C. At these temperatures, a long photoperiod (16 h) caused no improvement in leaf yields compared to a short photoperiod (12 h). Growth rates decreased at temperatures below 24°C, and the 12-h photoperiod further decreased growth compared to 16 h.[77] The temperature during forcing must be at least 18°C to eliminate flowering.[35]

C. PRODUCTION OF BULBS FOR FORCING

Bulbs grown for forcing are sown in late April or early May at a depth of 5 to 10 mm. Seeds are sown at a rate of approximately 30 kg/ha in 6 double rows in beds (130 cm wide

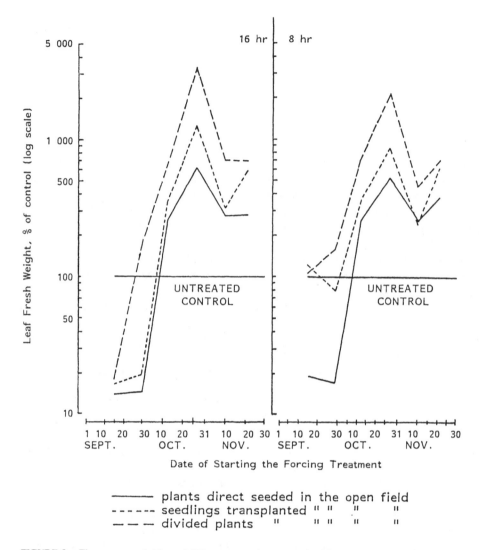

FIGURE 5.   The response of chives of different age and pregrown in different ways to forcing treatments. The forcing treatments were either warm-water immersion for 16 h (left) or 8 h (right), during which interval the temperature decreased gradually from 40 to 25°C. The 16-h treatment is termed the "standard warm-water treatment". After treatment the plants were grown at 20°C for 20 d, and leaf fresh weight was determined. For the untreated control plants, the warm-water treatment was omitted. (From Fölster, E. and Krug, H., *Sci. Hortic.*, 7, 213, 1977. With permission.)

plus 50 cm for tractor wheels), when the mean soil temperature reaches about 10°C. The plant population density is approximately 15 million plants per hectare. Leaf harvest during the summer is not recommended, since this depletes the reserves used to form the bulb, and consequently regrowth upon forcing is lessened. Only bulbs of fairly small diameter (approximately 0.5 cm) are used for forcing as the consumers prefer many leaves per 5-cm pot. The grower can control bulb diameter to some degree by varying the seed rate. By using a high seed rate, and the consequent high plant density, nearly all the bulbs produced are suitably small.

Until very recently most forcing was carried out in the autumn and winter. However, there is a year-round demand for the product, and by cold storing the bulbs forcing is possible throughout the year, or even from bulbs stored for 2 years.

Method 1. Plants treated with hot air after potting the plants.

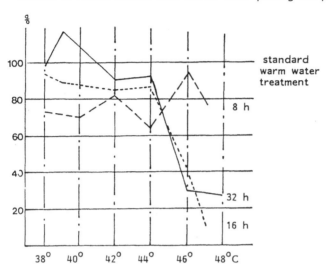

Method 2. Plants treated with hot air before potting the plants.

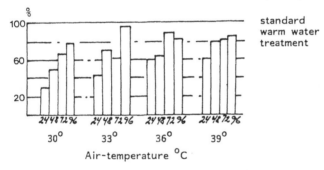

Air-temperature °C

FIGURE 6. The effect of different warm-air treatments on the yield of forced chives. Leaf yields after growth at 20°C for 20 d are given as a percentage of those from plants exposed to the "standard warm-water treatment" which was considered as 100%. The numbers 24, 48, 72, and 96 under the histograms refer to the duration of the warm-air treatment. (From Fölster, E. and Krug, H., *Sci. Hortic.*, 7, 213, 1977. With permission.)

D. PRODUCTION OF LEAVES FOR BUNCHING FROM PERENNIAL FIELDS

The production of bunches of fresh leaves is of limited economic importance. Such bunches are sold in fresh vegetable markets and to a lesser extent under contract to caterers.[94] The most economic production method is to harvest leaves from perennial fields. Harvesting is thus limited to the summer months only. Since the flower stalks contain much sclerenchyma, they are not palatable. Therefore, flower stalks must be removed and discarded at the end of the flowering period in May. Leaves may be hand harvested, and, in this case, it is most convenient if plants are grown as small clumps, although in time these usually coalesce into rows. For machine harvesting, plants are grown in rows.

E. PRODUCTION FOR PROCESSING

The processing industry produces a variety of products to supply the market throughout the year. Frozen leaves have been sold for many years. These retain their green color, but lose their turgidity when thawed.

Hot-air drying has been used for many years. In this case quality is reduced because of

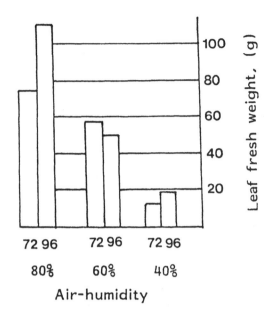

FIGURE 7.   The influence of air humidity during a warm-air treatment of 35°C for 72 or 96 h on the subsequent regrowth after 20 d at 20°C. (From Fölster, E. and Krug, H., *Sci. Hortic.*, 7, 213, 1977. With permission.)

loss of aroma and some decrease in green color, as well as some loss of structure. The best method of conservation is freeze drying. This retains most of the aroma, the structure, and the green color. It was concluded that a rapid freezing rate was acceptable only when drying was carried out at a pressure of about 1.5 mmHg.[95] This relatively new technique is carried out in a few countries only, namely, the U.S., Taiwan, Peru, and Denmark. About 3000 to 4000 t of freeze-dried leaves are produced annually worldwide. The method of production depends on local labor costs. In Denmark, the leaves were cut by hand for the first few years of freeze-drying production. In recent years, mechanical harvesting of single rows has been used, resulting in heavy traffic over these fields.

Fields for the perennial plants must be weed-free, and no other *Allium* species should have been grown in the previous year. Because of the risk of infection by downy mildew (*Peronospora destructor*), chive fields should be isolated from fields of other alliums. Many kinds of soils are suitable, but those of low available water capacity should be avoided. A soil pH of approximately 7 is optimal.

Seed is sown at a rate of 7 kg/ha in late April or early May. Alternatively, sowings can be made in early August. Seed is sown in double rows, 5 cm apart. The distance between these double rows is 50 to 65 cm, but 56 cm is most common in Denmark.

Irrigation is essential throughout the growing season; otherwise, water stress will promote chlorosis and necrosis. As for fresh leaves, harvesting must be carried out in dry weather; otherwise, necrosis develops at the leaf tips.[94]

After a spring sowing, the field can be harvested about three times in the first season. It is important to avoid overfrequent harvest in the first year; otherwise, productivity in the following year is reduced. From the second season onward fields can be first harvested in early May. When the plants are flowering in late May or early June, the cut material must be discarded because flowers and flower stalks are not palatable. The second harvest is taken in late June, but the quality is usually low because of chlorosis (yellowing) of the older (lower) leaves. Often this chlorosis is only on the adaxial side of the leaves. Maximum productivity occurs in July and August, when about four harvests can be taken. In early September the productivity declines and the plants enter the rest phase (see Section II.G). Unless the field is to be discontinued at the end of the season, only one harvest should be taken after this date. Otherwise, late harvesting will reduce early growth in the next season. In September the quality is also reduced due to chlorosis of the older leaves. Again, the chlorosis is often unilateral, but at this time it is abaxial. About 6 to 7 harvests annually are typical in Denmark, and the annual yield is about 7.5 t/ha fresh leaves. About 100 ha are grown for freeze drying in Denmark. In New Zealand about four harvests can be taken annually giving a yield of about 4.5 t/ha. Also, here regrowth after the first cutting produces flower heads, these are left for seed production (see Section III.F).[96] By cutting off flower heads at an early stage of development, it may be possible to increase the yield of leaves.

F. SEED PRODUCTION

Most seeds are produced by open-pollination of selected plant material, but high rates of self-pollination are common. Consequently, some inbreeding depression occurs. Only a limited amount of F_1 hybrid seed is produced, using male sterile lines. Fields for seed production are commonly grown from seeds rather than transplants. Each field is normally used for 3 or 4 years; hence, they must be weed-free. Many soil types are suitable, but soils of low available water capacity, heavy clay loams, and water-logged soils should be avoided. The fields must be well drained and have neutral pH.[97]

Chives are insect pollinated. The honeybee is especially important for pollination.[98,99] An isolation distance of 800 m or more should be strictly maintained to avoid cross-contamination of pollen.

Seeds are sown 5 to 10 mm deep. After emergence, weeds must be controlled using herbicides and hoeing. Seeds are harvested at maturation when the capsules start to open, and black seeds can be seen. This is normally at the beginning of July in Denmark. Umbels are cut, then dried on racks for at least 3 weeks. After threshing, the seed is further dried. Yield is typically 300 kg/ha in Denmark.[97] In New Zealand a seed yield of 60 kg/ha is reported.[96] Seeds can become dormant.[35]

Diseases and pests are rarely a problem, and since seeds are not used for food, chemical control can be freely used.[97]

G. FERTILIZER APPLICATION AND MINERAL NUTRITION

In the spring, perennial fields must be supplied with about 25 kg P_2O_5 and 100 kg K_2O per hectare. A high soil magnesium content must be maintained. Nitrogen must be supplied throughout the growing season after every harvest. A weekly application of about 10 to 12 kg/ha N is recommended, totalling 220 kg/ha per year.

When chives became a commercial crop, fertilizer trials were carried out. The maximum yield in Germany occurred with 300 kg/ha nitrogen.[100] The yield increase was correlated with a larger leaf number and diameter. However, the high solute levels resulting from high doses of N fertilizer can adversely affect yields.[100] A single application of N resulted in lower yields than the same total amount given as four consecutive dressings, probably because of the sensitivity of chives to high soil-solute concentration.[101,102] In Florida, a dressing of about 15 kg N and 30 kg K_2O per ha per week was the optimum fertilizer treatment. When 62 kg N per ha per week was applied, yield was reduced and excess solutes accumulated in the soil.[102]

Young plants are particularly responsive to nitrogen. Division is inhibited in young plants by high nitrogen levels and promoted by low levels.[73] Inhibition of division at this stage is undesirable. When the plants grow older, around mid-June, this response is reversed. The reason for this is not known.

In young plants, dry-matter content is highest with the highest levels of nitrogen fertilization. Toward the end of the growing season in early October, the highest dry matter percentage is found in plants given the lowest levels of N fertilization. In plants used for forcing, yields increase in parallel with nitrogen application rates. Several equally distributed applications during the growing season give a moderate yield increase over a single large nitrogen application. The increase in yield from nitrogen application is correlated with an increased leaf weight and number.

In studies concerning forcing and harvesting, it was found that the yield decreases caused by over-frequent harvesting could be counteracted by nitrogen fertilization.[103]

Deficiency symptoms caused by lack of N, P, K, Ca, Mg, and B have been studied using sand cultures. Plants were supplied periodically with complete nutrient solution for 30 d, and in the following 35 d they were given solutions deficient in one of the above elements. Following leaf harvest, plants were left to regrow for a further 43 d before a second leaf harvest. Deficiency symptoms were hardly noticeable. However, some decrease in growth occurred with nitrogen shortage. The quality of the leaves was affected by both nitrogen and calcium deficiency. Nitrogen deficiency caused chlorosis of the leaves, and nitrogen as well as calcium deficiency promoted senescence.[104] The lack of symptoms found in this work might be explained by the accumulation of sufficient amounts of the elements during the initial phase, when a complete nutrient solution was provided, to supply the plants during subsequent deprivation.

Chives, like all other cultivated *Allium* species, live in symbiosis with mycorrhizal fungi[105] (see chapter on "Mycorrhyzal Associations and Their Significance"). These may have an important influence on their nutrition in field conditions.

H. WEED CONTROL, MAJOR DISEASES, AND PESTS
1. Weeds

Weed control, especially of grasses should start in the preceding year. The field must be as free as possible from weeds. At sowing, chloropropham, diquat-dibromide, or other preemergence herbicides are applied. Weed control is more difficult in perennial crops than in the crop for forcing, since the latter is annual. In Denmark, only mechanical weed control is permitted if leaves are to be harvested.[94]

2. Major Diseases
a. Rust

Puccinia allii Rud. infects plants all year round and causes serious damage, especially during August and September. The extent of disease spread depends on the relative humidity of the air. It is difficult to control this disease in chives because fungicidal spraying is restricted since the leaves are for human consumption. Details on the fungus biology can be found in several references[106-109] (see chapter on "Onion Leaf Diseases").

b. Other Fungi

White rot (*Sclerotium cepivorum*) sometimes infects chives. Chemical control is little used and soil disinfection is uneconomic.[110] Some control is possible with crop hygiene, long rotations, adjustment of soil pH, altering the growing period by varying planting and lifting times, rotation with gladiolus, and mulching moist soil with polyethylene sheeting in summer (soil solarization) (see chapter on "Root Diseases").

Downy mildew (*Peronospora destructor*) can be a serious disease in some years. The infection usually occurs when leaves exceed 15 cm in length. It cannot easily be detected until the weather becomes wet. Then, elongated white to purplish fruiting lesions show on the foliage. The later become pale green and finally white or tan, unless a secondary infection by other fungi occurs, causing a black mold. Old leaves are more susceptible than young ones, and the former normally die when infected, whereas young ones survive, though these may become twisted and distorted. If the weather becomes dry before the plants die, new leaves develop. The infection can spread a long distance (many kilometers).[111] In Brazil, the disease has been controlled by the experimental fungicide CGA 48988 25W.[112]

Foot rot (*Pythium* ssp.) may be a serious disease in some years in newly sown fields.[113]

3. Pests

Onion fly (*Delia antiqua*) can cause serious damage, especially to young seedlings. Use of trichloronate and thiram at sowing gives some protection. Application of parathion between the rows has given good control.[114] Thrips (*Thrips tabaci*) are not a serious problem when the fields are irrigated.[94] Stem and bulb nematodes (*Ditylenchus dipsaci*) rarely attack chives.[94]

REFERENCES

1. **Poulsen, N.**, Purløg (*Allium schoenoprasum L.*). Et litteraturstudium, *Tidsskr. Planteavl Spec. Ser. S*, 1656, 39, 1983.
2. **Dahlgren, R.**, *Alliaceae*, in *Angiospermernes taxonomi*, Vol. 4, Dahlgren, R., Ed., Akademisk Forlag, Copenhagen, 1976, 98.
3. **Warming, E.**, Liliales; Lilieordenen, in *Frøplanterne (Spermatofyter)*, Warming, E., Ed., Gyldendal, Copenhagen, 1912, 136.
4. **Levan, A.**, Zytologische Studien an *Allium schoenoprasum*, *Hereditas*, 22, 1, 1936.
5. **Tatlioglu, T.**, Cytoplasmic male sterility in chives (*Allium schoenoprasum L.*), *Z. Pflanzenzuecht.*, 89, 251, 1982.
6. **Turesson, G.**, Über verschiedene Chromosomenzahlen in *Allium schoenoprasum L.*, *Bot. Not. (Lund.)*, 1931, 15, 1931.
7. **Bernhard, R. A.**, Chemotaxonomy: distribution studies of sulfur compounds in *Allium*, *Phytochemistry*, 9, 2019, 1970.
8. **Klozová, E., Turková, V., Pitterová, K., and Hadačová, V.**, Serological comparisons of seed proteins of some representatives of the genus *Allium*, *Biol. Plant.*, 23, 9, 1981.
9. **Klozová, E., Turková, V., Hadačová, V., and Švachulová, J.**, Serological comparisons of seed proteins of some *Allium* (L.) species belonging to the subgenus *Rhizirideum* (G. Don ex Koch) Wendelbo, *Biol. Plant.*, 23, 376, 1981.
10. **Hadačová, V., Klozová, E., Hadač, E., Turková, V., and Pitterova, K.**, Comparison of esterase isoenzyme patterns in seeds of some *Allium* species and in cultivars of *Allium cepa L.*, *Biol. Plant.*, 23, 174, 1981.
11. **Hadačová, V., Švachulová, J., Klozová, E., Hadač, E., and Pitterová, K.**, Use of esterase isoenzymes revealed by gel isoelectric focusing as an aid in chemotaxonomical study of the genus *Allium*, *Biol. Plant.*, 25, 36, 1983.
12. **Hadačová, V., Vacková, K., Klozová, E., Kutáček, M., and Pitterová, K.**, Cholinesterase activity in some species of the *Allium* genus, *Biol. Plant.*, 25, 209, 1983.
13. **El-Gadi, A. and Elkington, T. T.**, Numerical taxonomic studies on species in *Allium* subgenus *Rhizirideum*, *New Phytol.*, 79, 183, 1977.
14. **Kurita, M.**, Karyotypes of some species in *Allium*, *Mem. Ehime Univ. Sect II. Biol.*, 2, 11, 1956.
15. **Brat, S. V.**, Genetic systems in *Allium*. I. Chromosome variation, *Chromosoma*, 16, 486, 1965.
16. **Vosa, C. G.**, Heterochromatic patterns in *Allium*. I. The relationship between the species of the *cepa* group and its allies, *Heredity*, 36, 383, 1976.
17. **Hussain, L. A. Al-Sheikh, and Elkington, T. T.**, Giemsa C-band karyotypes of diploid and triploid *Allium caeruleum* and their genomic relationship, *Cytologia*, 43, 405, 1978.
18. **Löwe, Á**, Chromosome number reports. LXIX, *Taxon*, 29, 703, 1980.
19. **Pandita, T. K. and Mehra, P. N.**, Cytology of *Alliums* of Kashmir Himalayas. I. Wild species, *Nucleus*, 24, 5, 1981.
20. **Rudat, K. D.**, Vergleichende Untersuchungen über die Antibakterielle Wirksamkeit verschiedener Lauchgewächse und Cruciferen-Arten, *Qual. Plant. Mater. Veg.*, 18, 29, 1969.
21. **Anon.**, Allium L., in *Hortus Third*, Bailey, L. H., and Bailey, E. Z., Eds., Macmillan, New York, 1976, 47.
22. **Ueno, Y.**, *Allium schoenoprasum var. Foliosum, J. Jpn. Bot.*, 44. 348. 1969.

23. **Weber, E.,** Entwicklingsgeschichtliche Untersuchungen über die Gattung *Allium, Bot. Arch.*, 25, 1, 1929.

24. **Jones, H. A. and Mann, L. K.,** *Allium schoenoprasum* L., in *Onions and their Allies,* Jones, H. A. and Mann, L. K., Leonard Hill, London, 1963, 43.

25. **Ownbey, M.,** The genus *Allium* in Idaho, *Wash. Res. Study State Coll.,* 18, 3, 1950.

26. **Becker-Dillingen, J.,** Der Schnittlauch *Allium schoenoprasum* L., in *Handbuch des Gesamten Gemüsebaues,* Becker-Dillingen, J., Ed., Parey, Berlin, 1956, 706.

27. **Gillet, H.,** Ciboule et ciboulette, *J. Agric. Bot. Appl.,* 24, 55, 1977.

28. **Bougourd, S. M. and Parker, J. S.,** The B-chromosome system of *Allium schoenoprasum.* I. B-distribution, *Chromosoma,* 53, 273, 1975.

29. **Bougourd, S. M. and Parker, J. S.,** Nucleolar-organiser polymorphism in natural populations of *Allium schoenoprasum, Chromosoma,* 56, 301, 1976.

30. **Bougourd, S. M. and Parker, J. S.** The B-chromosome system of *Allium schoenoprasum.* II. Stability, inheritance and phenotypic effects, *Chromosoma,* 75, 369, 1979.

31. **Bougourd, S. M. and Parker, J. S.,** The B-chromosome system of *Allium schoenoprasum* III. An abrupt change in B-frequency, *Chromosoma,* 75, 385, 1979.

32. **Herklots, G. A. C.,** *Allium schoenoprasum* L., in *Vegetables in South-East Asia,* Herklots, G. A. C., Ed., Allen and Unwin, London, 1972, 397.

33. **Kurosawa, S.,** Notes on chromosome numbers of spermatophytes, *J. Jpn. Bot.,* 54, 155, 1979.

34. **Traub, H. P.,** The subgenera, sections and subsections of *Allium* L., *Plant Life,* 24, 147, 1968.

35. **Krug, H.,** Schnittlauch *(Allium schoenoprasum* L.), in *Gemüseproduktion,* Krug, H., Ed., Parey, Berlin, 1986, 406.

36. **Stearn, W. T.,** European species of *Allium* and allied genera of *Alliaceae:* a synonymic enumeration, *Ann. Musei Goulandris,* 4, 83, 1978.

37. **Lid, J. and Lid, D. T.,** *Norsk og svensk flora,* Det Norske Samlaget, Oslo, 1974, 808.

38. **Christiansen, M. S.,** Purløg *Allium schoenoprasum,* in *Danmarks Vilde Planter,* Vol. 1, Christiansen, M. S. and Anthon, H., Eds., Politiken, Copenhagen, 1970, 46.

39. **Helm, J.,** *Allium schoenoprasum* L., *Kulturpflanze,* 4, 154, 1956.

40. **Ranjekar, P. K., Pallotta, D., and Lafontaine, J. G.,** Analysis of plant genomes. V. Comparative study of molecular properties of DNAs of seven *Allium* species, *Biochem. Genet.,* 16, 957, 1978.

41. **Levan, A.,** Cytological studies in *Allium.* A preliminary note, *Hereditas,* 15, 347, 1931.

42. **Levan, A.,** Distribution of chromosome numbers in a progeny of triploid *Allium schoenoprasum, Nature (London),* 134, 254, 1934.

43. **Levan, A.,** Cytological studies in *Allium.* VI. The chromosome morphology of some diploid species of *Allium, Hereditas,* 20, 289, 1935.

44. **Levan, A.,** Different results in reciprocal crosses between diploid and triploid *Allium schoenoprasum,* L. *Nature (London),* 138, 508, 1936.

45. **Jones, R. N. and Rees, H.,** Nuclear DNA variation in *Allium, Heredity,* 23, 591, 1968.

46. **Ermakov, I. P., Morozova, E. M., and Karpova, L. V.,** DNA content in nuclei of male gametophytes of some flowering plants, *Dokl. Akad. Nauk SSSR Ser. Biol.,* 250-252, (32-34), 1980.

47. **Persson, N. E.,** Resultat från några växtanalyser av köksväxter, *Aktuel. Lantbrukshoegsk.,* 85, 1, 1966.

48. **Gorini, F.,** Erba cipollina o aglio cipollino, *Inf. Ortiflorofrutticoltura,* 19, 5, 1978.

49. **Rosenfeld, H. J.,** Ascorbic acid in vegetables grown at different temperatures, *Acta Hortic.,* 93, 425, 1979.

50. **Rinno, G.,** Die Beurteilung des ernährungsphysiologischen Wertes von Gemüse, *Arch. Gartenbau,* 13, 415, 1965.

51. **Franke, W.,** On the contents of vitamin c and thiamin during the vegetation period in leaves of three spice plants *(Allium schoenoprasum* L., *Melissa officinalis* L. and *Petroselinum crispum* (Mill.) Nym. ssp. *crispum, Acta Hortic.,* 73, 205, 1978.

52. **Ruhl, I. and Herrmann, K.,** Organische Säuren der Gemüsearten. I. Kohlarten, Blatt- und Zwiebelgemüse sowie Möhren und Sellerie, *Z. Lebensm. Unters. Forsch.,* 180, 215, 1985.

53. **Fujiwara, M., Yoshimura, M., Tsuno, S., and Murakami, F.,** "Allithiamine", a newly found derivative of vitamin B$_1$. IV. On the alliin homologues in the vegetables, *J. Biochem.,* 45, 141, 1958.

54. **Whitaker, J. R.,** Development of flavor, odor, and pungency in onion and garlic, *Adv. Food Res.,* 22, 73, 1976.

55. **Virtanen, A. I.,** Antimikrobielle und Antithyreoide Stoffe in einigen Nahrungspflanzen, *Qual. Plant. Mater. Veg.,* 18, 8, 1969.

56. **Saghir, A. R., Mann, L. K., Bernhard, A., and Jacobsen, J. V.,** Determination of aliphatic mono- and disulfides in *Allium* by gas chromatography and their distribution in the common food species, *Proc. Am. Soc. Hortic. Sci.,* 84, 386, 1964.

57. **Shankaranarayana, M. L., Raghaven, B., Abraham, K. O., and Natarajan, C. P.,** Volatile sulfur compounds in food flavors, *CRC Crit. Rev. Food Technol.,* 4, 395, 1973.

58. **Wahlroos, Ö. and Virtanen, A. I.**, Volatiles from chives *(Allium schoenoprasum)*, *Acta Chem. Scand.*, 19, 1327, 1965.
59. **Matikkala, E. J. and Virtanen, A. I.**, A new gamma-glutamylpeptide, gamma-L-glutamyl-*S*-(prop-1-enyl)-L-cysteine in the seeds of chives *(Allium schoenoprasum)*, *Acta Chem. Scand.*, 16, 2461, 1962.
60. **Matikkala, E. J. and Virtanen, A. I.**, New gamma-glutamylpeptides isolated from the seeds of chives *(Allium schoenoprasum)* *N'*, *N*-bis-(gamma-glutamyl)-cystine, *N'*, *N*-bis-(gamma-glutamyl)-3', 3-(2-methylethylene-1, 2-dithio)-dialanine, gamma-glutamyl-*S*-propylcysteine, *Acta Chem. Scand.*, 17, 1799, 1963.
61. **Virtanen, A. I.**, On sulphur-containing amino acids and gamma-glutamyl peptides in the bulbs and seeds of *Allium* species, *Bot. Mag. (Tokyo)*, 79, 506, 1966.
62. **Granroth, B.**, Separation of *Allium* sulfur amino acids and peptides by thin-layer electrophoresis and thin-layer chromatography, *Acta Chem. Scand.*, 22, 3333, 1968.
63. **Granroth, B.**, Biosynthesis and decomposition of cysteine derivatives in onion and other *Allium* species, *Ann. Acad. Sci. Fenn. Ser. A2*, 154, 1, 1970.
64. **Kameoka, H. and Hashimoto, S.**, Two sulfur constituents from *Allium schoenoprasum*, *Phytochemistry*, 22, 294, 1983.
65. **Hashimoto, S., Miyazawa, M., and Kameoka, H.**, Volatile flavor components of chive *(Allium schoenoprasum L.)*, *J. Food Sci.*, 48, 1858, 1983.
66. **Starke, H.**, Über die Flavonole der Zwiebel des Porrees, des Schnittlauchs und der Schwarzen Johannisbeeren, *Diss. Tech. Univ. Hannover*, p. 65, 1975.
67. **Starke, H. and Herrmann, K.**, Flavonole und Flavone der Gemüsearten, *Z. Lebensm. Unters.- Forsch.*, 161, 25, 1976.
68. **Schmidtlein, H. and Herrmann, K.**, Originalarbeiten über die Phenolsäuren des Gemüses. IV. Hydroxyzimtsäuren und Hydroxbenzoesäuren weiterer Gemüsearten und der Kartoffeln, *Z. Lebensm. Unters.- Forsch.*, 159, 255, 1975.
69. **Herrmann, K.**, Üaabersicht über nichtessentielle Inhaltsstoffe der Gemüsearten, *Z. Lebensm. Unters.- Forsch.*, 165, 151, 1977.
70. **Lerch, B.**, Phosphodiesterase. I. Spezifischer Nachweis nach Disk-Elektrophorese und Vorkommen in Pflanzen, *Experimentia*, 24, 889, 1968.
71. **Kelly, G. J. and Latzko, E.**, Chloroplast phosphofructokinase. I. Proof of phosphofructokinase activity in chloroplasts, *Plant Physiol.*, 60, 290, 1977.
72. **Fölster, E.**, Treiben von Schnittlauch, *Gemüse*, 19, 256, 1983.
73. **Hartmann, H. D.**, Stickstoffdüngungsversuche zu Schnittlauch, *Gartenbauwissenschaft*, 31, 51, 1966.
74. **Brewster, J. L.**, The response of growth rate to temperature in seedlings of several *Allium* crop species, *Ann. Appl. Biol.*, 93, 351, 1979.
75. **Takagi, H.**, Dormancy and seasonal changes of growth activities in asatsuki *(Allium schoenoprasum L. var. schoenoprasum)*, *J. Jpn. Soc. Hortic. Sci.*, 56, 60, 1987.
76. **Krug, H. and Fölster, E.**, Influence of the environment on growth and development of chives *(Allium schoenoprasum L.)*. I. Induction of the rest period, *Sci. Hortic.*, 4, 211, 1976.
77. **Fölster, E. and Krug, H.**, Influence of the environment on growth and development of chives *(Allium schoenoprasum L.)*. II. Breaking of the rest period and forcing, *Sci. Hortic.*, 7, 213, 1977.
78. **Tychsen, K. and Skytt Andersen, A.**, Abscisic acid and dormancy of chives *(Allium schoenoprasum L.)*, *Yearb. R. Vet. Agric. Univ.*, 1973, 39, 1973.
79. **Pollack, R.**, Untersuchungen über die Beziehungen zwischen Auxin-Gibberellin- und Inhibitorgehalten und der Wachstumsaktivität von Schnittlauch *(Allium schoenoprasum)*, Dissertation, Technical University, Hannover, 1975.
80. **Waris, H., Simola, L. K., and Granö, A.**, Aseptic cultures of seed plants at various sucrose concentrations with and without giberellin, *Ann. Acad. Sci. Fenn. Ser. A4*, 188, 1, 1972.
81. **Yamane, Y.**, Chives, *Jpn. J. Genet.*, 58, 698, 1983.
82. **Kobabe, G.**, Möglichkeiten zur züchterischen Verbesserung von Schnittlauchsorten, *Gemüse*, 1, 171, 1965.
83. **Tatlioglu, T. and Wricke, G.**, Genetisch-züchterische Untersuchungen am Schnittlauch *(Allium schoenoprasum L.)*, *Gartenbauwissenschaft*, 45, 278, 1980.
84. **Holm, E.**, Bestøvning af purløg, *Dansk Frøavl.*, 67, 359, 1984.
85. **Tatlioglu, T. and Wricke, G.**, Stand und Möglichkeiten der Schnittlauchzüchtung, *Gemüse*, 16, 392, 1980.
86. **Tatlioglu, T.**, Influence of temperature on the expression of cytoplasmic male sterility in chives *(Allium schoenoprasum L.)*. *Z. Pflanzenzuecht.*, 94, 156, 1985.
87. **Tatlioglu, T.**, Influence of tetracycline on the expression of cytoplasmic male sterility (cms) in chives *(Allium schoenoprasum L.)*, *Plant Breed.*, 97, 46, 1986.
88. **Singh, V. P. and Kobabe, G.**, Cyto-morphological investigation on male-sterility in *Allium schoenoprasum L.*, *Indian J. Genet. Plant Breed.*, 29, 241, 1969.
89. **Gvaladze, G. E.**, Forms of apomixis in the genus *Allium L.*, in *Apomixis and Breeding*, Khokhlov, S. S., Ed., Amerind, New York, 1976, 160.

90. **McCollum, J. P.,** A study of the rest period of chives, *Proc. Am. Soc. Hortic. Sci.*, 33, 491, 1935.
91. **Fölster, E.,** Zur schnittlauchtreiberei im Herbst, *Gartenbauwissenschaft*, 32, 503, 1967.
92. **Dunkel, K.-H.,** Zu jeder Jahreszeit frischen Schnittlauch, *Gemüse*, 3, 271, 1967.
93. **Heinze, W. and Werner, H.,** Frühtreiberei von Schnittlauch, *Gemüse*, 7, 245, 1971.
94. **Larsen, J. J.,** Purløg (*Allium schoenoprasum* L.), in *Grønsager pa Friland*, Jørgensen, M. Blangstrup, Ed., Gartnerinfo, Copenhagen, 1982, 315.
95. **Porsdal Poulsen, K., and Nielsen, P.,** Freeze-drying of chives and parsley- optimization attempts, *Prog. Refrig. Sci. Tech.*, 3, 275, 1979.
96. **Thompson, W. A. and Lammerink, J.,** Chives cuttings for processing, *N. Z. Commer. Grower*, 38, 23, 1983.
97. **Anon.,** Vejledning i froavl af purløg, from the seed company, *Daehnfeldt*, 1, 1985.
98. **Kropačová, S., Kropáč, A., and Nedbalová, V.,** Studie vztahů mezi opylenim květů a tvorbou semene pažitky, *Acta Univ. Agric. Brno.*, 17, 103, 1969.
99. **Nordestgaard, A.,** Frøavl af purløg, *Meddelelse Statens Planteavlsforsøg*, 85(1733), 3, 1983.
100. **Hartmann, H. D. and Frenz, F. W.,** Stickstoff-Düngungsversuche bei Schnittlauch, *Jahresschrift Hess. Lehr. u. Forsch. Inst. Wein-Obst-u. Gartenbau* 1963, 83, 1964.
101. **Hartmann, H. D.,** Zum Anbau von Schnittlauch, *Gemüse*, 5, 239, 1969.
102. **Woltz, S. S. and Waters, W. E.,** Chives production as affected by fertilizer practices, soil mixes and methyl bromide soil residues, *Proc. Fl. State Hortic. Soc.*, 88, 133, 1976.
103. **Hartmann, H. D.,** Anbaumethodische Versuche zu Schnittlauch, *Gartenbauwissenschaft*, 32, 263, 1967.
104. **Belfort, C. C., and Haag, H. P.,** Nutrição mineral de hortaliças -LVI- Carência de macronutrientes em cebolinha (*Allium schoenoprasum*), *An. Esc. Super. Agric. Luiz de Queiroz Univ Sao Paulo*, 40, 221, 1983.
105. **Sievers, E.,** Untersuchungen über die Mycorrhizen von *Allium*- und *Solanum*-Arten, *Arch Mikrobiol.*, 18, 289, 1953.
106. **Dale, W. T.,** Aecidia of *Puccinia allii* Rud. in chives in Britain, *Plant Pathol.*, 19, 149, 1970.
107. **Gjærum, H. B.,** Additional Norwegian finds of Uredinales and Ustilaginales. III, *Norw. J. Bot.*, 19, 17, 1972.
108. **Werner, A. and Krzan, Z.,** Etiology of *Puccinia porri* (sow.) Winter and behaviour of its urediniospores on poplar leaves, *Arbor. Kornickie*, 24, 201, 1979.
109. **Sørensen, N. H.,** Rust (*Puccinia allii* Rud.) on Chive (*Allium schoenoprasum* L.) Epidemiology and Resistance Biology, Dissertation, Department of Plant Pathology, Royal Veterinary and Agricultural University, 1984.
110. **Mordue, J. E. M.,** *Sclerotium cepivorum*, CMI Descriptions of pathogenic fungi and bacteria, 52, 2, 1976.
111. **Chupp, C. and Sherf, A. F.,** Onion diseases, in *Vegetable Diseases and Their Control*, Ronald Press Co., New York, 1960, 375.
112. **Reifschneider, J. B.,** Chives (*Allium schoenoprasum*) Downy mildew (*Peronospora destructor*), *Am. Phytopathol. Soc.*, 36, 58, 1981.
113. **Hansen, P. O.,** personal communication, 1987.
114. **Nøddegaard, E. and Hansen, K. E.,** Forsøg med plantebeskyttelsesmidler i landbrugs- og specialafgrøder 1970, *Tidsskr. Planteavl*, 76, 63, 1972.

INDEX

AUTHOR INDEX